OPTICAL METHODS OF MEASUREMENT

Wholefield Techniques

SECOND EDITION

OPTICAL SCIENCE AND ENGINEERING

1. Electron and Ion Microscopy and Microanalysis: Principles and Applications, *Lawrence E. Murr*
2. Acousto-Optic Signal Processing: Theory and Implementation, *edited by Norman J. Berg and John N. Lee*
3. Electro-Optic and Acousto-Optic Scanning and Deflection, *Milton Gottlieb, Clive L. M. Ireland, and John Martin Ley*
4. Single-Mode Fiber Optics: Principles and Applications, *Luc B. Jeunhomme*
5. Pulse Code Formats for Fiber Optical Data Communication: Basic Principles and Applications, *David J. Morris*
6. Optical Materials: An Introduction to Selection and Application, *Solomon Musikant*
7. Infrared Methods for Gaseous Measurements: Theory and Practice, *edited by Joda Wormhoudt*
8. Laser Beam Scanning: Opto-Mechanical Devices, Systems, and Data Storage Optics, *edited by Gerald F. Marshall*
9. Opto-Mechanical Systems Design, *Paul R. Yoder, Jr.*
10. Optical Fiber Splices and Connectors: Theory and Methods, *Calvin M. Miller with Stephen C. Mettler and Ian A. White*
11. Laser Spectroscopy and Its Applications, *edited by Leon J. Radziemski, Richard W. Solarz, and Jeffrey A. Paisner*
12. Infrared Optoelectronics: Devices and Applications, *William Nunley and J. Scott Bechtel*
13. Integrated Optical Circuits and Components: Design and Applications, *edited by Lynn D. Hutcheson*
14. Handbook of Molecular Lasers, *edited by Peter K. Cheo*
15. Handbook of Optical Fibers and Cables, *Hiroshi Murata*
16. Acousto-Optics, *Adrian Korpel*
17. Procedures in Applied Optics, *John Strong*
18. Handbook of Solid-State Lasers, *edited by Peter K. Cheo*
19. Optical Computing: Digital and Symbolic, *edited by Raymond Arrathoon*
20. Laser Applications in Physical Chemistry, *edited by D. K. Evans*
21. Laser-Induced Plasmas and Applications, *edited by Leon J. Radziemski and David A. Cremers*
22. Infrared Technology Fundamentals, *Irving J. Spiro and Monroe Schlessinger*

23. Single-Mode Fiber Optics: Principles and Applications, Second Edition, Revised and Expanded, *Luc B. Jeunhomme*
24. Image Analysis Applications, *edited by Rangachar Kasturi and Mohan M. Trivedi*
25. Photoconductivity: Art, Science, and Technology, *N. V. Joshi*
26. Principles of Optical Circuit Engineering, *Mark A. Mentzer*
27. Lens Design, *Milton Laikin*
28. Optical Components, Systems, and Measurement Techniques, *Rajpal S. Sirohi and M. P. Kothiyal*
29. Electron and Ion Microscopy and Microanalysis: Principles and Applications, Second Edition, Revised and Expanded, *Lawrence E. Murr*
30. Handbook of Infrared Optical Materials, *edited by Paul Klocek*
31. Optical Scanning, *edited by Gerald F. Marshall*
32. Polymers for Lightwave and Integrated Optics: Technology and Applications, *edited by Lawrence A. Hornak*
33. Electro-Optical Displays, *edited by Mohammad A. Karim*
34. Mathematical Morphology in Image Processing, *edited by Edward R. Dougherty*
35. Opto-Mechanical Systems Design: Second Edition, Revised and Expanded, *Paul R. Yoder, Jr.*
36. Polarized Light: Fundamentals and Applications, *Edward Collett*
37. Rare Earth Doped Fiber Lasers and Amplifiers, *edited by Michel J. F. Digonnet*
38. Speckle Metrology, *edited by Rajpal S. Sirohi*
39. Organic Photoreceptors for Imaging Systems, *Paul M. Borsenberger and David S. Weiss*
40. Photonic Switching and Interconnects, *edited by Abdellatif Marrakchi*
41. Design and Fabrication of Acousto-Optic Devices, *edited by Akis P. Goutzoulis and Dennis R. Pape*
42. Digital Image Processing Methods, *edited by Edward R. Dougherty*
43. Visual Science and Engineering: Models and Applications, *edited by D. H. Kelly*
44. Handbook of Lens Design, *Daniel Malacara and Zacarias Malacara*
45. Photonic Devices and Systems, *edited by Robert G. Hunsberger*
46. Infrared Technology Fundamentals: Second Edition, Revised and Expanded, *edited by Monroe Schlessinger*
47. Spatial Light Modulator Technology: Materials, Devices, and Applications, *edited by Uzi Efron*
48. Lens Design: Second Edition, Revised and Expanded, *Milton Laikin*
49. Thin Films for Optical Systems, *edited by Francoise R. Flory*
50. Tunable Laser Applications, *edited by F. J. Duarte*
51. Acousto-Optic Signal Processing: Theory and Implementation, Second Edition, *edited by Norman J. Berg and John M. Pellegrino*
52. Handbook of Nonlinear Optics, *Richard L. Sutherland*
53. Handbook of Optical Fibers and Cables: Second Edition, *Hiroshi Murata*

54. Optical Storage and Retrieval: Memory, Neural Networks, and Fractals, *edited by Francis T. S. Yu and Suganda Jutamulia*
55. Devices for Optoelectronics, *Wallace B. Leigh*
56. Practical Design and Production of Optical Thin Films, *Ronald R. Willey*
57. Acousto-Optics: Second Edition, *Adrian Korpel*
58. Diffraction Gratings and Applications, *Erwin G. Loewen and Evgeny Popov*
59. Organic Photoreceptors for Xerography, *Paul M. Borsenberger and David S. Weiss*
60. Characterization Techniques and Tabulations for Organic Nonlinear Optical Materials, *edited by Mark G. Kuzyk and Carl W. Dirk*
61. Interferogram Analysis for Optical Testing, *Daniel Malacara, Manuel Servin, and Zacarias Malacara*
62. Computational Modeling of Vision: The Role of Combination, *William R. Uttal, Ramakrishna Kakarala, Spiram Dayanand, Thomas Shepherd, Jagadeesh Kalki, Charles F. Lunskis, Jr., and Ning Liu*
63. Microoptics Technology: Fabrication and Applications of Lens Arrays and Devices, *Nicholas Borrelli*
64. Visual Information Representation, Communication, and Image Processing, *edited by Chang Wen Chen and Ya-Qin Zhang*
65. Optical Methods of Measurement, *Rajpal S. Sirohi and F. S. Chau*
66. Integrated Optical Circuits and Components: Design and Applications, *edited by Edmond J. Murphy*
67. Adaptive Optics Engineering Handbook, *edited by Robert K. Tyson*
68. Entropy and Information Optics, *Francis T. S. Yu*
69. Computational Methods for Electromagnetic and Optical Systems, *John M. Jarem and Partha P. Banerjee*
70. Laser Beam Shaping, *Fred M. Dickey and Scott C. Holswade*
71. Rare-Earth-Doped Fiber Lasers and Amplifiers: Second Edition, Revised and Expanded, *edited by Michel J. F. Digonnet*
72. Lens Design: Third Edition, Revised and Expanded, *Milton Laikin*
73. Handbook of Optical Engineering, *edited by Daniel Malacara and Brian J. Thompson*
74. Handbook of Imaging Materials: Second Edition, Revised and Expanded, *edited by Arthur S. Diamond and David S. Weiss*
75. Handbook of Image Quality: Characterization and Prediction, *Brian W. Keelan*
76. Fiber Optic Sensors, *edited by Francis T. S. Yu and Shizhuo Yin*
77. Optical Switching/Networking and Computing for Multimedia Systems, *edited by Mohsen Guizani and Abdella Battou*
78. Image Recognition and Classification: Algorithms, Systems, and Applications, *edited by Bahram Javidi*
79. Practical Design and Production of Optical Thin Films: Second Edition, Revised and Expanded, *Ronald R. Willey*
80. Ultrafast Lasers: Technology and Applications, *edited by Martin E. Fermann, Almantas Galvanauskas, and Gregg Sucha*
81. Light Propagation in Periodic Media: Differential Theory and Design, *Michel Nevière and Evgeny Popov*

82. Handbook of Nonlinear Optics, Second Edition, Revised and Expanded, *Richard L. Sutherland*
83. Polarized Light: Second Edition, Revised and Expanded, *Dennis Goldstein*
84. Optical Remote Sensing: Science and Technology, *Walter Egan*
85. Handbook of Optical Design: Second Edition, *Daniel Malacara and Zacarias Malacara*
86. Nonlinear Optics: Theory, Numerical Modeling, and Applications, *Partha P. Banerjee*
87. Semiconductor and Metal Nanocrystals: Synthesis and Electronic and Optical Properties, *edited by Victor I. Klimov*
88. High-Performance Backbone Network Technology, *edited by Naoaki Yamanaka*
89. Semiconductor Laser Fundamentals, *Toshiaki Suhara*
90. Handbook of Optical and Laser Scanning, *edited by Gerald F. Marshall*
91. Organic Light-Emitting Diodes: Principles, Characteristics, and Processes, *Jan Kalinowski*
92. Micro-Optomechatronics, *Hiroshi Hosaka, Yoshitada Katagiri, Terunao Hirota, and Kiyoshi Itao*
93. Microoptics Technology: Second Edition, *Nicholas F. Borrelli*
94. Organic Electroluminescence, *edited by Zakya Kafafi*
95. Engineering Thin Films and Nanostructures with Ion Beams, *Emile Knystautas*
96. Interferogram Analysis for Optical Testing, Second Edition, *Daniel Malacara, Manuel Sercin, and Zacarias Malacara*
97. Laser Remote Sensing, *edited by Takashi Fujii and Tetsuo Fukuchi*
98. Passive Micro-Optical Alignment Methods, *edited by Robert A. Boudreau and Sharon M. Boudreau*
99. Organic Photovoltaics: Mechanism, Materials, and Devices, *edited by Sam-Shajing Sun and Niyazi Serdar Saracftci*
100. Handbook of Optical Interconnects, *edited by Shigeru Kawai*
101. GMPLS Technologies: Broadband Backbone Networks and Systems, *Naoaki Yamanaka, Kohei Shiomoto, and Eiji Oki*
102. Laser Beam Shaping Applications, *edited by Fred M. Dickey, Scott C. Holswade and David L. Shealy*
103. Electromagnetic Theory and Applications for Photonic Crystals, *Kiyotoshi Yasumoto*
104. Physics of Optoelectronics, *Michael A. Parker*
105. Opto-Mechanical Systems Design: Third Edition, *Paul R. Yoder, Jr.*
106. Color Desktop Printer Technology, *edited by Mitchell Rosen and Noboru Ohta*
107. Laser Safety Management, *Ken Barat*
108. Optics in Magnetic Multilayers and Nanostructures, *Štefan Višňovský*
109. Optical Inspection of Microsystems, *edited by Wolfgang Osten*
110. Applied Microphotonics, *edited by Wes R. Jamroz, Roman Kruzelecky, and Emile I. Haddad*
111. Organic Light-Emitting Materials and Devices, *edited by Zhigang Li and Hong Meng*
112. Silicon Nanoelectronics, *edited by Shunri Oda and David Ferry*

113. Image Sensors and Signal Processor for Digital Still Cameras, *Junichi Nakamura*
114. Encyclopedic Handbook of Integrated Circuits, *edited by Kenichi Iga and Yasuo Kokubun*
115. Quantum Communications and Cryptography, *edited by Alexander V. Sergienko*
116. Optical Code Division Multiple Access: Fundamentals and Applications, *edited by Paul R. Prucnal*
117. Polymer Fiber Optics: Materials, Physics, and Applications, *Mark G. Kuzyk*
118. Smart Biosensor Technology, *edited by George K. Knopf and Amarjeet S. Bassi*
119. Solid-State Lasers and Applications, *edited by Alphan Sennaroglu*
120. Optical Waveguides: From Theory to Applied Technologies, *edited by Maria L. Calvo and Vasudevan Lakshiminarayanan*
121. Gas Lasers, *edited by Masamori Endo and Robert F. Walker*
122. Lens Design, Fourth Edition, *Milton Laikin*
123. Photonics: Principles and Practices, *Abdul Al-Azzawi*
124. Microwave Photonics, *edited by Chi H. Lee*
125. Physical Properties and Data of Optical Materials, *Moriaki Wakaki, Keiei Kudo, and Takehisa Shibuya*
126. Microlithography: Science and Technology, Second Edition, *edited by Kazuaki Suzuki and Bruce W. Smith*
127. Coarse Wavelength Division Multiplexing: Technologies and Applications, *edited by Hans Joerg Thiele and Marcus Nebeling*
128. Organic Field-Effect Transistors, *Zhenan Bao and Jason Locklin*
129. Smart CMOS Image Sensors and Applications, *Jun Ohta*
130. Photonic Signal Processing: Techniques and Applications, *Le Nguyen Binh*
131. Terahertz Spectroscopy: Principles and Applications, *edited by Susan L. Dexheimer*
132. Fiber Optic Sensors, Second Edition, *edited by Shizhuo Yin, Paul B. Ruffin, and Francis T. S. Yu*
133. Introduction to Organic Electronic and Optoelectronic Materials and Devices, *edited by Sam-Shajing Sun and Larry R. Dalton*
134. Introduction to Nonimaging Optics, *Julio Chaves*
135. The Nature of Light: What Is a Photon?, *edited by Chandrasekhar Roychoudhuri, A. F. Kracklauer, and Katherine Creath*
136. Optical and Photonic MEMS Devices: Design, Fabrication and Control, *edited by Ai-Qun Liu*
137. Tunable Laser Applications, Second Edition, *edited by F. J. Duarte*
138. Biochemical Applications of Nonlinear Optical Spectroscopy, *edited by Vladislav Yakovlev*
139. Dynamic Laser Speckle and Applications, *edited by Hector J. Rabal and Roberto A. Braga Jr.*
140. Slow Light: Science and Applications, *edited by Jacob B. Khurgin and Rodney S. Tucker*
141. Laser Safety: Tools and Training, *edited by Ken Barat*
142. Near-Earth Laser Communications, *edited by Hamid Hemmati*

143. Polarimetric Radar Imaging: From Basics to Applications, *Jong-Sen Lee and Eric Pottier*
144. Photoacoustic Imaging and Spectroscopy, *edited by Lihong V. Wang*
145. Optoelectronics: Infrared-Visible-Ultraviolet Devices and Applications, Second Edition, *edited by Dave Birtalan and William Nunley*
146. Optical Methods of Measurement: Wholefield Techniques, Second Edition, *Rajpal S. Sirohi*

OPTICAL METHODS OF MEASUREMENT
Wholefield Techniques

SECOND EDITION

RAJPAL S. SIROHI

CRC Press
Taylor & Francis Group
Boca Raton London New York

CRC Press is an imprint of the
Taylor & Francis Group, an **informa** business

CRC Press
Taylor & Francis Group
6000 Broken Sound Parkway NW, Suite 300
Boca Raton, FL 33487-2742

First issued in paperback 2017

ISBN 13: 978-1-138-11549-1 (pbk)
ISBN 13: 978-1-57444-697-5 (hbk)

Library of Congress Cataloging-in-Publication Data

Sirohi, R. S.
Optical methods of measurement : wholefield techniques / Rajpal S. Sirohi. -- 2nd ed.
p. cm. -- (Optical science and engineering)
Includes bibliographical references and index.
ISBN 978-1-57444-697-5 (hardcover : alk. paper)
1. Optical measurements. I. Title. II. Series.

QC367.S57 2009
681'.25--dc22 2009019252

Visit the Taylor & Francis Web site at
http://www.taylorandfrancis.com

and the CRC Press Web site at
http://www.crcpress.com

Contents

From the Series Editor

Over the last 20 years, optical science and engineering has emerged as a discipline in its own right. This has occurred with the realization that we are dealing with an integrated body of knowledge that has resulted in optical science, engineering, and technology becoming an enabling discipline that can be applied to a variety of scientific, industrial, commercial, and military problems to produce operating devices and hardware systems. This book series is a testament to the truth of the preceding statement.

Quality control and the testing have become essential tools in modern industry and modern laboratory processes. Optical methods of measurement have provided many of the essential techniques of current practice. This current volume on *Optical Methods of Measurement* emphasizes wholefield measurement methods as opposed to point measurement, that is, sensing a field all at once and then mapping that field for the parameter or parameters under consideration as contrasted to building up that field information by a time series of point-by-point determinations. The initial output of these wholefield systems of measurement is often a fringe pattern that is then processed. The required fringes are formed by direct interferometry, holographic interferometry, phase-shifting methods, heterodyning, speckle pattern interferometry, and moiré techniques.

The methods described here are applicable to many measurement scenarios, although the examples focus on modern experimental mechanics. Since this volume covers the variety of techniques available and their range of applicability together with their sensitivity and accuracy as determined by the underlying principle, the reader will find these pages an excellent practice guide of wholefield measurement as well as a desk reference volume.

Brian J. Thompson

Preface

Many good ideas originated from several of my professional colleagues that have considerably improved the book. I would like to thank Prof. Osten from Stuttgart University and Professors Hinsch and Helmers from the University of Oldenburg for providing numerous contributions to the contents of the book. I owe a great deal to my colleagues Prof. M. P. Kothiyal, Prof. Chandra Shakher, and Dr. N. Krishna Mohan for their help and support at various stages.

Since the phenomenon of interference is central to many techniques of measurement, I have introduced a chapter on "Optical Interference." There have been several additions like the non-diffracting beam and singular beam with their metrological applications. Bibliography and additional reading have been expanded. The book contains references to 103 books, 827 journal papers, and 130 figures.

Revision of this book was originally slated for late 2005; however, administrative responsibilities prevented this. It is only through the persistent efforts by the staff of Taylor & Francis that the revision finally materialized. I would therefore like to express my sincere thanks to Jessica Vakili and Catherine Giacari for keeping my interest in the book alive.

Rajpal S. Sirohi

Preface to First Edition

Optical techniques of measurement are among the most sensitive known today. In addition, they are noncontact, noninvasive, and fast. In recent years, the use of optical techniques for measurement have dramatically increased, and applications range from determining the topography of landscapes to checking the roughness of polished surfaces.

Any of the characteristics of a light wave—amplitude, phase, length, frequency, polarization, and direction of propagation—can be modulated by the measurand. On demodulation, the value of the measurand at a spatial point and at a particular time instant can be obtained. Optical methods can effect measurement at discrete points or over the wholefield with extremely fine spatial resolution. For many applications, wholefield measurements are preferred.

This book contains a wealth of information on wholefield measurement methods, particularly those employed frequently on modern experimental mechanics since the variable that is normally monitored is displacement. Thus, the methods can be used to determine surface deformation, strains, and stresses. By extension, they can be applied in the nondestructive evaluation of components. There is no doubt that the methods described are applicable to other fields as well, such as the determination of surface contours and surface roughness. With the appropriate setup, these wholefield optical methods can be used to obtain material properties and temperature and pressure gradients. Throughout the book, emphasis is placed on the physics of the techniques, with demonstrative examples of how they can be applied to tackle particular measurement problems.

Any optical technique has to involve a light source, beam-handling optics, a detector, and a data-handling system. In many of the wholefield measurement techniques, a laser is the source of light. Since much information on lasers is available, this aspect is not discussed in the book. Instead, we have included an introductory chapter on the concept of waves and associated phenomena. The propagation of waves is discussed in Chapter 2.

Chapter 3 deals with current phase evaluation techniques, since almost all the techniques described here display the information in the form of a fringe pattern. This fringe pattern is evaluated to a high degree of accuracy using currently available methods to extract the information of interest. The various detectors available for recording wholefield information are described in Chapter 4. Thus the first four chapters provide the background for the rest of the book.

Chapter 5 is on holographic interferometry. A number of techniques have been included to give readers a good idea of how various applications may be handled.

The speckle phenomenon, although a bane of holographers, has emerged as a very good measurement technique. Several techniques are now available to measure the in-plane and out-of-plane displacement components, tilts, or slopes. In Chapter 6, these techniques are contrasted with those based on holographic interferometry, with discussion on their relative advantages and areas of applicability. In recent times, electronic detection in conjunction with phase evaluation methods has been increasingly used. This makes all these speckle techniques almost real-time approaches, and they have consequently attracted more attention from industry. Chapter 6 includes descriptions of various measurement techniques in speckle metrology.

Photoelasticity is another wholefield measurement technique that has existed for some time. Experiments are conducted on transparent models that become birefringent when subjected to load. Unlike holographic interferometry and speckle-based methods, which measure displacement or strain, photoelasticity gives the stress distribution directly. Photoelasticity, including the technique of holophotoelasticy, is covered in Chapter 7.

The moiré technique, along with the Talbot phenomenon, is a powerful measurement technique and covers a very wide range of sensitivity. Chapter 8 presents various techniques, spanning the measurement range from geometrical moiré with low-frequency diffraction gratings.

The book has evolved as a result of the authors' involvement with research and teaching of these techniques for more than two decades. It puts together the major wholefield methods of measurement in one text, and, therefore should be useful to students taking a graduate course in optical experimental mechanics and to experimentalists who are interested in investigating the techniques available and the physics behind them. It will also serve as a useful reference for researchers in the field.

We acknowledge the strong support given by Professor Brian J. Thompson and the excellent work of the staff of Marcel Dekker, Inc. This book would not have been possible without the constant support and encouragement of our wives, Vijayalaxmi Sirohi and Chau Swee Har, and our families, to whom we express our deepest thanks.

Rajpal S. Sirohi
Fook Siong Chau

Author

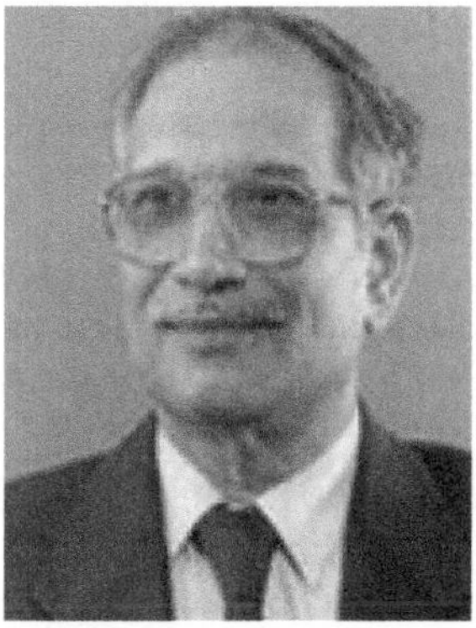

R. S. Sirohi, Ph.D. is presently the vice chancellor of Amity University, Rajasthan, India. Prior to this, he was the vice chancellor of Barkatullah University, Bhopal, and director, Indian Institute of Technology Delhi, Delhi. He has also served at the Indian Institute of Science, Bangalore and in various capacities at the Indian Institute of Technology Madras, Chennai.

He holds a postgraduate degree in applied optics and a Ph.D. in physics from IIT Delhi, India.

Prof. Sirohi worked in Germany as a Humboldt Fellow and Awardee. He was senior research associate at Case Western Reserve University, Cleveland, Ohio, and associate professor at the Rose Hulman Institute of Technology, Terre Haute Indiana. He was an ICTP (International Center for Theoretical Physics, Trieste, Italy) visiting scholar to the Institute for Advanced Studies, University of Malaya, Malaysia, and visiting professor at the National University of Singapore. Currently he is an ICTP visiting scientist to University of Namibia.

Dr. Sirohi is a Fellow of several important academies in India and abroad including the Indian National Academy of Engineering, National Academy of Sciences, Optical Society of America, Optical Society of India, SPIE (The International Society for Optical Engineering) and honorary fellow of ISTE (Indian Society for Technical Education) and Metrology Society of India. He is a member of several other scientific societies, and founding member of the India Laser Association. He was also the chair for SPIE-INDIA Chapter, which he established with cooperation from SPIE in 1995 at IIT Madras. He was invited as a JSPS (Japan Society for the Promotion of Science) Fellow to Japan. He was a member of the Education Committee of SPIE.

Dr. Sirohi has received the following awards from various organizations: Humboldt Research Award (1995) by the Alexander von Humboldt Foundation, Germany; Galileo Galilei Award of International Commission for Optics (1995); Amita De Memorial Award of the Optical Society of India (1998); 13th Khwarizmi International Award, IROST (Iranian Research Organisation for Science and Technology) (2000); Albert Einstein Silver Medal, UNESCO (2000); Dr. Y.T. Thathachari Prestigious Award for Science by Thathachari Foundation, Mysore (2001); Pt. Jawaharlal Nehru Award in Engineering & Technology (2000) by the MP Council of Science and Technology; Giant's Gaurav Samman (2001); NAFEN's Annual Corporate Excellence Award (2001); NRDC Technology Invention Award (2003); Sir C.V. Raman

Award: Physical Sciences (2002) by UGC; Padma Shri, a National Civilian Award (2004); Sir C.V. Raman Birth Centenary Award (2005) by the Indian Science Congress Association, Kolkata; Inducted into Order of Holo-Knights during the International Conference Fringe 05 held at Stuttgart, Germany, 2005; Centenarian Seva Ratna Award (2004) by The Centenarian Trust, Chennai; Instrument Society of India Award (2007); and the Lifetime Achievement Award (2007) by the Optical Society of India. SPIE—The International Society for Optical Engineering—has bestowed on him the SPIE Gabor Award 2009 for his work on holography, speckle metrology, interferometry, and confocal microscopy.

Dr. Sirohi was the president of the Optical Society of India during 1994–1996. He was also president of the Instrument Society of India during 2003–2006 and reelected for another term beginning 2008. He is on the international advisory board of the *Journal of Modern Optics*, UK, on the editorial boards of the *Journal of Optics (India)* and *Optik (Germany)*. Currently, he is associate editor of the international journal *Optical Engineering* and regional editor of the *Journal of Holography and Speckle*. He was also guest editor for the journals *Optics and Lasers in Engineering* and *Optical Engineering*.

Dr. Sirohi has 438 papers to his credit with 240 published in national and international journals, 60 in proceedings of the conferences and 140 presented in conferences. He has authored/co-authored/edited thirteen books including five Milestones for SPIE, was principal coordinator for 26 projects sponsored by government funding agencies and industries, and supervised 24 Ph.D., 7 M.S. and numerous B.Tech., M.Sc. and M.Tech. theses.

Dr. Sirohi's research areas are optical metrology, optical instrumentation, holography and speckle phenomenon.

1 Waves and Beams

Optics pertains to the generation, amplification, propagation, detection, modification, and modulation of light. Light is considered a very tiny portion of the electromagnetic spectrum responsible for vision. In short, optics is science, technology, and engineering with light. Measurement techniques based on light fall into the subject of optical metrology. Thus, optical metrology comprises varieties of measurement techniques, including dimensional measurements, measurement of process variables, and measurement of electrical quantities such as current and voltages.

Optical methods are noncontact and noninvasive. Indeed, the light does interact with the measurand, but does not influence its value in the majority of cases, and hence such measurements are termed noncontact. Further, information about the whole object can be obtained simultaneously, thereby giving optical methods the capability of wholefield measurement. The measurements are not influenced by electromagnetic fields. Further, light wavelength is a measurement stick in length metrology, and hence high accuracy in measurement of length/displacement is attained. Some of these attributes make optical methods indispensable for several different kinds of measurements.

1.1 THE WAVE EQUATION

Light is an electromagnetic wave, and therefore satisfies the wave equation

$$\nabla^2\mathbf{E}(\mathbf{r},t) - \frac{1}{c^2}\frac{\partial^2\mathbf{E}(\mathbf{r},t)}{\partial t^2} = 0 \tag{1.1}$$

where $\mathbf{E}(\mathbf{r},t)$ is the instantaneous electric field of the light wave and c is the velocity of light. [The instantaneous magnetic field of the wave, $\mathbf{B}(\mathbf{r},t)$, satisfies a similar equation.] Equation 1.1 is a vector wave equation. We are interested in those of its solutions that represent monochromatic waves, that is,

$$\mathbf{E}(\mathbf{r},t) = \mathbf{E}(\mathbf{r})\exp(-i\omega t) \tag{1.2}$$

where $\mathbf{E}(\mathbf{r})$ is the amplitude and ω is the circular frequency of the wave. Substituting Equation 1.2 into Equation 1.1 leads to

$$\nabla^2\mathbf{E}(\mathbf{r}) + k^2\mathbf{E}(\mathbf{r}) = 0 \tag{1.3}$$

where $k^2 = \omega^2/c^2$. This is the well-known Helmholtz equation. A solution of the Helmholtz equation for $\mathbf{E}(\mathbf{r})$ will provide a monochromatic solution of the wave equation.

1.2 PLANE WAVES

There are several solutions of the wave equation; one of these is of the form

$$\mathbf{E}(\mathbf{r}) = \mathbf{E}_0 \exp(i\mathbf{k} \cdot \mathbf{r}) \tag{1.4}$$

This represents a plane wave, which has constant amplitude $\mathbf{E}_0$ and is of infinite cross-section. A plane monochromatic wave spans the whole of space for all times, and is a mathematical idealization. A real plane wave, called a collimated wave, is limited in its transverse dimensions. The limitation may be imposed by optical elements/systems or by physical stops. This restriction on the transverse dimensions of the wave leads to the phenomenon of diffraction, which is discussed in more detail in Chapter 3. A plane wave is shown schematically in Figure 1.1a.

1.3 SPHERICAL WAVES

Another solution of Equation 1.3 is

$$\mathbf{E}(\mathbf{r}) = \frac{A}{r} \exp(i\mathbf{k} \cdot \mathbf{r}) \tag{1.5}$$

where A is a constant representing the amplitude of the wave at unit distance. This solution represents a spherical wave. A plane wave propagates in a particular direction, whereas a spherical wave is not unidirectional. A quadratic approximation to the spherical wave, in rectangular Cartesian coordinates, is

$$\mathbf{E}(x, y, z) = \frac{A}{z} \exp(ikz) \exp\left[\frac{ik}{2z}(x^2 + y^2)\right] \tag{1.6}$$

This represents the amplitude at any point (x, y, z) on a plane distant z from a point source. This expression is very often used when discussing propagation through optical systems. A collimated wave is obtained by placing a point source (source size diffraction-limited) at the focal point of a lens. This is, in fact, a truncated plane wave. A spherical wave emanates from a point source, or it may converge to a point. Diverging and converging spherical waves are shown schematically in Figure 1.1b.

1.4 CYLINDRICAL WAVES

Often, a fine slit is illuminated by a broad source. The waves emanating from such line sources are called cylindrical waves. The surfaces of constant phase, far away from the source, are cylindrical. A cylindrical lens placed in a collimated beam will

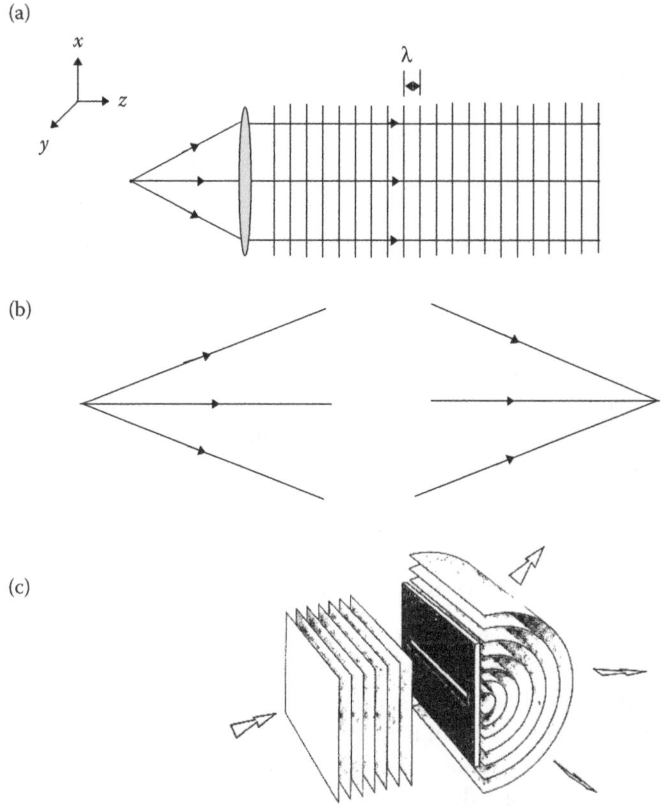

FIGURE 1.1 (a) Plane wave. (b) Diverging and converging spherical waves. (c) Cylindrical wave. (Figure 1.1c from E. Hecht and A. Zajac, *Optics*, Addison-Wesley, Reading, MA, 1974. With permission.)

generate a cylindrical wave that would converge to a line focus. The amplitude of a cylindrical wave, far from the narrow slit, can be written as

$$\mathbf{E}(\mathbf{r}) = \frac{\mathrm{A}}{\sqrt{r}} \exp(i\mathbf{k} \cdot \mathbf{r}) \tag{1.7}$$

A cylindrical wave is shown schematically in Figure 1.1c.

1.5 WAVES AS INFORMATION CARRIERS

Light is used for sensing a variety of parameters, and its domain of applications is so vast that it pervades all branches of science, engineering, and technology, biomedicine, agriculture, etc. Devices that use light for sensing, measurement, and control are termed optical sensors. Optical sensing is generally noncontact and noninvasive, and provides very accurate measurements. In many cases, accuracy can be varied over a wide range. In these sensors, an optical wave is an information sensor and carrier of information.

Any one of the following characteristics of a wave can be modulated by the measured property (the measurand):

- Amplitude or intensity
- Phase
- Polarization
- Frequency
- Direction of propagation

However, the detected quantity is always intensity, as the detectors cannot follow the optical frequency. The measured property modifies the characteristics of the wave in such a way that, on demodulation, a change in intensity results. This change in intensity is related to the measured property. In some measurements, the wave intensity is modulated directly, and hence no demodulation is used before detection.

1.5.1 Amplitude/Intensity-Based Sensors

There are many sensors that measure the change in intensity immediately after the wave propagates through the region of interest. The measurand changes the intensity or attenuates the wave—the simplest case being measurement of opacity or density. In another application, rotation of the plane of polarization of linearly polarized light is measured by invoking Malus's law. The refractive index of a dielectric medium is obtained by monitoring reflected light when a p-polarized wave is incident on the dielectric–air interface. Measurement of absorption coefficients using Beer's law and measurement of very high reflectivities of surfaces are other examples. Fiber-optic sensors based on attenuation are used for force/pressure measurement, displacement measurement, etc. The measurement process may lead to absolute measurements or relative measurements as the need arises.

1.5.2 Sensors Based on Phase Measurement

A variety of measurements are performed using phase measurement over a range of accuracies. A light wave can be modulated, and the phase of the modulated wave with respect to a certain reference can be measured to extract information about the measurand. Alternatively, the phase of the light wave itself can be measured. Phase can be influenced by distance, refractive index, and wavelength of the source. A variety of phase measuring instruments, generally known as interferometers, are available that have accuracies from nanometers to millimeters. Interferometers can also measure derivatives of displacement or, in general, of a measurand. Heterodyne interferometry is used for measuring distance, absolute distance, and very high velocities, among other things. Interferometry can also be performed on real objects without surface treatment, and their response to external perturbations can be monitored.

1.5.3 Sensors Based on Polarization

Malus's law, optical activity, the stress-optic law, Faraday rotation, and so on, all based on changes of polarization by the measurand, have been exploited for the measurement of a number of quantities. Current flowing in a wire is measured using Faraday rotation, and electrically induced birefringence is used for measuring voltage, thereby measurement of power is accomplished with ease. Force is measured using the stress-optic law. Measurement of the angle of rotation of the plane of polarization can be done by application of Malus's law.

1.5.4 Sensors Based on Frequency Measurement

Reflection of light from a moving object results in a shift in the frequency (Doppler shift) of the reflected light. This frequency shift, known as the Doppler shift, is directly related to the velocity of the object. It is measured by heterodyning the received signal with the unshifted light signal. Heterodyne interferometry is used to measure displacement. The Hewlett-Packard interferometer, a commercial instrument, is based on this principle. Flow measurement also utilizes measurements of Doppler shift. Laser–Doppler anemometers/velocimeters (LDA/LDV) are common instruments used to measure all three components of the velocity vector simultaneously. Turbulence can also be monitored with an LDA. Measurement of velocities has also been performed routinely by measuring Doppler shift. However, for very high velocities, the Doppler shift becomes very large, and hence unmanageable by electronics. Using Doppler interferometry, in which Doppler-shifted light is fed into an interferometer, a low-frequency signal is obtained, and this is then related to the velocity.

1.5.5 Sensors Based on Change of Direction

Optical pointers are devices based on change of direction, and can be used to monitor a number of variables, such as displacement, pressure, and temperature. Wholefield optical techniques utilizing this effect include shadowgraphy, schlieren photography, and speckle photography.

Sensing can be performed at a point or over the wholefield. Interferometry, which converts phase variations into intensity variations, is in general a wholefield technique. Interference phenomena is discussed in detail in Chapter 2.

1.6 THE LASER BEAM

A laser beam propagates as a nearly unidirectional wave with little divergence and with finite cross-section. Let us now seek a solution of the Helmholtz equation (Equation 1.3) that represents such a beam. We therefore write a solution in the form

$$\mathbf{E}(\mathbf{r}) = \mathbf{E}_0(\mathbf{r})\exp(ikz) \tag{1.8}$$

This solution differs from a plane wave propagating along the z-direction in that its amplitude $\mathbf{E}_0(\mathbf{r})$ is not constant. Further, the solution should represent a beam, that is it should be unidirectional and of finite cross-section. The variations of $\mathbf{E}_0(\mathbf{r})$ and $\partial\mathbf{E}_0(\mathbf{r})/\partial z$ over a distance of the order of the wavelength along z-direction are assumed to be negligible. This implies that the field varies approximately as e^{ikz} over a distance of a few wavelengths. When the field $\mathbf{E}_0(\mathbf{r})$ satisfies these conditions, the Helmholtz equation takes the form

$$\nabla_{\mathrm{T}}^2\mathbf{E}_0 + 2ik\frac{\partial\mathbf{E}_0}{\partial z} = 0 \tag{1.9}$$

where ∇_{T}^2 is the transverse Laplacian. Equation 1.9 is known as the paraxial wave equation, and is a consequence of the weak factorization represented by Equation 1.8 and other assumptions mentioned earlier.

1.7 THE GAUSSIAN BEAM

The intensity distribution in a beam from a laser oscillating in the TEM_{00} mode is given by

$$I(r) = I_0\exp(-2r^2/w^2) \tag{1.10}$$

where $r = (x^2+y^2)^{1/2}$. The intensity $I(r)$ drops to $1/e^2$ of the value of the peak intensity I_0 at a distance $r = w$. The parameter w is called the spot size of the beam. w depends on the z-coordinate. The intensity profile is Gaussian; hence the beam is called a Gaussian beam. Such a beam maintains its profile as it propagates. The Gaussian beam has a very special place in optics, and is often used for measurement.

The solution of Equation 1.9 that represents the Gaussian beam can be written as

$$\begin{aligned} E(r,z) &= A\frac{\exp[-r^2/w^2(z)]}{[1+(\lambda z/\pi w_0^2)^2]^{1/2}}\exp\left[\frac{ikr^2}{2R(z)}\right]\exp\left[-i\tan^{-1}\left(\frac{\lambda z}{\pi w_0^2}\right)\right]\exp(ikz)\\ &= A\frac{w_0}{w(z)}\exp\left[-\frac{r^2}{w^2(z)}\right]\exp\left[\frac{ikr^2}{2R(z)}\right]\exp\left[-i\tan^{-1}\left(\frac{\lambda z}{\pi w_0^2}\right)\right]\exp(ikz)\\ &= A\frac{w_0}{w(z)}\exp\left\{-r^2\left[\frac{1}{w^2(z)}-\frac{ik}{2R(z)}\right]\right\}\exp[i(kz-\varphi)] \end{aligned} \tag{1.11}$$

where A and w_0 are constants and $\varphi = \tan^{-1}(\lambda z/w_0^2)$. The constant w_0 is called the beam waist size and $w(z)$ is called the spot size. We introduce another constant, called the Rayleigh range z_0, which is related to the beam waist size by $z_0 = \pi w_0^2/\lambda$. The radius of curvature $R(z)$ and spot size $w(z)$ of the beam at an arbitrary z-plane are given by

$$R(z) = z\left[1+\left(\frac{\pi w_0^2}{\lambda z}\right)^2\right] = z\left[1+\left(\frac{z_0}{z}\right)^2\right] \tag{1.12a}$$

$$w(z) = w_0 \left[1 + \left(\frac{\lambda z}{\pi w_0^2}\right)^2\right]^{1/2} = w_0 \left[1 + \left(\frac{z}{z_0}\right)^2\right]^{1/2} \tag{1.12b}$$

The amplitude of the wave is expressed as

$$A \frac{w_0}{w(z)} \exp\left[-\frac{r^2}{w^2(z)}\right] \tag{1.13}$$

The amplitude decreases as $\exp[-r^2/w^2(z)]$ (Gaussian beam). The spot size $w(z)$ increases as the beam propagates, and consequently the amplitude on axis decreases. In other words, the Gaussian distribution flattens. The spot size $w(z)$ at very large distances is

$$w(z)\Big|_{z\to\infty} = \frac{\lambda z}{\pi w_0} \tag{1.14}$$

The half divergence angle θ is obtained as

$$\theta = \frac{dw}{dz} = \frac{\lambda}{\pi w_0} \tag{1.15}$$

The exponential factor $\exp[ikr^2/2R(z)]$ in Equation 1.11 is the radial phase factor, with $R(z)$ as the radius of curvature of the beam. At large distances, that is, $z \gg z_0, R(z) \to z$, the beam appears to originate at $z = 0$. However, the wavefront is plane at the $z = 0$ plane. In general, a $z =$ constant plane is not an equiphasal plane. The second exponential factor in Equation 1.11 represents the dependence on the propagation phase, which comprises the phase of a plane wave, kz, and Guoy's phase φ. These can be summarized for near-field and far-field situations as follows. For the near field

- Phase fronts are nearly planar (collimated beam).
- The beam radius is nearly constant.
- The wave is a plane wave with Gaussian envelope.
- For $z \ll z_0, w(z) \to w_0, R(z) \to \infty$, and $\tan^{-1}(z/z_0) = 0$.
- There is no radial phase variation if $R \to \infty$.

For the far field

- The curvature increases linearly with z.
- The beam radius increases linearly with z.
- For $z \gg z_0, \omega(z) = \lambda z/\pi\omega_0, R(z) = z$, and $\tan^{-1}(z/z_0) \approx \pi/2$.

Figure 1.2 shows the variation of the amplitude as the beam propagates.

The intensity distribution in a Gaussian beam is

$$I(r,z) = |A|^2 \frac{w_0^2}{w^2(z)} \exp\left[-\frac{2r^2}{w^2(z)}\right] \tag{1.16}$$

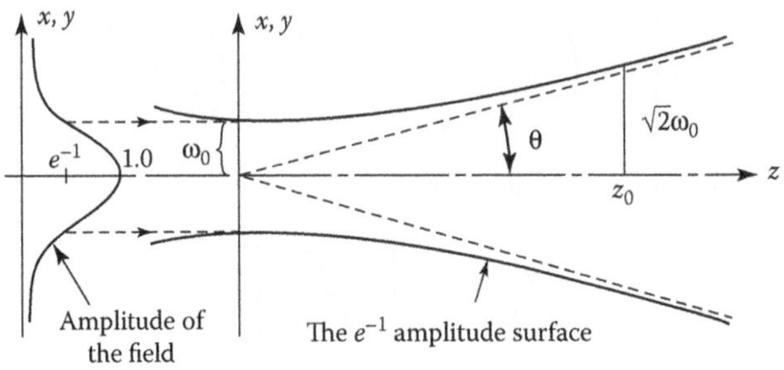

FIGURE 1.2 A Gaussian beam.

On the beam axis ($r = 0$), the intensity varies with distance z as follows:

$$I(0, z) = I_0 \frac{w_0^2}{w^2(z)} = \frac{I_0}{1 + (z/z_0)^2} \tag{1.17}$$

It has its peak value I_0 at $z = 0$, and keeps decreasing with increasing z. The value drops to 50% of the peak value at $z = \pm z_0$.

1.8 *ABCD* MATRIX APPLIED TO GAUSSIAN BEAMS

The propagation of a ray through optical elements under the paraxial approximation (linear optics) can be described by 2 × 2 matrices. The height of the output ray and its slope are related to the input ray and its slope through a 2 × 2 matrix characteristic of the optical element. This is described by

$$\begin{bmatrix} r_2 \\ r_2' \end{bmatrix} = \begin{bmatrix} A & B \\ C & D \end{bmatrix} \begin{bmatrix} r_1 \\ r_1' \end{bmatrix} \tag{1.18}$$

or

$$r_2 = Ar_1 + Br_1'$$
$$r_2' = Cr_1 + Dr_1'$$

where r_2 and r_2' are the height and slope of the output ray. Consider a ray emanating from a point source, the wavefront being spherical. Its radius of curvature $R_2 (= r_2/r_2')$ at the output side is related to the radius of curvature at the input side as follows:

$$R_2 = \frac{r_2}{r_2'} = \frac{Ar_1 + Br_1'}{Cr_1 + Dr_1'} = \frac{AR_1 + B}{CR_1 + D}. \tag{1.19}$$

The propagation of a spherical wave can easily be handled by this approach. It is also known that the propagation of Gaussian beams can also be described by these matrices. Since in any measurement process, a lens is an important component of the experimental setup, we shall discuss how the propagation of a Gaussian beam through a lens can be handled by a 2×2 matrix.

ABCD matrices for free space and many optical elements are given in standard textbooks. For a thin lens of focal length f, the matrix is

$$\begin{bmatrix} 1 & 0 \\ -f^{-1} & 1 \end{bmatrix}$$

The amplitude of a Gaussian beam as described by Equation 1.11 can be rewritten as

$$E(r,z) = A(z)\exp[i(kz - \varphi)]\exp\left(i\frac{k}{2}\right)(x^2 + y^2)\left[\frac{1}{R(z)} - \frac{i\lambda}{\pi\omega^2(z)}\right] \tag{1.20}$$

This expression is similar to the paraxial approximation to the spherical wave, with the radius of curvature R replaced by a complex Gaussian wave parameter q, given by

$$\frac{1}{q} = \frac{1}{R} - \frac{i\lambda}{\pi\omega^2} \tag{1.21}$$

A Gaussian beam is described completely by w, R, and λ. These are contained in the complex parameter q, which transforms as follows:

$$q_2 = \frac{Aq_1 + B}{Cq_1 + D}. \tag{1.22}$$

The proof of this equality is rather tedious, but, in analogy with the case of a spherical wave, we may accept that it is correct. We shall now study the propagation of a Gaussian beam through free space and then through a lens.

1.8.1 Propagation in Free Space

The *ABCD* matrix for free space is $\begin{bmatrix} 1 & z \\ 0 & 1 \end{bmatrix}$, where z is the propagation distance. Therefore, $q_2 = q_1 + z$. In order to study the propagation of a Gaussian beam in free space, we will assume that the beam waist lies at $z = 0$ where the spot size and radius of curvature are w_0 and $R_0 = \infty$, respectively. At this plane, $q_0 = i\pi w_0^2/\lambda$; the complex parameter at the waist is purely imaginary. Therefore, the complex parameter at any plane distant z from the beam waist is $q(z) = q_0 + z$. Thus,

$$\frac{1}{q(z)} = \frac{1}{R(z)} - \frac{i\lambda}{\pi w^2(z)} = \frac{1}{q_0 + z} = \frac{1}{z + \dfrac{i\pi w_0^2}{\lambda}}$$

$$= \frac{z}{z + \left(\dfrac{\pi w_0^2}{\lambda}\right)^2} - \frac{i\dfrac{\pi w_0^2}{\lambda}}{z^2 + \left(\dfrac{\pi w_0^2}{\lambda}\right)^2}$$

It can be seen that although the Gaussian parameter q was purely imaginary at the beam waist, it has now become complex at any plane z and has both real and imaginary components. On equating real and imaginary parts, we obtain, after some simplification,

$$R(z) = z\left[1 + \left(\frac{\pi w_0^2}{\lambda z}\right)^2\right] \quad (1.23a)$$

$$w(z) = w_0\left[1 + \left(\frac{\lambda z}{\pi w_0^2}\right)^2\right]^{1/2} \quad (1.23b)$$

These are the same expressions for the radius of curvature and spot size as were obtained earlier.

1.8.2 Propagation through a Thin Lens

Consider a Gaussian beam incident on a thin lens of focal length f. We will assume that the beam waists w_1 and w_2 of the incident and transmitted beams lie at distances d_1 and d_2 from the lens, respectively, as shown in Figure 1.3. The $ABCD$ matrix for the light to propagate from the plane at d_1 to the plane at d_2 is given by

$$\begin{bmatrix} A & B \\ C & D \end{bmatrix} = \begin{bmatrix} 1 & d_2 \\ 0 & 1 \end{bmatrix}\begin{bmatrix} 1 & 0 \\ -f^{-1} & 1 \end{bmatrix}\begin{bmatrix} 1 & d_1 \\ 0 & 1 \end{bmatrix} = \begin{bmatrix} 1 - \dfrac{d_2}{f} & d_1 + d_2 - \dfrac{d_1 d_2}{f} \\ -\dfrac{1}{f} & 1 - \dfrac{d_1}{f} \end{bmatrix} \quad (1.24)$$

Using the relation given in Equation 1.22, we obtain

$$q_2 = \frac{Aq_1 + B}{Cq_1 + D} = \frac{\left(1 - \dfrac{d_2}{f}\right)q_1 + \left(d_1 + d_2 - \dfrac{d_1 d_2}{f}\right)}{\left(-\dfrac{1}{f}\right)q_1 + \left(1 - \dfrac{d_1}{f}\right)} \quad (1.25)$$

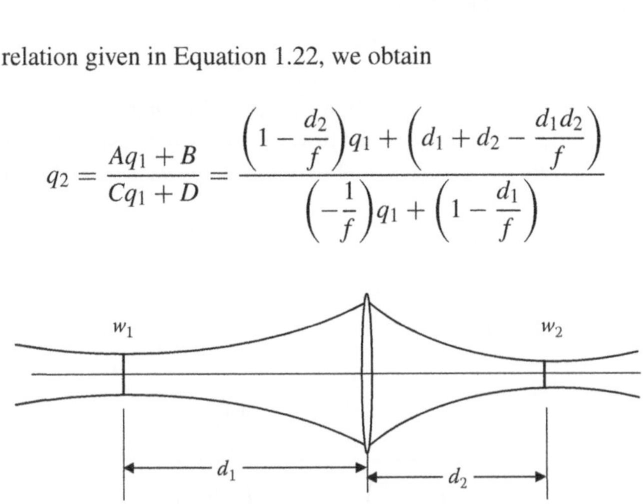

FIGURE 1.3 Propagation of a Gaussian beam through a lens.

The complex parameter q_1 is purely imaginary at the beam waist, and hence $q_1 = i\pi w_1^2/\lambda$. Therefore, q_2 can be expressed as follows:

$$q_2 = \frac{\left(1-\frac{d_2}{f}\right)(i\pi w_1^2/\lambda)+\left(d_1+d_2-\frac{d_1 d_2}{f}\right)}{\left(-\frac{1}{f}\right)(i\pi w_1^2/\lambda)+\left(1-\frac{d_1}{f}\right)}$$

Alternatively,

$$\frac{1}{q_2} = \frac{\left(-\frac{1}{f}\right)(i\pi w_1^2/\lambda)+\left(1-\frac{d_1}{f}\right)}{\left(1-\frac{d_2}{f}\right)(i\pi w_1^2/\lambda)+\left(d_1+d_2-\frac{d_1 d_2}{f}\right)} = \frac{1}{R_2} - \frac{i\lambda}{\pi w_2^2} \tag{1.26}$$

Equating real and imaginary parts, we obtain

$$\frac{1}{R_2} = \frac{(f-d_1)\left[(d_1+d_2)f - d_1 d_2\right] - (f-d_2)\left(\pi w_1^2/\lambda\right)^2}{\left[(d_1+d_2)f-d_1 d_2\right]^2 + (f-d_2)^2\left(\pi w_1^2/\lambda\right)^2} \tag{1.27a}$$

$$\frac{\lambda}{\pi w_2^2} = \frac{\pi w_1^2}{\lambda}\frac{f^2}{\left[(d_1+d_2)f-d_1 d_2\right]^2 + (f-d_2)^2\left(\pi w_1^2/\lambda\right)^2} \tag{1.27b}$$

We now consider that the beam waist lies at a distance d_2, implying that w_2 is purely imaginary, and hence $R_2 = \infty$. This occurs when

$$(f-d_1)\left[(d_1+d_2)f-d_1 d_2\right] - (f-d_2)\left(\pi w_1^2/\lambda\right)^2 = 0$$

In order to determine the location of the beam waist, we solve this equation for d_2 as follows:

$$d_2 = \frac{f + \left[d_1 f^{(d_1-f)}/z_1^2\right]}{1+\left[(d_1-f)/z_1\right]^2} \tag{1.28a}$$

where $z_1 = \pi w_1^2/\lambda$.

From Equation 1.27b, we obtain

$$\frac{w_2^2}{w_1^2} = \frac{(f/z_1)^2}{1+\left[(d_1-f)/z_1\right]^2} = \frac{d_2-f}{d_1-f} \tag{1.28b}$$

As a special case, when the beam waist w_1 lies at the lens plane, that is, $d_1 = 0$, Equations 1.28a and 1.28b can be expressed as

$$d_2 = \frac{f}{1+(f/z_1)^2} \tag{1.29a}$$

$$\frac{w_2}{w_1} = \frac{f/z_1}{[1+(f/z_1)^2]^{1/2}} \tag{1.29b}$$

1.8.3 Mode Matching

We assume that the beam waists lie at distances d_1 and d_2 from the lens. At the beam waists, the complex parameters are purely imaginary, and hence $q_1 = i\pi w_1^2/\lambda$ and $q_2 = i\pi w_2^2/\lambda$. Then Equation 1.26 can be rewritten as follows:

$$\frac{i\pi w_2^2}{\lambda} = \frac{\left(1-\frac{d_2}{f}\right)(i\pi w_1^2/\lambda) + f\left(d_1 + d_2 - \frac{d_1 d_2}{f}\right)}{\left(-\frac{1}{f}\right)(i\pi w_1^2/\lambda) + \left(1-\frac{d_1}{f}\right)} \tag{1.26a}$$

On equating real and imaginary parts, we obtain

$$(d_1 - f)(d_2 - f) = f^2 - f_g^2 \tag{1.30a}$$

$$\frac{d_2 - f}{d_1 - f} = \frac{w_2^2}{w_1^2} \tag{1.30b}$$

where $f_g = \pi w_1 w_2/\lambda$ is defined by the products of the waists. The right-hand side of Equation 1.30b is always positive, and hence $d_1 - f$ and $d_2 - f$ have to be either both positive or both negative. Hence, $(d_1 - f)(d_2 - f)$ is always positive, and so $f > f_g$. From Equations 1.30a and 1.30b, we can obtain the following two relations:

$$d_1 = f \pm \frac{w_1}{w_2}(f^2 - f_g^2)^{1/2} \tag{1.31a}$$

$$d_2 = f \pm \frac{w_2}{w_1}(f^2 - f_g^2)^{1/2} \tag{1.31b}$$

Propagation of a Gaussian beam in the presence of other optical elements can be studied in a similar way.

1.9 NONDIFFRACTING BEAMS—BESSEL BEAMS

There is a class of beams that propagate over a certain distance without diffraction. These beams can be generated using either axicons or holograms.

Diffractionless solutions of the Helmholtz wave equation in the form of Bessel functions constitute Bessel beams. A Bessel beam propagates over a considerable distance without diffraction. It also has a "healing" property in that it exhibits self-reconstruction after encountering an obstacle. A true Bessel beam, being unbounded, cannot be created. However, there are several practical ways to create a beam that is a close approximation to a true Bessel beam, including diffraction at an annular aperture, focusing by an axicon, and the use of diffractive elements (holograms). Figure 1.4 shows a schematic for producing Bessel beam using an axicon. A Gaussian beam is incident on the axicon, which produces two plane waves in conical geometry. These plane waves interfere to compensate for the spread of the beam. A Bessel beam exists in the region of interference.

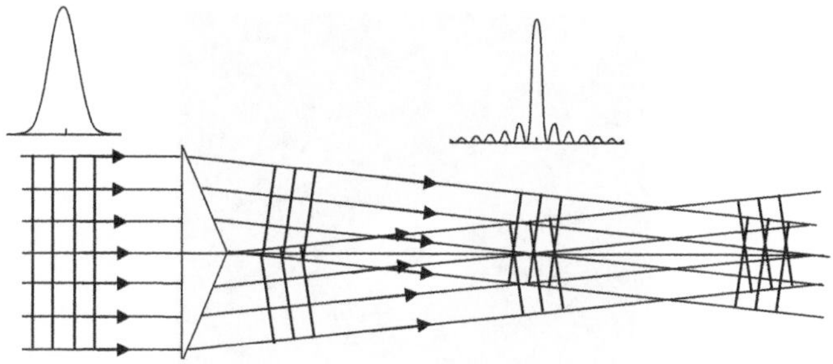

FIGURE 1.4 Generation of a Bessel beam with an axicon.

On the other hand, a hologram with transmittance function $t(\rho, \phi)$ can generate Bessel beams. The transmittance function is given by

$$t(\rho, \phi) = \begin{cases} A(\phi)\exp(-2\pi i\rho/\rho_0), & \rho \leq D \\ 0, & \rho > D \end{cases} \tag{1.32}$$

where D is the size of the hologram. The diffraction pattern of the Bessel beam shows a very strong central spot surrounded by large number of rings. The rings can be suppressed. However, the intensity in the central spot is weaker than that of a Gaussian beam of comparable size.

Only first-order Bessel beams can be generated with an axicon, while a hologram generates beams of several orders.

Because of several interesting properties of Bessel beams, they find applications in several areas of optical research, such as optical trapping and micromanipulation.

1.10 SINGULAR BEAMS

An optical beam possessing isolated zero-amplitude points with indeterminate phase and helical phase structure is called a singular beam; such points are called singular points. Both real and imaginary parts of the complex amplitude are simultaneously zero at a singular point. A helical phase front of the singular beam is described by the wave field $\exp(il\varphi)$, where l is the topological charge. The value of the topological charge of the vortex field determines the total phase that the wavefront accumulates in one complete rotation around the singular point. The sign of l determines the sense of helicity of the singular beam. The phase gradient of a wave field with phase singularity is nonconservative in nature. The line integral of the phase gradient over any closed path surrounding the point of singularity is nonzero, that is, $\oint \nabla\phi \cdot dl = m2\pi$. At an isolated point where a phase singularity occurs, the phase is uncertain and the amplitude is zero, thereby creating an intensity null at the point of singularity. The intensity profile of a phase-singular beam with unit topological charge ($l = 1$) is shown in Figure 1.5. It is now well established that such beams possess orbital angular momentum equal to $l\hbar$ per photon, where $\hbar$ is Planck's constant divided by 2π.

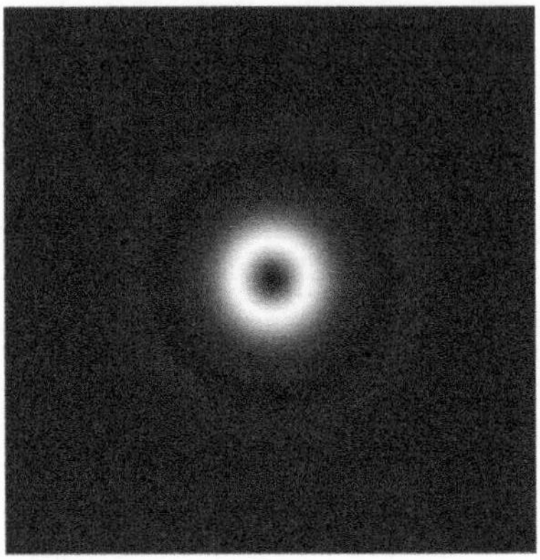

FIGURE 1.5 Intensity profile of a phase-singular beam with topological charge $l = 1$. (Courtesy of D. P. Ghai.)

According to terminology introduced by Nye and Berry, wavefront dislocation in a monochromatic wave can be divided into three main categories: (i) screw dislocation; (ii) edge dislocation; and (iii) mixed screw edge dislocation. Screw dislocation, also called optical vortex, is the most common type of phase defect. The equiphase surfaces of an optical field exhibiting screw dislocation are helicoids or groups of helicoids, nested around the dislocation axis. One round trip on the continuous surface around the dislocation axis will lead to the next coil with a pitch $l\lambda$, where l is the topological charge and λ is the wavelength of operation. Today, "screw-type wavefront dislocation" has become synonymous with "phase singularity." The laser TEM_{01}^{*} mode, also called the doughnut mode, is the most common example of a screw dislocation of topological charge ± 1.

Properties of vortices have been studied in both linear and nonlinear regimes. It has been shown that a unit-charge optical vortex is structurally stable. On the other hand, multiple-charge optical vortices are structurally unstable. Optical vortices with the same topological charge rotate about each other on propagation, with the rotation rate depending on the inverse square of the separation between the two vortices. In contrast, vortices with equal magnitude but opposite polarity of their topological charges drift away from each other. Vortices with small cores, called vortex filaments, exhibit fluid-like rotation similar to vortices in liquids. An optical vortex propagating in a nonlinear medium generates a soliton. The optical vortex soliton is a three-dimensional, robust spatial structure that propagates without changing its size. Optical vortex solitons are formed in a self-defocusing medium when the effects of diffraction are offset by the refractive index variations in the nonlinear medium.

One of the possible solutions of the wave equation, as suggested by Nye and Berry, is

$$E(r, \phi, z, t) \propto \begin{pmatrix} r^{|l|} \exp(il\phi + ikz - iwt) \\ r^{-|l|} \exp(il\phi + ikz - iwt) \end{pmatrix} \tag{1.33}$$

where $E(r, \phi, z, t)$ is the complex amplitude of a monochromatic light wave with frequency w and wavelength λ propagating along the z axis, and $k = 2\pi/\lambda$ is the propagation vector. Both solutions have azimuthal phase dependence. As the wave propagates, the equiphase surfaces (wavefront) trace a helicoid given by

$$l\phi + kz = \text{constant} \tag{1.34}$$

After one round trip of the wavefront around the z axis, there is a continuous transition into the next wavefront sheet, separated by $l\lambda$, which results in a continuous helicoidal wave surface. The sign of the topological charge is positive or negative, depending upon whether the wave surface has right- or left-handed helicity.

The solutions given by Equation 1.33 do not describe a real wave, because of the radial dependence of the wave amplitude. Therefore, it is necessary to take an appropriate wave as the host beam. We may consider, for example, a phase-singular beam embedded in a Gaussian beam. Using the paraxial approximation to the scalar wave equation,

$$\frac{1}{r}\frac{\partial}{\partial r}\left(r\frac{\partial E}{\partial r}\right)\frac{1}{r^2} + \frac{1}{r^2}\frac{\partial^2 E}{\partial r^2} + 2ik\frac{\partial E}{\partial z} = 0 \tag{1.35}$$

we obtain the solution of a vortex Gaussian envelope as

$$E(r,\phi,z) = E_0\left(\frac{r}{w_0}\right)^l \left\{\frac{\frac{1}{2}kw_0^2}{\left[z^2 + \left(\frac{1}{2}kw_0^2\right)^2\right]^{1/2}}\right\}^{l+1}$$

$$\times \exp\left[-r^2\frac{\frac{1}{4}k^2w_0^4}{z^2 + \left(\frac{1}{2}kw_0^2\right)^2}\right]\exp[i\Phi(r,\phi,z)] \tag{1.36}$$

where E_0 is the real amplitude and w_0 is the beam waist. The phase Φ is given by

$$\Phi(r,\phi,z) = (|l|+1)\tan^{-1}\left(\frac{z}{\frac{1}{2}kw_0^2}\right) - \frac{kr^2}{2z + k^2w_0^2/2z} - l\phi - kz \tag{1.37}$$

Optical vortices find many applications in interferometry, such as in the study of fractally rough surfaces, beam collimation testing, optical vortex metrology, and microscopy. An optical vortex interferometer employing three-wave interference can be used for tilt and displacement measurement, wavefront reconstruction, 3D scanning, and super-resolution microscopy. Spiral interferometry, where a spiral phase element is used as a spatial filter, removes the ambiguity between elevation and depression in the surface height of a test object. Lateral shear interferometers have been used to study gradients of phase-singular beams. It has been shown that shearograms do not represent true phase gradients when vortices are presents in the optical fields. However, lateral shear interferometry is a potential technique for the detection of both an isolated vortex and randomly distributed vortices in a speckle field. It is a simple, robust, and self-referencing technique, which is insensitive to vibrations. Unlike the existing interferometric techniques of vortex detection, it does not require any high-quality plane or spherical reference wavefront to form an interference pattern.

Vortex generation using interferometric methods is based on the superposition of linearly varying phase distributions in the plane of observation that arise from interference of three, four, or more plane or spherical waves. Interferometric techniques have been used universally to study the vortex structure of singular beams in both linear and nonlinear media. Depending on whether a plane or a spherical wave interferes with the singular beam, fork-type (bifurcation of fringes at the vortex points) or spiral-type fringes are observed in the interference pattern, as shown in Figure 1.6. Fork fringes, which are a signature of optical vortices, give information about both the magnitude and the sign of the topological charge.

Important methods of vortex generation make use of wedge plates, spatial light modulators, laser-etched mirrors, and adaptive mirrors. Most common techniques use interference of three or four plane waves for generating vortex arrays. Computer-generated holograms (CGHs) are now widely used for the generation of phase-singular beams. Spiral phase plates (SPPs) have been used for the generation of phase singularities for millimeter waves. A random distribution of optical vortices is also observed in the speckle field resulting from laser light scattered from a diffuse surface or transmitted through a multimode fiber. There is a finite possibility of the presence of an optical vortex in any given patch of a laser speckle field. On average, there is one vortex per speckle. There is little probability of the presence of optical vortices of topological charge greater than one in a speckle field.

The magnitude and sign of the topological charge of the vortex beam can be determined from the interference pattern of the beam with a plane or a spherical reference wave. The interference of a vortex beam of finite curvature with a spherical reference wave gives spiral fringes. The number of fringes originating from the center

(a) (b)

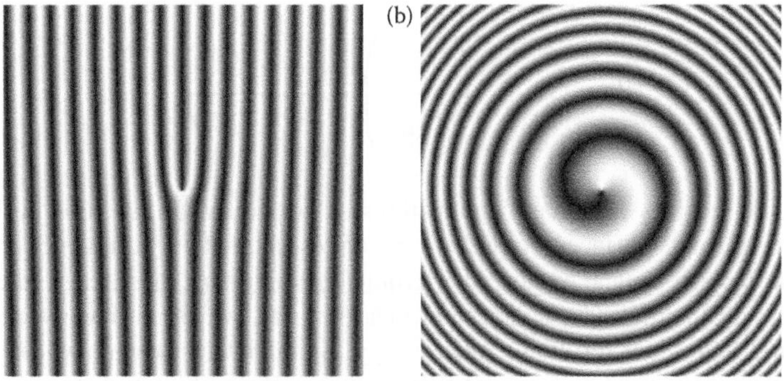

FIGURE 1.6 Simulated interferograms between (a) a tilted plane wave and singular beam; (b) a spherical wave and a singular beam of topological charge $l = 1$. (Courtesy of D. P. Ghai.)

equals the magnitude of the topological charge of the vortex beam. The circulation of the fringes depends upon the sign of the topological charge and the relation between the radius of curvature of the vortex and the reference wave.

Referring to Equation 1.37, the transverse phase dependence of the vortex beam is given by

$$\Phi(r,\phi) = -\frac{kr^2}{2R(z)} - l\phi \tag{1.38}$$

where $R\,(z)$ is the radius of curvature of the phase-singular beam. The condition for fringe maxima in the interference pattern is

$$\cos\left[-\frac{kr^2}{2R(z)} + \frac{kr^2}{2R(0)} - l\phi\right] = 1 \tag{1.39}$$

that is,

$$\left(kr^2\frac{R(z)-R(0)}{2R(z)R(0)} - l\phi\right) = 2\pi l \tag{1.40}$$

When the radius of curvature of the reference beam is larger than that of the vortex beam, that is, $R_0 > R(z)$, clockwise rotation of fringes corresponds to a positive value of the topological charge and anticlockwise rotation to a negative value. For $R_0 < R(z)$, clockwise rotation corresponds to a negative value of the topological charge and anticlockwise rotation to a positive value. When $R_0 = R(z)$, instead of spiral fringes, cross-type fringes are observed.

BIBLIOGRAPHY

1. W. T. Welford, *Geometrical Optics*, North-Holland, Amsterdam, 1962.
2. W. J. Smith, *Modern Optical Engineering*, McGraw-Hill, New York, 1966.
3. R. S. Longhurst, *Geometrical and Physical Optics*, Longmans, London, 1967.
4. F. A. Jenkins and H. E. White, *Fundamentals of Optics*, McGraw-Hill, New York, 1976.
5. A. E. Siegman, *Lasers*, University Science Books, California, 1986.
6. M. Born and E. Wolf, *Principles of Optics*, 7th edn, Cambridge University Press, Cambridge, 1999.
7. O. Svelto, *Principles of Lasers*, Plenum Press, New York, 1989.
8. R. S. Sirohi, *Wave Optics and Applications*, Orient Longmans, Hyderabad, 1993.
9. E. Hecht and A. R. Ganesan, *Optics*, 4th edn, Dorling Kindersley, Delhi, 2008.

ADDITIONAL READING

1. H. Kogelnik and T. Li, Laser beams and resonators, *Proc. IEEE*, 54, 1312–1392, 1966.
2. J. F. Nye and M. V. Berry, Dislocations in wave trains, *Proc. R. Soc. London Ser. A*, 336, 165–190, 1974.
3. J. M. Vaughan and D. V. Willetts, Interference properties of a light beam having a helical wave surface, *Opt. Commun.*, 30, 263–267, 1979.
4. J Durnin, Exact solutions for nondiffracting beams. I. The scalar theory, *J. Opt. Soc. Am.*, 4, 651–654, 1987.

5. P. Coullet, L. Gil, and F. Rocca, Optical vortices, *Opt. Commun.*, 73, 403–408, 1989.
6. V. Yu Bazhenov, M. S. Soskin, and M. V. Vasnetsov, Screw dislocations in light wavefronts, *J. Mod. Opt.*, 39, 985–990, 1992.
7. M. Harris, C. A. Hill, and J. M. Vaughan, Optical helices and spiral interference fringes, *Opt. Commun.*, 106, 161–166, 1994.
8. M. Padgett and L. Allen, Light with a twist in its tail, *Contemp. Phys.*, 41, 275–285, 2000.
9. P. Muys and E. Vandamme, Direct generation of Bessel beams, *Appl. Opt.*, 41, 6375–6379, 2002.
10. V. Garcés-Chávez, D. McGloin, H. Melville, W. Sibbett, and K. Dholakia, Simultaneous micromanipulation in multiple planes using a self-reconstructing light beam, *Nature* 419, 145–147, 2002.
11. Miguel A. Bandres, Julio C. Gutiérrez-Vega, and S. Chávez-Cerda, Parabolic non-diffracting optical wavefields, *Opt. Lett.*, 29, 44–46, 2004.
12. Carlos López-Mariscal, Julio C. Gutiérrez-Vega, and S. Chávez-Cerda, Production of high-order Bessel beams with a Mach–Zehnder interferometer, *Appl. Opt.*, 43, 5060–5063, 2004.
13. D. McGloin and K. Dholakia, Bessel beams: Diffraction in a new light, *Contemp. Phys.*, 46, 15–28, 2005.
14. S. Fürhapter, A. Jesacher, S. Bernet, and M. Ritsch-Marte, Spiral interferometry, *Opt. Lett.*, 30, 1953–1955, 2005.
15. A. Jesacher, S. Fürhapter, S. Bernet, and M. Ritsch-Marte, Spiral interferogram analysis, *J. Opt. Soc. Am. A*, 23, 1400–1408, 2006.
16. Carlos López-Mariscal and Julio C. Gutiérrez-Vega, The generation of nondiffracting beams using inexpensive computer-generated holograms, *Am. J. Phys*, 75, 36–42, 2007.
17. D. P. Ghai, S. Vyas, P. Senthilkumaran, and R. S. Sirohi, Detection of phase singularity using a lateral shear interferometer, *Opt. Lasers Eng.*, 46, 419–423, 2008.

2 Optical Interference

2.1 INTRODUCTION

Light waves are electromagnetic waves. They can be spherical, cylindrical, or plane, as has been explained in Chapter 1. By modulating wave properties such as amplitude, phase, and polarization, it is possible for light waves to carry information. In a homogenous, isotropic medium, there is usually no interaction between the wave and the medium. However, at very high intensities, some nonlinear effects begin to come into play. Our interest for now is to study the superposition of two or more waves. If the waves are traveling in the same direction, they will have a very large region of superposition. In contrast, if the angle between their directions of propagation is large, the region of superposition will be small. As the angle between the two waves increases, the region of superposition continues to decrease, and reaches a minimum when they propagate at right angles to each other. As the angle is further increased, the region of superposition also increases, reaching a maximum when they are antipropagating.

Let us consider the superposition of two coherent monochromatic waves. The amplitude distribution of these waves can be written as follows:

$$\mathbf{E}_1(x, y, z; t) = \mathbf{E}_{01}(x, y, z)\exp[i(\omega t + \mathbf{k}_1 \cdot \mathbf{r} + \delta_1)] \tag{2.1}$$

$$\mathbf{E}_2(x, y, z; t) = \mathbf{E}_{02}(x, y, z)\exp[i(\omega t + \mathbf{k}_2 \cdot \mathbf{r} + \delta_2)] \tag{2.2}$$

The total amplitude is the sum of the amplitudes of the individual waves. This can be expressed as follows:

$$\mathbf{E}(x, y, z; t) = \mathbf{E}_1(x, y, z; t) + \mathbf{E}_2(x, y, z; t) \tag{2.3}$$

The intensity distribution at a point (x, y, z) can be obtained as follows:

$$\begin{aligned} I(x, y, z) &= \langle \mathbf{E}(x, y, z; t) \cdot \mathbf{E}(x, y, z; t) \rangle \\ &= \mathbf{E}_{01}^2 + \mathbf{E}_{02}^2 + 2\mathbf{E}_{01} \cdot \mathbf{E}_{02} \cos[(\mathbf{k}_2 - \mathbf{k}_1) \cdot \mathbf{r} + \phi] \end{aligned} \tag{2.4}$$

Here $(\mathbf{k}_2 - \mathbf{k}_1) \cdot \mathbf{r} + \phi$ is the phase difference between the waves and $\phi = \delta_2 - \delta_1$ is the initial phase difference, which can be set equal to zero. It is seen that the intensity at a point depends on the phase difference between the waves. There does exist a varying intensity distribution in space, which is a consequence of superposition and hence interference of waves. When taken on a plane, this intensity pattern is known as an interference pattern. In standard books on optics, the conditions for obtaining a stationary interference pattern are as follows:

1. Waves must be monochromatic and of the same frequency.
2. They must be coherent.
3. They should enclose a small angle between them.
4. They must have the same state of polarization.

These conditions were stated at a time when fast detectors were not available and to realize a monochromatic wave was extremely difficult. It is therefore worth commenting that interference does take place when two monochromatic waves of slightly different frequencies are superposed. However, this interference pattern is not stationary but moving: heterodyne interferometry is an example. In order to observe interference, the waves should have time-independent phases. In the light regime, therefore, these waves are derived from the same source either by amplitude division or wavefront division. This is an important condition for observing a stable interference pattern. It is not necessary that these waves enclose a small angle between them. Previously, when the observations were made visually with the naked eye or under low magnification, the fringe width had to be large to be resolved. However, with the availability of high-resolution recording media, an interference pattern can be recorded even between oppositely traveling waves (with the enclosed angle approaching 180°). Further interference between two linearly polarized waves is observed except when they are orthogonally polarized, albeit with reduced contrast.

2.2 GENERATION OF COHERENT WAVES

There are only two methods that are currently used for obtaining two or more coherent beams from the parent beam: (1) division of wavefront and (2) division of amplitude.

2.2.1 Interference by Division of Wavefront

Two or more portions of the parent wavefront are created and then superposed to obtain an interference pattern. Simple devices such as the Fresnel biprism, the Fresnel bi-mirror, and the split lens sample the incident wavefront and superpose the sampled portions in space, where the interference pattern is observed. This is the well-known two-beam interference. A simple method that has recently been used employs an aperture plate, and samples two or more portions of the wavefront. If the sampling aperture is very fine, diffraction will result in the superposition of waves at some distance from the aperture plane. The Smartt interferometer and the Rayleigh interferometer are examples of interference by wavefront division. The use of multiple apertures such as a grating gives rise to multiple-beam interference based on wavefront division.

2.2.2 Interference by Division of Amplitude

A wave incident on a plane parallel or wedge plate is split into a transmitted and a reflected wave, which can be superposed by judicious application of mirrors, resulting in various kinds of interferometers. Michelson and Mach–Zehnder interferometers are two well-known examples. All polarization-based interferometers use division of amplitude. The Fabry–Perot interferometer, the Fabry–Perot etalon, and the Lummer–Gerchke plate are examples of multiple-beam interference by division of amplitude.

2.3 INTERFERENCE BETWEEN TWO PLANE MONOCHROMATIC WAVES

As seen from Equation 2.4, the intensity distribution in the interference pattern when two monochromatic plane waves are propagating in the directions of propagation vectors $\mathbf{k}_1$ and $\mathbf{k}_2$, respectively, is given by

$$I(x, y, z) = \mathbf{E}_{01}^2 + \mathbf{E}_{02}^2 + 2\mathbf{E}_{01} \cdot \mathbf{E}_{02} \cos[(\mathbf{k}_2 - \mathbf{k}_1) \cdot \mathbf{r} + \phi] \tag{2.5}$$

This equation can also be written as

$$I(x, y, z) = I_1 + I_2 + 2\mathbf{E}_{01} \cdot \mathbf{E}_{02} \cos[(\mathbf{k}_2 - \mathbf{k}_1) \cdot \mathbf{r} + \phi] \tag{2.6}$$

where I_1 and I_2 are the intensities of the individual beams. It is thus seen that the resultant intensity varies in space as the phase changes. This variation in intensity is a consequence of the interference between two beams. The intensity will be maximum wherever the argument of the cosine function takes values equal to $2m\pi$ $(m = 0, 1, 2, \ldots)$ and will be minimum wherever the phase difference is $(2m + 1)\pi$. The loci of constant phase $2m\pi$ or $(2m + 1)\pi$ are bright or dark fringes, respectively. The phase difference between two bright fringes or dark fringes is always 2π. Following Michelson, we define the contrast of the fringes as follows:

$$C = \frac{I_{max} - I_{min}}{I_{max} + I_{min}} = \frac{2\mathbf{E}_{01} \cdot \mathbf{E}_{02}}{I_1 + I_2} \tag{2.7}$$

It can be seen that there are no fringes when the contrast is zero. This can occur when the two light beams are orthogonally polarized. The fringes are always formed, albeit with low contrast, if the two beams are not in the same state of polarization. Assuming the beams to be in the same state of polarization, the contrast is given by

$$C = \frac{2\sqrt{I_1 I_2}}{I_1 + I_2} = \frac{2\sqrt{\frac{I_1}{I_2}}}{1 + \frac{I_1}{I_2}} \tag{2.8}$$

Fringes of appreciable contrast are formed even if the beams differ considerably in intensity.

Assuming the plane waves to be confined to the (x, z) plane as shown in Figure 2.1, the phase difference is given by

$$\delta = [(\mathbf{k}_2 - \mathbf{k}_1) \cdot \mathbf{r} + \phi] = \frac{2\pi}{\lambda} 2x \sin\theta$$

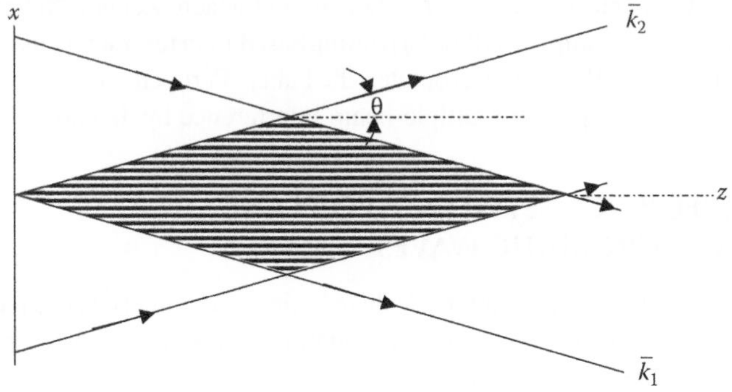

FIGURE 2.1 Interference pattern between two plane waves.

where we have taken $\phi = 0$. Here the beams are symmetric with respect to the z axis, making an angle θ with the axis. It is thus seen that straight-line fringes are formed that run parallel to the z axis, with spacing $\bar{x} = \lambda/(2\sin\theta)$. The fringes have a larger width for smaller enclosed angles.

2.3.1 Young's Double-Slit Experiment

This is a basic experiment to explain the phenomenon of interference. Radiation from an extremely narrow slit illuminates a pair of slits situated on a plane a distance p from the plane of the first slit. The slits are parallel to each other. A cylindrical wave from the first slit is incident on the pair of slits, which thus sample the wave at two places. The incident wave is diffracted by each slit of the pair, and the interference pattern between the diffracted waves is observed at a plane a distance z from the plane of the pair of slits, as shown in Figure 2.2. The amplitude distribution at an observation point P on the screen is given by

$$I(x, z) = 2I_{0s}\,\mathrm{sinc}^2\left(\frac{1}{2}kb\sin\theta\right)[1 + \cos(ka\sin\theta)] \tag{2.9}$$

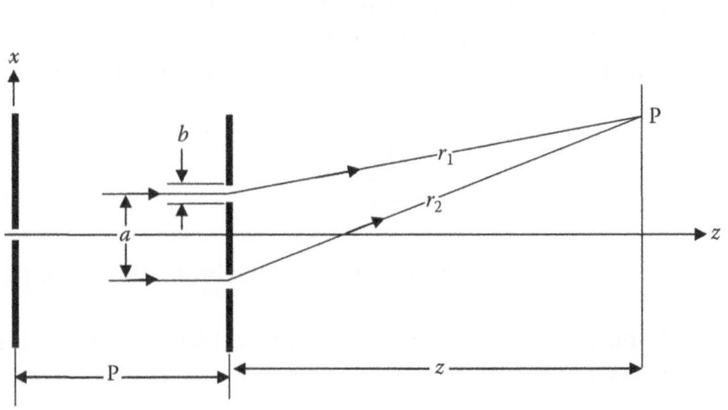

FIGURE 2.2 Young's double-slit experiment.

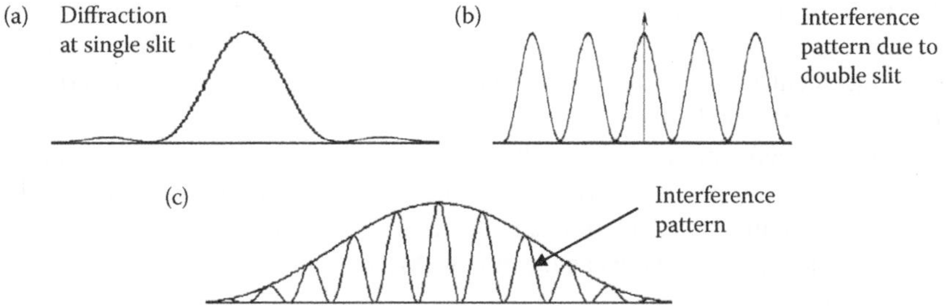

FIGURE 2.3 (a) Diffraction pattern of a single slit. (b) Interference pattern due to a double slit. (c) Interference pattern as observed in Young's double-slit experiment (with diffraction envelope shown).

where $I_{0s}\,\mathrm{sinc}^2(kb\sin\theta/2)$ is the intensity distribution on the observation plane due to a single slit and $ka\sin\theta$ is the phase difference between two beams diffracted by the pair of slits. It can be seen that the intensity distribution in the fringe pattern is governed by the diffraction pattern of a single slit, and the fringe spacing is governed by the slit separation. This is shown in Figure 2.3.

Let us now broaden the illuminating slit, thereby decreasing the region of coherence. The width of the region of coherence can be calculated by invoking the van Cittert–Zernike theorem.

Another interesting interferometer based on the division of wavefronts is the Michelson stellar interferometer, which was devised to measure the angular sizes of distant stars. A schematic of the Michelson stellar interferometer is shown in Figure 2.4.

The combination of mirrors M_3 and M_1 samples a portion of the wavefront from a distant source and directs it to one of the apertures, while the other mirror combination M_4 and M_2 samples another portion of the same wavefront and directs it

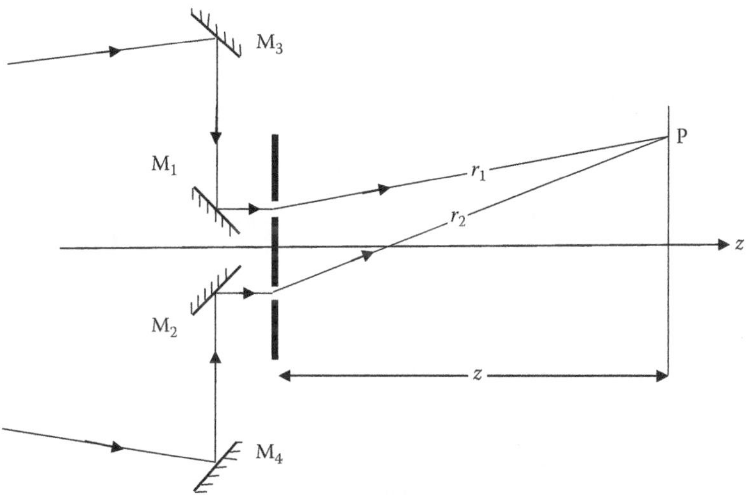

FIGURE 2.4 Michelson stellar interferometer for angular size measurement.

to the other aperture. Diffraction takes place at these apertures, and an interference pattern is observed at the observation plane. In the Michelson stellar interferometer, this arrangement is placed in front of the lens of a telescope, and the interference pattern is observed at the focal plane.

In this arrangement, the width of the interference fringes is governed by the separation between the pair of apertures, and the intensity distribution is governed by diffraction at the individual apertures. For a broad source, the visibility of the fringes depends on the separation between the mirrors M_3 and M_4. This is also the case when a binary object is considered.

2.3.2 Michelson Interferometer

The Michelson interferometer consists of a pair of mirrors M_1 and M_2, a beam-splitter BS, and a compensating plate CP identical to the BS, as shown in Figure 2.5. Light from a source S is divided into two parts by amplitude division at the beam-splitter. The beams usually travel in orthogonal directions and are reflected back to the beam-splitter, where they recombine to form an interference pattern. Reflection at BS produces a virtual image M_2' of the mirror M_2. The interference pattern observed is therefore the same as would be observed in an air layer bounded by M_1 and M_2'.

With a monochromatic source, there is no need to introduce the compensating plate CP. It is required, however, when a quasi-monochromatic source is used, to compensate for dispersion in glass. It is also interesting to note that nonlocalized

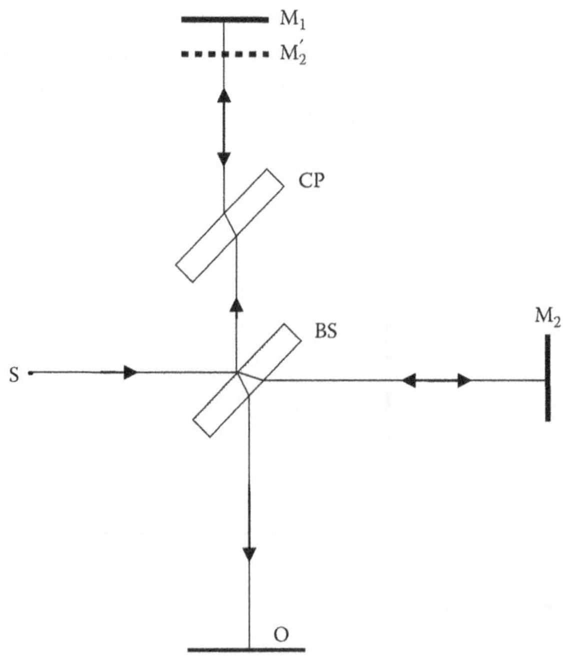

FIGURE 2.5 Michelson interferometer.

fringes are formed when a monochromatic point source is used. These fringes can be circular or straight-line, depending on the alignment.

With an extended monochromatic source, circular fringes localized at infinity are formed when M_1 and M_2' are parallel. If M_1 and M_2' enclose a small angle and are very close to each other (a thin air wedge), then fringes of equal thickness are formed that run parallel to the apex of the wedge and are localized in the wedge itself.

2.4 MULTIPLE-BEAM INTERFERENCE

Multiple-beam interference can be observed either with division of wavefront or with division of amplitude. There are, however, two important differences:

1. Beams of equal amplitudes participate in multiple-beam interference by division of wavefront, while the amplitude of successive beams continues to decrease when division of amplitude is used.
2. The number of beams participating in interference is finite owing to the finite size of the elements generating multiple beams by division of wavefront, and hence coherence effects due to the finite size of the source may also come into play. The number of beams participating in interference can be infinite when division of amplitude is employed.

Except for these differences, the theory of fringe formation is the same.

2.4.1 Multiple-Beam Interference: Division of Wavefront

An obvious example of the generation of multiple beams is a grating. Let us consider a grating that has N equispaced slits. Let the slit width be b and let the period of the slits be p. An incident beam is diffracted by each slit. We consider beams propagating in the direction θ as shown in Figure 2.6.

The total amplitude at any point on the observation screen in the direction θ is obtained by the summation of the amplitudes from each slit along with the appropriate phase differences:

$$A = a + ae^{i\delta} + ae^{2i\delta} + \cdots + ae^{i(N-1)\delta} \tag{2.10}$$

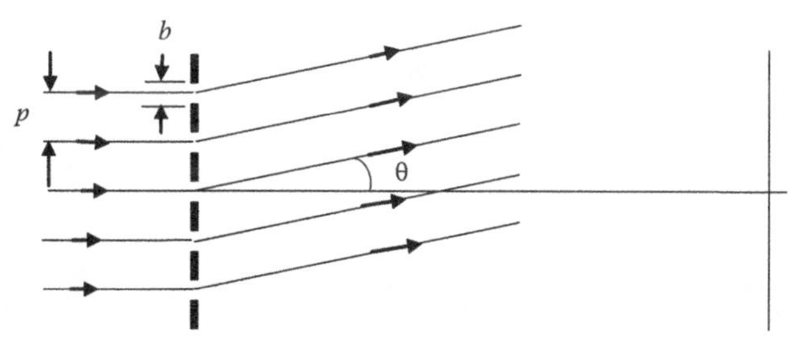

FIGURE 2.6 Diffraction at a grating: multiple-beam interference by division of wavefront.

where $\delta = kp \sin\theta$ is the phase difference between two waves diffracted in the direction θ by two consecutive slits and a is the amplitude of the diffracted wave from any slit. The intensity distribution is obtained by summing this series of N terms as follows:

$$I = AA^* = a^2 \frac{\sin^2 \frac{1}{2}N\delta}{\sin^2 \frac{1}{2}\delta} \tag{2.11}$$

Substituting for $a^2 = I_{0s}\,\mathrm{sinc}^2(kb \sin\theta/2)$ as the intensity distribution due to a single slit, the intensity distribution in the interference pattern due to the grating is

$$I = I_{0s}\,\mathrm{sinc}^2\left(\frac{1}{2}kb \sin\theta\right) \frac{\sin^2 \frac{1}{2}N\delta}{\sin^2 \frac{1}{2}\delta} \tag{2.12}$$

The first term is due to diffraction at the single slit and the second is due to interference. For a Ronchi grating, $2b = p$ (the transparent and opaque parts are equal), the intensity distribution is given by

$$I = \frac{I_{0s}}{4} \frac{\sin^2 \frac{1}{2}N\delta}{\cos^2 \frac{1}{4}\delta}$$

The principal maxima are formed wherever $\delta = 2m\pi$. Figure 2.7 shows the intensity distribution due to $N = 10$ slits. It can be seen that the positions of the principal maxima remain the same as would be obtained with a two-beam interference, but their width narrows considerably. In addition to the principal maxima, there are secondary maxima in between two consecutive principal maxima. With an increasing number of slits, the intensity of the secondary maxima falls off.

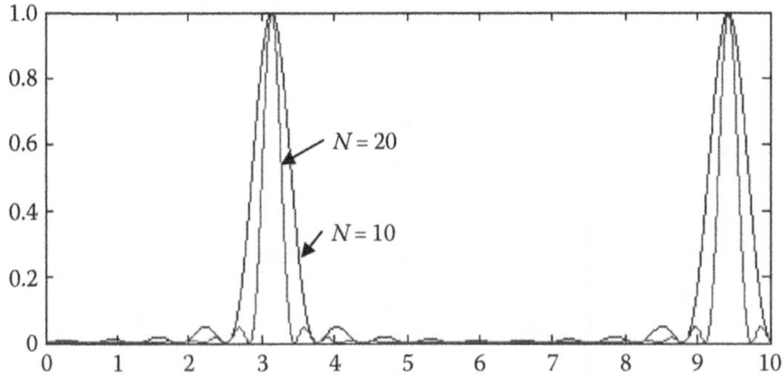

FIGURE 2.7 Intensity distribution in the diffraction pattern of a grating with $N = 10$ and 20.

2.4.2 Multiple-Beam Interference: Division of Amplitude

Consider a plane parallel plate of a dielectric material of refractive index μ and thickness d in air, as shown in Figure 2.8. A plane monochromatic wave of amplitude a is incident at an angle θ_1. A ray belonging to this wave, as shown in Figure 2.8, gives rise to a series of reflected and transmitted rays owing to multiple reflections. Let the complex reflection and transmission coefficients for the rays incident from the side of the surrounding medium on the two surface of the plate be r_1, t_1 and r_2, t_2, respectively, and let those for the rays incident from the plate side be r_1', t_1' and r_2', t_2', respectively. The amplitudes of the waves reflected back into the first medium and transmitted are given in Figure 2.8. The phase difference between any two consecutives rays is given by

$$\delta = \frac{4\pi}{\lambda}\mu d \cos\theta_2 \tag{2.13}$$

It is obvious that we can observe an interference pattern in both reflection and transmission. If the angle of incidence is very small and the plate is fairly large, an infinite number of beams take part in the interference.

2.4.2.1 Interference Pattern in Transmission

The resultant amplitude when all the beams in transmission are superposed is given by

$$A_t(\delta) = at_1t_2'e^{-i\delta/2}(1 + r_1'r_2'e^{-i\delta} + r_1'^2r_2'^2e^{-i2\delta} + r_1'^3r_2'^3e^{-i3\delta} + \cdots) \tag{2.14}$$

The infinite series can be summed to yield

$$A_t(\delta) = \frac{at_1t_2'e^{-i\delta/2}}{1 - r_1'r_2'e^{-i\delta}} \tag{2.15}$$

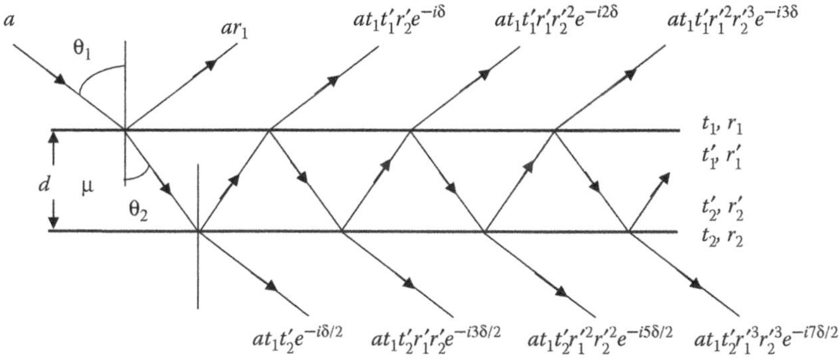

FIGURE 2.8 Multiple-beam formation in a plane parallel plate.

The intensity distribution in transmission is then given by

$$I(\delta) = I_{\text{in}} \frac{T_1 T_2}{1 + R_1 R_2 - 2\sqrt{R_1 R_2}\cos(\delta + \psi_1 + \psi_2)} \tag{2.16}$$

where $T_1 = |t_1|^2$, $T_2 = |t_2'|^2$, and $R_1 = |r_1'|^2$, $R_2 = |r_2'|^2$ are the transmittances and reflectances of the two surfaces of the plate; ψ_1 and ψ_2 are the phases acquired by the wave on reflection; and $I_{\text{in}} = |a|^2$ is the intensity of the incident wave. In the special case when the surrounding medium is the same on both sides of the plate (as here, where it is assumed that the plate is in air), $R_1 = R_2 = R$, $T_1 = T_2 = T$, and $\psi_1 + \psi_2 = 0$ or 2π. Therefore, the intensity distribution in the interference pattern in transmission takes the form

$$\begin{aligned} I_t(\delta) &= I_{\text{in}} \frac{T^2}{1 + R^2 - 2R\cos\delta} \\ &= I_{\text{in}} \frac{T^2}{(1-R)^2 + 4R\sin^2\frac{1}{2}\delta} \end{aligned} \tag{2.17}$$

This is the well-known Airy formula. The maximum intensity in the interference pattern occurs when $\sin\frac{1}{2}\delta = 0$. The maximum intensity is given by $I_{\max} = I_{\text{in}}T^2/(I-R)^2$. Similarly, the minimum intensity occurs when $\sin\frac{1}{2}\delta = 1$. The minimum intensity is given by $I_{\min} = I_{\text{in}}T^2/(I+R)^2$. The intensity distribution can now be expressed in terms of $I_{\max}$ as follows:

$$I_t(\delta) = \frac{I_{\max}}{1 + \dfrac{4R}{(1-R)^2}\sin^2\dfrac{1}{2}\delta} \tag{2.18}$$

Figure 2.9 shows a plot of the intensity distribution for two values of R. This distribution does not display any secondary maxima, completely in contrast with that obtained in a multiple-beam interference by division of wavefront.

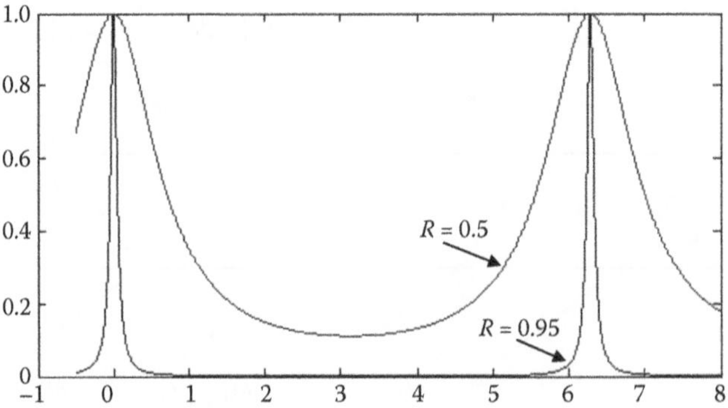

FIGURE 2.9 Intensity distribution in transmission due to multiple-beam interference.

2.4.2.2 Interference Pattern in Reflection

The resultant amplitude in reflection is obtained by the summation of the amplitudes of all the reflected waves as follows:

$$\begin{aligned} A_r(\delta) &= ar_1 + at_1t_1'r_2'e^{-i\delta}(1 + r_1'r_2'e^{-i\delta} + r_1'^2r_2'^2e^{-i2\delta} + \cdots) \\ &= ar_1 + \frac{at_1t_1'r_2'e^{-i\delta}}{1 - r_1'r_2'e^{-i\delta}} \end{aligned} \tag{2.19}$$

Using the Stokes relations $r_1 = -r_1'$ and $r_1^2 + t_1t_1' = 1$, the above relation can be expressed as follows:

$$A_r(\delta) = a\frac{-r_1' + r_2'e^{-i\delta}}{1 - r_1'r_2'e^{-i\delta}} \tag{2.20}$$

The intensity distribution in the interference pattern in reflection, when the plate is in air, is thus given by

$$\begin{aligned} I_r(\delta) &= I_{in}\frac{2R(1 - \cos\delta)}{1 + R^2 - 2R\cos\delta} \\ &= I_{in}\frac{\dfrac{4R}{(1-R)^2}\sin^2\dfrac{1}{2}\delta}{1 + \dfrac{4R}{(1-R)^2}\sin^2\dfrac{1}{2}\delta} \end{aligned} \tag{2.21}$$

The intensity maxima occur when $\sin\frac{1}{2}\delta = 1$. The intensity distribution is complementary to the intensity distribution in transmission.

2.5 INTERFEROMETRY

Interferometry is a technique of measurement that employs the interference of light waves, and the devices using this technique are known as interferometers. These use an arrangement to generate two beams, one of which acts as a reference and the other is a test beam. The test beam gathers information about the process to be measured or monitored. These two beams are later combined to produce an interference pattern that arises from the acquired phase difference. Interferometers based on division of amplitude use beam-splitters for splitting the beam into two and later combining them for interference. Ingenuity lies in designing beam-splitters and beam-combiners. The interference pattern is either recorded on a photographic plate or sensed by a photodetector or array-detector device. Currently charge-coupled device (CCD) arrays are used along with phase-shifting to display the desired information such as surface profile, height variation, and refractive index variation. Observation of the fringe pattern is completely hidden in the process.

Two-beam interferometers using division of wavefront have been used for the determination of the refractive index (Rayleigh interferometer), determination of the angular size of distant objects (Michelson stellar interferometer), deformation studies (speckle interferometers), and so on. Interferometers based on division of amplitude are used for the determination of wavelength (Michelson interferometer), the testing of optical components (Twyman–Green interferometer, Zygo interferometer, Wyco interferometer, and shear interferometers), the study of microscopic objects (Mirau interferometer), the study of birefringent objects (polarization interferometers), distance measurement (modified Michelson interferometer), and so on. The Michelson interferometer is also used in spectroscopy, particularly in the infrared region, and offers Fellgett's advantage. The Michelson–Morley experiment, which used a Michelson interferometer to show the absence of the ether, played a great part in the advancement of science. Gratings are also used as beam-splitters and beam-combiners. Because of their dispersive nature, they are used in the construction of achromatic interferometers.

Multiple-beam interferometers are essentially dispersive, and find applications in spectroscopy. They offer high sensitivity.

2.5.1 Dual-Wavelength Interferometry

The interferometer is illuminated simultaneously with two monochromatic waves of wavelengths λ_1 and λ_2. Each wave produces its own interference pattern. The two interference patterns produce a moiré, which is characterized by a synthetic wavelength λ_s. If the path difference variation is large, thereby producing an interference pattern with narrow fringes, then only the pattern due to the synthetic wavelength will be visible. Under the assumption that the two wavelengths are very close to each other and the waves are of equal amplitude, the intensity distribution can be expressed as follows:

$$\begin{aligned} I_{tw} &= I_0(a + b\cos\delta_1 + b\cos\delta_2) \\ &= I_0\left\{a + 2b\cos\left[\frac{1}{2}(\delta_1+\delta_2)\right]\cos\left[\frac{1}{2}(\delta_1-\delta_2)\right]\right\} \end{aligned} \tag{2.22}$$

where a and b are constants and the phase differences δ_1 and δ_2 are given by

$$\delta_1 = \frac{2\pi}{\lambda_1}\Delta, \quad \delta_2 = \frac{2\pi}{\lambda_2}\Delta$$

Δ is the optical path difference and is assumed to be constant for the two wavelengths under consideration. The second term in the intensity distribution has a modulation term; the argument of the cosine function can be expressed as

$$\frac{\delta_1-\delta_2}{2} = \pi\left(\frac{1}{\lambda_1} - \frac{1}{\lambda_2}\right)\Delta = \frac{\pi}{\lambda_s}\Delta \tag{2.23}$$

The synthetic wavelength λ_s is thus given by

$$\lambda_s = \frac{\lambda_1\lambda_2}{|\lambda_1-\lambda_2|} \tag{2.24}$$

Dual-wavelength interferometry has been used to measure large distances with high accuracy. Phase-shifting can be incorporated in dual-wavelength interferometry. There are several other techniques, such as dual-wavelength sinusoidal phase-modulating interferometry, dual-wavelength phase-locked interferometry, and dual-wavelength heterodyne interferometry.

2.5.2 White Light Interferometry

Each wavelength in white light produces an interference pattern, which adds up on an intensity basis at any point on the observation plane. The fringe width of these patterns is different. However, at a point on the observation plane where the two interfering beams corresponding to each wavelength have zero phase difference, the interference results in a bright fringe. At this point, all wavelengths in white light will interfere constructively, thereby producing a white fringe. As one moves away from this point, thereby increasing the path difference, an interference color fringe corresponding to the shortest wavelength is seen first, followed by color fringes of increasing wavelengths. With increasing path difference, these colors become less and less saturated, and finally the interference pattern is lost. Figure 2.10 shows the intensity distribution in an interference pattern of white light.

White light interferometry has been used to measure thicknesses and air-gaps between two dielectric interfaces. The problem of 2π phase ambiguity can also be overcome by using white light interferometry and scanning the object in depth when the object profile is to be obtained. Instead of scanning the object, spectrally resolved interferometry can be used for profiling.

2.5.3 Heterodyne Interferometry

Let us consider the interference between two waves of slightly different frequencies traveling along the same direction. (In dual-wavelength interferometry, the two wavelengths are such that their frequency difference is in the terahertz region, which the optical detector cannot follow, while in heterodyne interferometry, the frequency

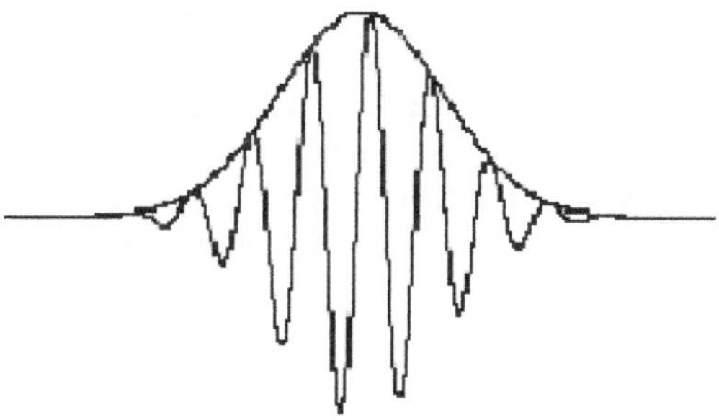

FIGURE 2.10 White light interferogram.

difference is in megahertz, which the detector can follow.) The two real waves can be described by

$$E_1(t) = a_1 \cos(2\pi i \nu_1 t + \delta_1)$$

$$E_2(t) = a_2 \cos(2\pi i \nu_2 t + \delta_2)$$

where ν_1, ν_2 are the frequencies and δ_1, δ_2 are the phases of the two waves. These two waves are superposed on a square-law detector. The resultant amplitude on the detector is thus the sum of the two; that is,

$$E(t) = E_1(t) + E_2(t)$$

The output current $i(t)$ of the detector is proportional to $|E(t)|^2$. Hence,

$$\begin{aligned} i(t) &\propto a_1^2 \cos^2(2\pi i \nu_1 t + \delta_1) + a_2^2 \cos^2(2\pi i \nu_2 t + \delta_2) \\ &\quad + 2a_1 a_2 \cos(2\pi i \nu_1 t + \delta_1)\cos(2\pi i \nu_2 t + \delta_2) \\ &= \frac{1}{2}(a_1^2 + a_2^2) + \frac{1}{2}[a_1^2 \cos 2(2\pi i \nu_1 t + \delta_1) + a_2^2 \cos 2(2\pi i \nu_2 t + \delta_2)] \\ &\quad + a_1 a_2 \cos[2\pi i(\nu_1 + \nu_2)t + \delta_1 + \delta_2] + a_1 a_2 \cos[2\pi i(\nu_1 - \nu_2)t + \delta_1 - \delta_2] \end{aligned} \tag{2.25}$$

The expression for the output current contains oscillatory components at frequencies $2\nu_1, 2\nu_2$, and $\nu_1 + \nu_2$ that are too high for the detector to follow, and thus the terms containing these frequencies are averaged to zero. The output current is then given by

$$i(t) = \frac{1}{2}(a_1^2 + a_2^2) + a_1 a_2 \cos[2\pi i(\nu_1 - \nu_2)t + \delta_1 - \delta_2] \tag{2.26}$$

The output current thus consists of a DC component and an oscillating component at the beat frequency $\nu_1 - \nu_2$. The beat frequency usually lies in the radiofrequency range, and hence can be observed. Thus superposition of two waves of slightly different frequencies on a square-law detector results in a moving interference pattern; the number of fringes passing any point on the detector in unit time is equal to the beat frequency. Obviously, the size of the detector must be much smaller than the fringe width for the beat frequency to be observed. The phase of the moving interference pattern is determined by the phase difference between the two waves. The phase can be measured electronically with respect to a reference signal derived either from a second detector or from the source deriving the modulator. Heterodyne interferometry is usually performed with a frequency-stabilized single-mode laser and offers extremely high sensitivity at the cost of system complexity.

2.5.4 Shear Interferometry

In shear interferometry, interference between a wave and its sheared version is observed. A sheared wave can be obtained by lateral shear, rotational shear, radial

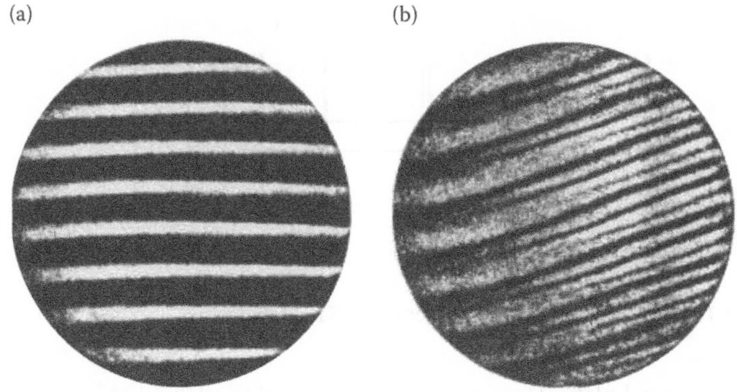

FIGURE 2.11 Multiple-beam shear interferograms: (a) at collimation; (b) off collimation.

shear, folding shear, or inversion shear. Shear interferometers have certain advantages. They do not require a reference wave for comparison. Both waves travel along the same path, and thus the interferometer is less susceptible to vibrations and thermal fluctuations. However, the interpretation of shear fringes is more tedious than that of fringes obtained with a Michelson or Twyman–Green interferometer. Fringes represent the gradient of path difference in the direction of shear. Fringe interferometry has found applications where one is interested in slope and curvature measurements rather than in displacement alone. Multiple-beam shear interferometry using a coated shear plate has been used for collimation testing. It can be shown that strong satellite fringes are seen in transmission when the beam is not collimated, as shown in Figure 2.11.

2.5.5 Polarization Interferometers

These are basically used to study birefringent specimens. A polarizer is used to polarize the incident collimated beam, which passes through a birefringent material. The polarized beam then splits into ordinary and extraordinary beams, which travel through the same object but cannot interfere, because they are in orthogonal polarization states. These beams are brought to interfere by another polarizer. Some of these interferometers are described in Chapter 8.

There is another class of polarization interferometers, which use beam-splitters and beam-combiners such as a Savart plate or a Wollaston prism made of birefringent material. Since the two beams generated in the beam-splitter have a very small separation, these constitute shear interferometers. Figure 2.12 shows a polarization shear interferometer using Savart plates as a beam-splitter (S_1) and a beam-combiner (S_2). These are two identical plates of birefringent material cut from a uniaxial crystal, either from calcite or quartz, with the optic axis at approximately 45° to the entrance and exit faces and put together with their principal sections crossed. A polarizer P_1 is placed before the Savart plate and oriented such as to produce a ray linearly polarized at 45°. A ray from the polarizer P_1 is split into an e-ray and an o-ray in the first plate. Since the principal sections of the two plates are orthogonal, the e-ray in the first plate becomes an o-ray in the second plate, while the o-ray in the first plate becomes an

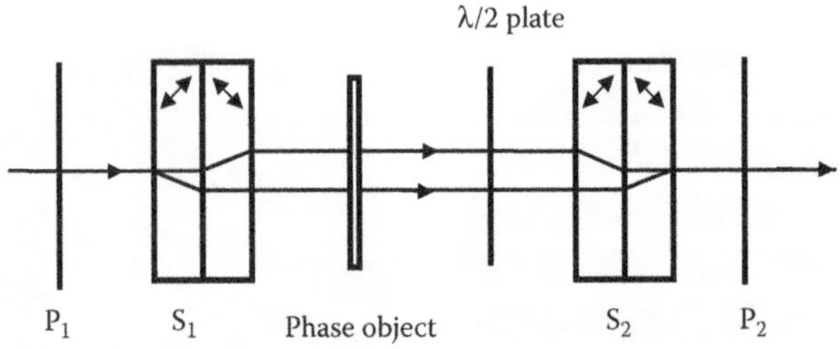

FIGURE 2.12 Polarization interferometer using Savart plates.

e-ray in the second plate. These rays emerge parallel to each other but have a linear shear (displacement) given by

$$\Delta_{\text{sa}} = \sqrt{2}d \left|(n_{\text{e}}^2 - n_{\text{o}}^2)\right| (n_{\text{e}}^2 + n_{\text{o}}^2) \tag{2.27}$$

where d is the thickness of the plate and n_{e} and n_{o} are the extraordinary and ordinary refractive indices of the uniaxial crystal. These two rays are combined using another Savart plate. Since the two rays are orthogonally polarized, another polarizer P_2 is used to make them interfere. The interference pattern consists of equally spaced straight fringes localized at infinity. The object is placed in the beam, and its refractive index and thickness variations introduce phase variations, which distort the fringes.

A Savart plate introduces linear shear in a collimated beam. Usually, the incident beam has a small divergence. Savart plates can be modified to accept wider fields. However, with a spherical beam, linear shear can be introduced by a Wollaston prism.

2.5.6 Interference Microscopy

In order to study small objects, it is necessary to magnify them. The response of micromechanical components to an external agency can be studied under magnification. Therefore, microinterferometers, particularly Michelson interferometers, have been built that can be placed in front of a microscope objective. Obviously, one requires microscope objectives with long working distances. The Mirau interferometer employs a very compact optical system that can be incorporated in a microscope objective. It is more compact than the micro-Michelson interferometer and fully compensated. Figure 2.13 shows a schematic view of the Mirau interferometer. Light from the illuminator through the microscope objective illuminates the object. The beam from the illuminator is split into two parts by the beam-splitter. One part illuminates the object, while the other is directed toward a reference surface. On reflection from the reference surface as well as from the object, the beams combine at the beam-splitter, which directs these beams to the microscope objective. An interference pattern is thus observed between the reference beam and the object beam. The pattern represents the path variations on the surface of the object. White light can be used with a Mirau interferometer. Phase-shifting is introduced by mounting the beam-splitter on PZT (lead zirconate titanate).

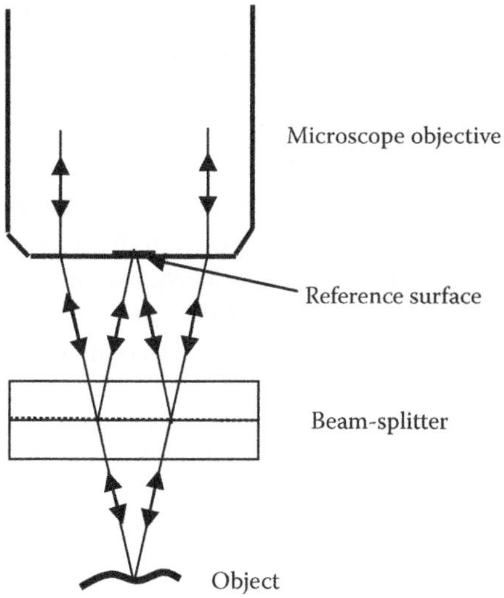

FIGURE 2.13 The Mirau interferometer.

Shearing interferometry can also be carried out under high magnification. The Nomarski interferometer is commonly used within a microscope, as shown in Figure 2.14. It uses two modified Wollaston prisms as a beam-splitter and a beam-combiner. The modified Wollaston prism offers the wider field of view that is required in microscopy. The angular shear γ between the two rays exiting from the Wollaston prism is given by

$$\gamma = 2\,|(n_e - n_o)|\,\sin\theta \qquad (2.28)$$

where θ is the angle of either wedge forming the Wollaston prism. The Nomarski interferometer can be employed in two distinctly different modes. With an isolated microscopic object, it is convenient to use a lateral shear that is larger than the dimensions of the object. Two images of the object are then seen, covered with fringes that contour the phase changes due to the object. Often, the Nomarski interferometer is used with smaller shear than the dimensions of a microscopic object. The interference pattern then shows the phase gradients. This mode of operation is known as differential interference contrast (DIC) microscopy. The Nomarski interferometer can

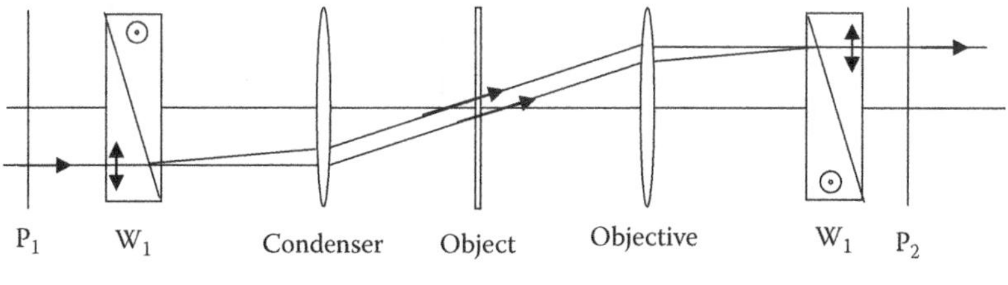

FIGURE 2.14 The Nomarski interferometer.

also be used with white light; the phase changes are then decoded as color changes. Biological objects, usually phase objects, can also be studied using microscopes.

2.5.7 Doppler Interferometry

An intense beam from a laser is expanded by a beam-expander and is directed onto a moving object. The light reflected/scattered from the moving object is redirected and then collimated before being fed into a Michelson interferometer as shown in Figure 2.15. The arms of the interferometer contain two $4f$ arrangements, but the focal lengths of the lenses in the two arms are different. This introduces a net path difference.

The light from a moving object is Doppler-shifted. The Doppler-shifted light illuminates the interferometer. For a retro-reflected beam, the Doppler shift $\Delta\lambda$ is given by

$$\Delta\lambda = \frac{2v}{c}\lambda$$

where v is the velocity of the object and c is the velocity of light. For objects moving with a speed of about 1 km/s, the Doppler-shift can be in tens of gigahertz, which is difficult to follow, and, to overcome this problem and reveal the time history of an object's motion, Doppler interferometers have been developed.

If the object is moving with constant velocity, the wavelength of the incident radiation remains constant, and hence the path difference between two beams in the interferometer is constant with time; the interference pattern remains stationary. However, if the object is accelerating, there is a change in phase difference, which is given by

$$d\delta = -\frac{2\pi}{\lambda^2}d\lambda\Delta = -\delta\frac{2v}{c} \tag{2.29}$$

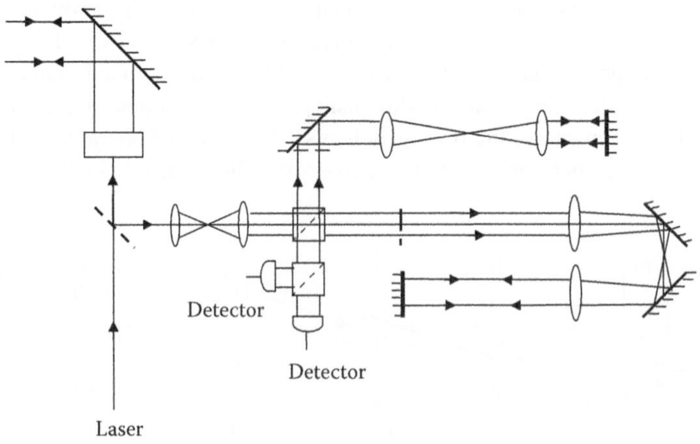

FIGURE 2.15 Michelson interferometer for observing the velocity history of a projectile.

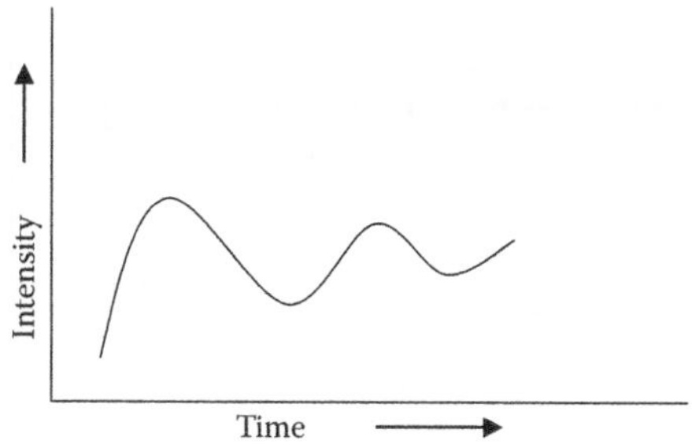

FIGURE 2.16 Variation of intensity with time in a Doppler interferometer.

The change in phase is directly proportional to the path difference Δ, which depends on the difference in the arm lengths of the interferometer. Further phase change is directly proportional to the speed of the object. If the interferometer is aligned for a single broad fringe in the field, the change of phase will result in a variation of intensity. Therefore, a detector will show a variation of intensity with time. The interference is governed by the equation $d\delta = 2m\pi$, where m is the order of the fringe. This gives $m = (2v/\lambda c)\Delta$. A full cycle change ($m = 1$) occurs when the velocity changes by $\lambda c/2\Delta$. A larger path difference between two arms thus gives higher sensitivity. Figure 2.16 shows an example of the variation of intensity with time. Usually, an event may occur only over a fraction of second. Therefore, a Doppler-shift of tens of gigahertz is converted into tens of hertz.

2.5.8 Fiber-Optic Interferometers

Two-beam interferometers can be easily constructed using optical fibers. Michelson and Fizeau interferometers have been realized. Electronic speckle pattern interferometry and its shear variant have been performed using optical fibers. A laser Doppler velocimetry setup can also be easily assembled.

The building blocks of fiber-optic interferometers are single-mode fibers, birefringent fibers, fiber couplers, fiber polarization elements, and micro-optics. Single-mode fibers allow only a single mode to propagate, and therefore a smooth wavefront is obtained at the output end. Very long path differences are achieved with fiber interferometers, and therefore they are inherently suited for certain measurements requiring high sensitivity. The fibers are compatible with optoelectronic devices such as semiconductor lasers and detectors. These interferometers have found applications as sensors with distributed sensing elements.

Figure 2.17 shows a fiber-optic Doppler velocimeter arrangement for measuring fluid velocities. Radiation from a semiconductor laser is coupled to a single-mode fiber. Part of the radiation is extracted into a second fiber by a directional coupler, thereby creating two beams. Graded-index rod lenses are attached to the fiber ends to obtain collimated beams. The fiber ends can be separated by a desired amount, and a large lens focuses the beams in the sample volume. An interference pattern is thus generated there. The beams can be aligned such that the interference fringes in the sample volume run perpendicular to the direction of flow. Alternatively, these beams

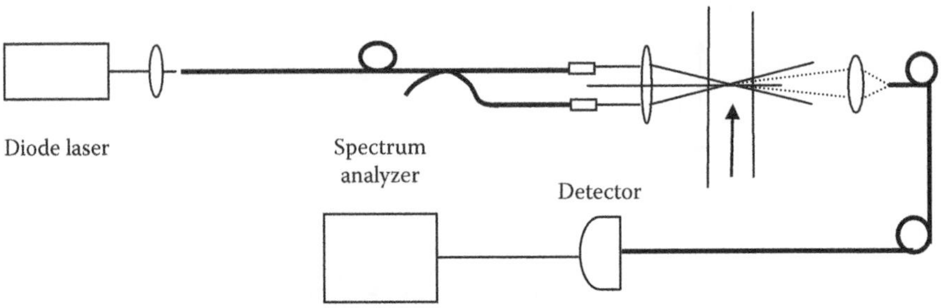

FIGURE 2.17 Fiber-optic Doppler velocimetry.

can be separated by any amount in space by simply displacing the fiber ends. The fiber ends can be imaged in the sample volume by individual lenses, and can be superposed by manipulating them. An interference pattern can thus be created in a small volume, with the fringe running normal to the direction of flow. The scattered light from the interference volume is collected and sent to a photodiode through a multimode fiber. The Doppler signal from the photodiode is analyzed by an RF spectrum analyzer. A frequency offset for direction determination can be introduced between the beams by using a piezoelectric phase-shifter, driven by a sawtooth waveform, in one arm.

2.5.9 Phase-Conjugation Interferometers

Interference between a signal wave or a test wave and its phase-conjugated version is observed in phase-conjugation interferometers. The phase-conjugated wave is realized by four-wave mixing in a nonlinear crystal such as $BaTiO_3$. The phase-conjugated wave carries phase variations of opposite signs and travels in the opposite direction to the signal or test wave. For example, a diverging wave emanating from a point source, when phase-conjugated, will become a converging wave and will converge to the same point source.

Since interference between the test wave and the phase-conjugated wave is observed in phase-conjugation interferometers, there is no need to have a reference

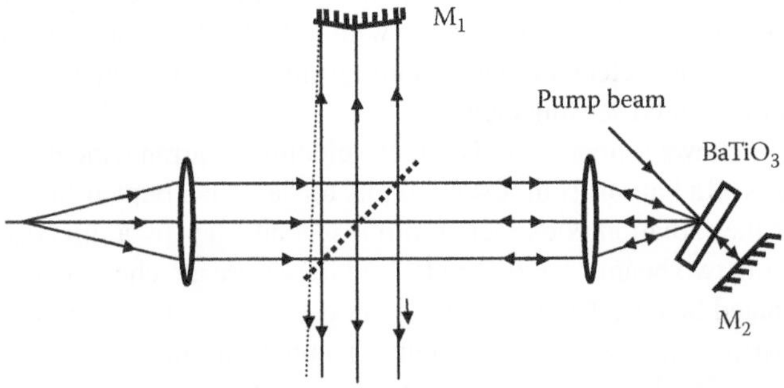

FIGURE 2.18 Phase-conjugated Michelson interferometer for collimation testing.

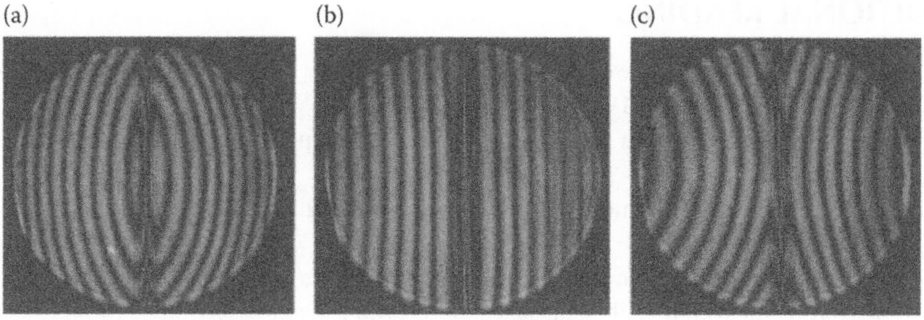

FIGURE 2.19 Interferograms (a) inside, (b) at, and (c) outside the collimation position.

wave, and the interference pattern exhibits twice the sensitivity. Phase-conjugate interferometers are robust and are less influenced by thermal and vibration effects than other interferometers.

Michelson, Twyman–Green, and Mach–Zehnder interferometers have been realized using phase-conjugate mirrors. We describe here an interferometer that is used for testing collimation. Figure 2.18 shows a schematic view of the interferometer. M_1 is a double mirror enclosing an angle close to 180°. A $BaTiO_3$ crystal is used for generating a phase-conjugated beam by four-wave mixing. At collimation, straight-line fringes are observed in both fields, while with a converging or diverging wave from the collimator, curved fringes with curvatures reversed in each field are observed. These interferograms are shown in Figure 2.19. The technique is self-referencing and allows sensitivity to be increased fourfold.

BIBLIOGRAPHY

1. C. Candler, *Modern Interferometers*, Hilger and Watts, London, 1951.
2. S. Tolansky, *An Introduction to Interferometry*, Longmans, London, 1955.
3. S. Tolansky, *Multiple-Beam Interferometry of Surfaces and Films*, Oxford University Press, London, 1948 (reprinted by Dover Publications, 1970).
4. M. Françon and S. Mallick, *Polarization Interferometers: Applications in Microscopy and Macroscopy*, Wiley-Interscience, London, 1971.
5. D. F. Melling and J. H. Whitelaw, *Principles and Practice of Laser–Doppler Anemometry*, Academic Press, London, 1976.
6. K. Leonhardt, *Optische Interferenzen*, Wissenschaftliche Vergesellschaft, Stuttgart, 1981.
7. W. H. Steel, *Interferometry*, Cambridge University Press, Cambridge, 1983.
8. B. Culshaw, *Optical Fibre Sensing and Signal Processing*, Peregrinus, London, 1984.
9. J. M. Vaughan, *The Fabry–Perot Interferometer*, Adam Hilger, Bristol, 1989.
10. Prasad L. Polavarapu (Ed.), *Principles and Applications of Polarization-Division Interferometry*, Wiley, New York, 1998.
11. M. Born and E. Wolf, *Principles of Optics*, 7th edn, Cambridge University Press, Cambridge, 1999.
12. P. Hariharan, *Optical Interferometry*, Academic Press, San Diego, 2003.

ADDITIONAL READING

1. G. Delaunay, Microscope interferential A. Mirau pour la measure du fini des surfaces, *Rev. Opt.*, 32, 610–614, 1953.
2. J. Dyson, Common-path interferometer for testing purposes, *J. Opt. Soc. Am.*, 47, 386–390, 1957.
3. M. V. R. K. Murty, Common path interferometer using Fresnel zone plates, *J. Opt. Soc. Am.*, 53, 568–570, 1963.
4. M. V. R. K. Murty, The use of a single plane parallel plate as a lateral shearing interferometer with a visible gas laser source, *Appl. Opt.*, 3, 531–534, 1964.
5. W. H. Steel, A polarization interferometer for the measurement of transfer functions, *Opt. Acta*, 11, 9–19, 1964.
6. W. H. Steel, On Möbius-band interferometers, *Opt. Acta*, 11, 211–217, 1964.
7. O. Bryngdahl, Applications of shearing interferometry, In *Progress in Optics* (ed. E. Wolf), *IV*, 37–83, North-Holland, Amsterdam, 1965.
8. R. Allen, G. David, and G. Nomarski, The Zeiss–Nomarski differential interference equipment for transmitted-light microscopy, *Z. Wiss. Mikroskop. Mikroskop. Tech.*, 69, 193–221, 1969.
9. A. K. Chakrabory, Polarization effect of beam-splitter on two-beam interferometry, *Nouv. Rev. Opt.*, 4, 331–335, 1973.
10. R. J. Clifton, Analysis of the laser velocity interferometer, *J. Appl. Phys.*, 41, 5335–5337 (1970)
11. O. Bryngdahl, Shearing interferometry with constant radial displacement, *J. Opt. Soc. Am.*, 61, 169–172, 1971.
12. L. M. Barker and R. E. Hollenbach, Laser interferometer for measuring high velocities of any reflecting surface, *J. Appl. Phys.*, 43, 4669–4675, 1972.
13. L. M. Barker and K. W. Schuler, Correction to the velocity per fringe relationship for the VISAR interferometer, *J. Appl. Phys.*, 45, 3692–3693, 1974.
14. R. N. Smartt, Zone plate interferometer, *Appl. Opt.*, 13, 1093–1099, 1974.
15. P. Hariharan, W. H. Steel, and J. C. Wyant, Double grating interferometer with variable shear, *Opt. Commun.*, 11, 317–320, 1974.
16. K. Leonhardt, Intensity and polarization of interference systems of a two-beam interferometer, *Opt. Commun.*, 11, 312–316, 1974.
17. D. R. Goosman, Analysis of the laser velocity interferometer, *J. Appl. Phys.*, 46, 3516–3525, 1975.
18. J. C. Wyant, A simple interferometric OTF instrument, *Opt. Commun.*, 19, 120–122, 1976.
19. W. F. Hemsing, Velocity sensing interferometer (VISAR) modification, *Rev. Sci. Instrum.*, 50, 73–78, 1979.
20. T. Yatagai and T. Kanou, Aspherical surface testing with shearing interferometer using fringe scanning detection method, *Opt. Eng.*, 23, 357–360, 1984.
21. D. Jackson and J. D. C. Jones, Fibre optic sensors, *J. Mod. Opt.*, 33, 1469–1503, 1986.
22. D. Z. Anderson, D. M. Lininger, and J. Feinberg, Optical tracking novelty filter, *Opt. Lett.*, 12, 123–125, 1987.
23. W. Zou, G. Yuan, Q. Z. Tian, B. Zhang, and C. Wu, In-bore projectile velocity measurement with laser interferometer, *Proc. SPIE*, 1032, 669–671, 1988.
24. R. Dändliker, R. Thalmann, and D. Prongué, Two-wavelength laser interferometry using superheterodyne detection, *Opt. Lett.*, 13, 339–341, 1988.
25. T. Suzuki, O. Sasaki, and T. Maruyama, Phase locked laser diode interferometry for surface profile measurement, *Appl. Opt.*, 28, 4407–4410, 1989.

26. M. Küchel, The new Zeiss interferometer, *Proc. SPIE*, 1332, 655–663, 1991.
27. Y. Ishii and R. Onodera, Two-wavelength laser-diode interferometry that uses phase-shifting techniques, *Opt. Lett.*, 16, 1523–1525, 1991.
28. D. J. Anderson and J. D. C. Jones, Optothermal frequency and power modulation of laser diodes, *J. Mod. Opt.*, 39, 1837–1847, 1992.
29. C. S. Vikram, W. K. Witherow, and J. D. Trolinger, Determination of refractive properties of fluid for dual-wavelength interferometry, *Appl. Opt.*, 31, 7249–7252, 1992.
30. R. Onodera and Y. Ishii, Two-wavelength laser-diode interferometer with fractional fringe techniques, *Appl. Opt.*, 34, 4740–4746, 1995.
31. K. Matsuda, T. Eiju, T. H. Barnes, and S. Kokaji, A novel shearing interferometer with direct display of lens lateral aberrations, *Jpn J. Appl. Phys.*, 34, 325–330, 1995.
32. I. Yamaguchi, J. Liu, and J. Kato, Active phase-shifting interferometers for shape and deformation measurements, *Opt. Eng.*, 35, 2930–2937, 1996.
33. T. Suzuki, T. Muto, O. Sasaki, and T. Maruyama, Wavelength-multiplexed phase-locked laser diode interferometry using a phase-shifting technique, *Appl. Opt.*, 36, 6196–6201, 1997.
34. L. Erdmann and R. Kowarschik, Testing of silicon micro-lenses by use of a lateral shearing interferometer in transmission, *Appl. Opt.*, 37, 676–682, 1998.
35. S. S. Helen, M. P. Kothiyal, and R. S. Sirohi, Achromatic phase shifting by a rotating polarizer, *Opt. Commun.*, 154, 249–254, 1998.
36. M. C. Park and S. W. Kim, Direct quadratic polynomial fitting for fringe peak detection of white light scanning interferograms, *Opt. Eng.*, 39, 952–959, 2000.
37. T. Yokoyama, T. Araki, S. Yokoyama, and N. Suzuki, A subnanometre heterodyne interferometric system with improved phase sensitivity using three-longitudinal-mode He–Ne Laser, *Meas. Sci. Technol.*, 12, 157–162, 2001.
38. DeVon W. Griffin, Phase-shifting shearing interferometer, *Opt. Lett.*, 26, 140–141, 2001.
39. H. Kihm and S. Kim, Fiber-diffraction interferometer for vibration desensitization, *Opt. Lett.*, 30, 2059–2061, 2005.
40. J. E. Millerd, N. J. Brock, J. B. Hayes, M. B. North-Morris, M. Novak, and J. C. Wyant, Pixelated phase-mask dynamic interferometer, *Proc. SPIE*, 5531, 304–314, 2004.
41. R. M. Neal and J. C. Wyant, Polarization phase-shifting point-diffraction interferometer, *Appl. Opt.*, 45, 3463–3476, 2006.
42. O. Ferhanoglu, M. F. Toy, and H. Urey, Two-wavelength grating interferometry for MEMS sensors, *Photonics Technol. Lett. IEEE*, 19, 1895–1897, 2007.
43. J. M. Flores, M. Cywiak, M. Servín, and L. P. Zuárez, Heterodyne two beam Gaussian microscope interferometer, *Opt. Exp.*, 15, 8346–8359, 2007.
44. J. Schmit and P. Hariharan, Improved polarization Mirau interference microscope, *Opt. Eng.*, 46, 077007, 2007.
45. Z. Ge, T. Saito, M. Kurose, H. Kanda, K. Arakawa, and M. Takeda, Precision interferometry for measuring wavefronts of multi-wavelength optical pickups, *Opt. Exp.*, 16, 133–143, 2008.

3 Diffraction

In an isotropic medium, a monochromatic wave propagates with its characteristic speed. For plane and spherical waves, if the phase front is known at time $t = t_1$, its location at a later time is obtained easily by multiplying the elapsed time by the velocity in the medium. The wave remains either plane or spherical. Mathematically, we can find the amplitude at any point by solving the Helmholtz equation (Equation 1.3). However, when the wavefront is restricted in its lateral dimension, diffraction of light takes place. Diffraction problems are rather complex in nature, and analytical solutions exist for only a few of them. The Kirchhoff theory of diffraction, although afflicted with inconsistency in the boundary conditions, yields predictions that are in close agreement with experiments.

The term "diffraction" has been conveniently described by Sommerfeld as "any deviation of light rays from rectilinear paths that cannot be interpreted as reflection or refraction." It is often stated in a different form as "bending of rays near the corners." Grimaldi first observed the presence of bands in the geometrical shadow region of an object. A satisfactory explanation of this observation was given by Huygens by introducing the concept of secondary sources on the wavefront at any instant and obtaining the subsequent wavefront as an envelope of the secondary waves. Fresnel improved upon the ideas of Huygens, and the theory is known as the Huygens–Fresnel theory of diffraction.

This theory was placed on firmer mathematical ground by Kirchhoff, who developed his mathematical theory using two assumptions about the boundary values of the light incident on the surface of an obstacle placed in its path of propagation. This theory is known as Fresnel–Kirchhoff diffraction theory. It was found that the two assumptions of Fresnel–Kirchhoff theory are mutually inconsistent. The theory, however, yields results that are in surprisingly good agreement with experiments in most situations. Sommerfeld modified Kirchhoff's theory by eliminating one of the assumptions, and the resulting theory is known as Rayleigh–Sommerfeld diffraction theory. Needless to say, there have been subsequent workers who have introduced several refinements to the diffraction theories.

3.1 FRESNEL DIFFRACTION

Let $u(x_o, y_o)$ be the field distribution at an aperture lying on the (x_o, y_o) plane located at $z = 0$. Under small-angle diffraction, the field at any point $P(x, y)$ (assumed to be

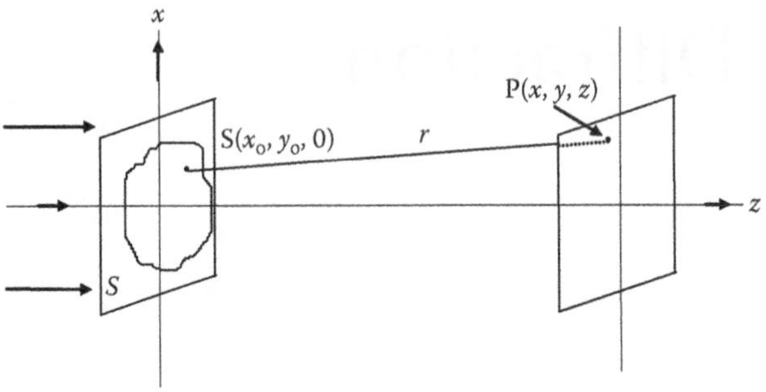

FIGURE 3.1 Diffraction at an aperture.

near the optical axis) on a plane a distance z from the obstacle or aperture plane can be expressed using Fresnel–Kirchhoff diffraction theory (Figure 3.1) as follows:

$$u(x, y) = \frac{1}{i\lambda z} \iint u(x_o, y_o) e^{ikr}\, dx_o\, dy_o \tag{3.1}$$

where the integral is over the area S of the aperture, and $k = 2\pi/\lambda$ is the propagation vector of the light. We can expand r in the phase term in a Taylor series as follows:

$$r = z + \frac{(x - x_o)^2 + (y - y_o)^2}{2z} - \frac{[(x - x_o)^2 + (y - y_o)^2]^2}{8z^3} + \cdots \tag{3.2}$$

If we impose the condition that

$$[(x - x_o)^2 + (y - y_o)^2]^2/8z^3 \ll \lambda \tag{3.3}$$

then we are in the Fresnel diffraction region. The Fresnel region may extend from very near the aperture to infinity. The condition 3.3 implies that the path variation at the point $P(x, y)$ as the point $S(x_o, y_o)$ spans the whole area S is much less than one wavelength. The amplitude $u(x, y)$ in the Fresnel region is given by

$$u(x, y) = \frac{e^{ikz}}{i\lambda z} \iint_S u(x_o, y_o) \exp\left\{\frac{ik}{2z}[(x - x_o)^2 + (y - y_o)^2]\right\} dx_o\, dy_o \tag{3.4}$$

Since we are invariably in the Fresnel region in most of the situations in optics, Equation 3.4 will be used frequently.

3.2 FRAUNHOFER DIFFRACTION

A more stringent constraint may be placed on the distance z and/or on the size of the obstacle or aperture by setting

$$\frac{x_o^2 + y_o^2}{2z} \ll \lambda. \tag{3.5}$$

This is the far-field condition, and when this condition is met, we are in the Fraunhofer region. The amplitude $u(x, y)$ is then given by

$$u(x,y) = \frac{e^{ikz}}{i\lambda z}\exp\left[\frac{ik}{2z}(x^2+y^2)\right]\iint_s u(x_o,y_o)\exp\left[-\frac{ik}{z}(xx_o+yy_o)\right]dx_o\,dy_o \quad (3.6)$$

Let us consider the illumination of the aperture by a plane wave of amplitude A. The transmittance function $t(x_o, y_o)$ of the aperture may be defined as

$$t(x_o, y_o) = u(x_o, y_o)/A$$

Therefore, the field $u(x, y)$ may be expressed as

$$\begin{aligned} u(x,y) &= \frac{Ae^{ikz}}{i\lambda z}\exp\left[\frac{ik}{2z}(x^2+y^2)\right]\int_{-\infty}^{\infty}\int_{-\infty}^{\infty} t(x_o,y_o)\exp\left[-\frac{ik}{z}(xx_o+yy_o)\right]dx_o\,dy_o \\ &= \frac{Ae^{ikz}}{i\lambda z}\exp\left[\frac{ik}{2z}(x^2+y^2)\right]\int_{-\infty}^{\infty}\int_{-\infty}^{\infty} t(x_o,y_o)\exp[-2\pi i(\mu x_o+\nu y_o)]\,dx_o\,dy_o \end{aligned} \quad (3.7)$$

The limits of integration have been changed to $\pm\infty$, since $t(x_o, y_o)$ is nonzero within the opening and zero outside. It can be seen seen that $u(x, y)$ can be expressed as the Fourier transform of the transmittance function. The Fourier transform is evaluated at the spatial frequencies $\mu = x/\lambda z$ and $\nu = y/\lambda z$.

3.3 ACTION OF A LENS

A plane wave incident on a lens is transformed to a converging/diverging spherical wave; a converging spherical wave will come to a focus. Therefore, a lens may be described by a transmittance function $\tau(x_L, y_L)$, which is represented as

$$\tau(x_L, y_L) = \exp(i\phi_0)\exp\left[\pm\frac{ik}{2f}(x_L^2+y_L^2)\right] \quad \text{for } x_L^2+y_L^2 \le R^2 \quad (3.8)$$

where φ_0 is a constant phase, $(k/2f)(x_L^2 + y_L^2)$ is a quadratic approximation to the phase of a spherical wave, and $2R$ is the diameter of the lens aperture. The minus sign corresponds to a positive lens. The lens is assumed to be perfectly transparent; that is, it does not introduce any attenuation. If a mask is placed in front of the lens, its transmittance function is obtained by multiplying the transmittance of the mask by that of the lens.

3.4 IMAGE FORMATION AND FOURIER TRANSFORMATION BY A LENS

We consider the geometry shown in Figure 3.2. A transparency of amplitude transmittance $t(x_o, y_o)$ is illuminated by a diverging spherical wave emanating from a point

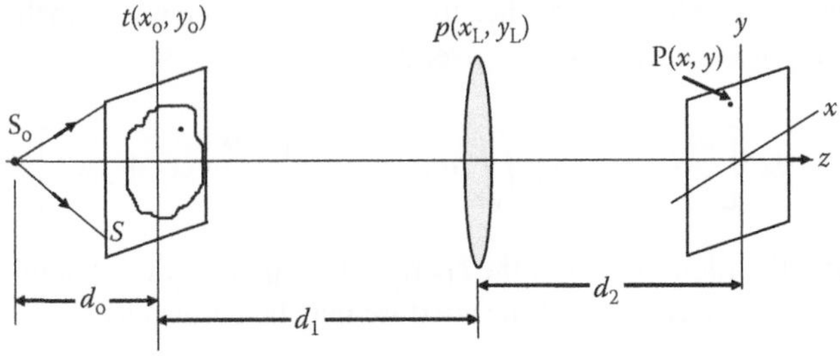

FIGURE 3.2 Imaging and Fourier transformation by a lens.

source S_o. We wish to obtain the amplitude distribution at a plane a distance d_2 from the lens. This is done by invoking Fresnel–Kirchhoff diffraction theory twice in succession: first at the transparency plane and then at the lens plane. The amplitude at any point $P(x, y)$ on the observation plane is expressed as

$$u(x,y) = -\frac{Ae^{ik(d_o+d_1+d_2)}}{\lambda^2 d_o d_1 d_2} \exp\left[\frac{ik(x^2+y^2)}{2d_2}\right]$$
$$\times \iiiint p(x_L, y_L) \exp\left[\frac{ik(x_L^2+y_L^2)}{2}\left(\frac{1}{d_1}+\frac{1}{d_2}-\frac{1}{f}\right)\right]$$
$$\times\, t(x_o, y_o) \exp\left[\frac{ik(x_o^2+y_o^2)}{2}\left(\frac{1}{d_o}+\frac{1}{d_1}\right)\right]$$
$$\times \exp\left\{-ik\left[x_L\left(\frac{x_o}{d_1}+\frac{x}{d_2}\right)+y_L\left(\frac{y_o}{d_1}+\frac{y}{d_2}\right)\right]\right\} dx_o\, dy_o\, dx_L\, dy_L \tag{3.9}$$

where we have introduced the lens pupil function

$$p(x_L, y_L) = \begin{cases} 1 & \text{for } (x_L^2+y_L^2)^{1/2} \le R \\ 0 & \text{otherwise} \end{cases}$$

The limits of integration on the lens plane are therefore taken as $\pm\infty$. Writing $1/\varepsilon = 1/d_1 + 1/d_2 - 1/f$, we evaluate the integral over the lens plane in Equation 3.9 assuming $p(x_L, y_L) = 1$ over the infinite plane. This holds valid provided that the entire diffracted field from the transparency has been accepted by the lens. In other words, there is no diffraction at the lens aperture. On substitution of the result thus obtained, we can obtain the amplitude $u(x, y)$ as

$$u(x,y) = -\frac{Ae^{ik(d_o+d_1+d_2)}}{\lambda d_o d_1 d_2}\varepsilon(1+i)^2 \exp\left[\frac{ik(x^2+y^2)}{2d_2}\left(1-\frac{\varepsilon}{d_2}\right)\right]$$

$$\times \iint t(x_o, y_o) \exp\left[\frac{ik(x_o^2 + y_o^2)}{2}\left(\frac{1}{d_o} + \frac{1}{d_1} - \frac{\varepsilon}{d_1^2}\right)\right]$$
$$\times \exp\left[-\frac{ik\varepsilon(xx_o + yy_o)}{d_1 d_2}\right] dx_o\, dy_o \tag{3.10}$$

We now examine both Equation 3.9 and Equation 3.10 to see under what conditions the (x, y) plane is an image plane or a Fourier transform plane of the (x_o, y_o) plane.

3.4.1 Image Formation

It is known in geometrical optics that an image is formed when the imaging condition $1/d_1 + 1/d_2 - 1/f = 0$ is satisfied. We invoke this condition and see whether the amplitude distribution at the (x, y) plane is functionally similar to that existing at the (x_o, y_o) plane. We begin with Equation 3.9, and invoke the imaging condition to give

$$u(x, y) = -\frac{Ae^{ik(d_o+d_1+d_2)}}{\lambda^2 d_o d_1 d_2} \exp\left[\frac{ik(x^2 + y^2)}{2d_2}\right]$$
$$\times \iiiint p(x_L, y_L) t(x_o, y_o) \exp\left[\frac{ik(x_o^2 + y_o^2)}{2}\left(\frac{1}{d_o} + \frac{1}{d_1}\right)\right]$$
$$\times \exp\left\{-ik\left[x_L\left(\frac{x_o}{d_1} + \frac{x}{d_2}\right) + y_L\left(\frac{y_o}{d_1} + \frac{y}{d_2}\right)\right]\right\} dx_o\, dy_o\, dx_L dy_L \tag{3.11}$$

We evaluate the integral over the lens plane assuming no diffraction. We thus obtain

$$u(x, y) = -\frac{Ae^{ik(d_o+d_1+d_2)}}{d_o M} \exp\left[\frac{ik(x^2 + y^2)}{2d_2}\right]$$
$$\times \exp\left[\frac{ik(x^2 + y^2)}{2M^2}\left(\frac{1}{d_o} + \frac{1}{d_1}\right)\right] t\left(-\frac{x}{M}, -\frac{y}{M}\right)$$
$$= \frac{C}{M} t\left(-\frac{x}{M}, -\frac{y}{M}\right) \tag{3.12}$$

where $M(= d_2/d_1)$ is the magnification of the system and C is a complex constant. The amplitude $u(x, y)$ at the observation plane is identical to that of the input transparency, except for the magnification. Hence, an image of the transparency is formed on the plane that satisfies the condition $1/d_1 + 1/d_2 = 1/f$.

3.4.2 Fourier Transformation

If the quadratic term within the integral in Equation 3.10 is zero, then the amplitude $u(x, y)$ is a two-dimensional Fourier transform of the function $t(x_o, y_o)$ multiplied by a quadratic phase factor, provided that the limits of integration extend from $-\infty$ to ∞. For the quadratic term to vanish, the following condition should be satisfied:

$$\frac{1}{d_o} + \frac{1}{d_1} - \frac{\varepsilon}{d_1^2} = 0 \tag{3.13}$$

This condition is satisfied when $d_o = \infty$ and $d_2 = f$, because then $\varepsilon = d_1$. Thus, the Fourier transform of the transparency $t(x_o, y_o)$ illuminated by a collimated beam is observed at the back focal plane of the lens. It should be noted that the lens does not take the Fourier transform. The Fourier transform is extracted by propagation, and the lens merely brings it from the far field to the back focal plane.

The amplitude $u(x, y)$ when the transparency is illuminated by a collimated beam of amplitude A_p is given by

$$u(x, y) = -\frac{A_p}{\lambda f} e^{ik(d_1+f)}(1+i)^2 \exp\left[\frac{ik}{2f}(x^2+y^2)\left(1-\frac{d_1}{f}\right)\right] \times \iint t(x_o, y_o) \exp\left[-\frac{ik}{2f}(xx_o + yy_o)\right] dx_o\, dy_o \tag{3.14}$$

The position-dependant quadratic phase term in front of the integral sign will also vanish if the transparency is placed at the front focal plane of the lens. The amplitude is then given by

$$u(x, y) = u_o \iint t(x_o, y_o) \exp\left[-\frac{ik}{2f}(xx_o + yy_o)\right] dx_o\, dy_o \tag{3.15}$$

where u_0 is a complex constant. Here $u(x, y)$ represents the pure Fourier transform of the function $t(x_o, y_o)$.

We thus arrive at the following conclusions:

1. The Fourier transform of a transparency illuminated by a collimated beam is obtained at the back focal plane of a lens. However, it includes a variable quadratic phase factor.
2. The pure Fourier transform (without the quadratic phase factor) is obtained when the transparency is placed at the front focal plane of the lens.

We have considered the very specific case of plane-wave illumination of the transparency. It can be shown that when the transparency is illuminated by a spherical wave, the Fourier transform is still obtained, but at a plane other than the focal plane. As mentioned earlier, the Fourier transform is obtained when Equation 3.13 is satisfied. This equation can also be written as follows:

$$\frac{1}{d_o} + \frac{1}{d_1} - \frac{\varepsilon}{d_1^2} = \frac{d_2 f + f(d_1 + d_o) - d_2(d_1 + d_o)}{d_2(d_2 + f + d_1 f - d_1 d_2)} = 0 \tag{3.16}$$

Equation 3.16 is satisfied when

$$\frac{1}{d_o + d_1} + \frac{1}{d_2} - \frac{1}{f} = 0 \tag{3.17}$$

This equation implies that the Fourier transform of an input is obtained at a plane that is conjugate to the point source plane. Indeed, the pure Fourier transform with non-unit scale factor is obtained when a transparency placed at the front focal plane is illuminated by a diverging spherical wave. It is also possible to obtain the Fourier transform with many other configurations.

3.5 OPTICAL FILTERING

The availability of the Fourier transform at a physical plane allows us to modify it using a transparency or a mask called a filter. In general, the Fourier transform of an object is complex; that is, it has both amplitude and phase. A filter should therefore act upon both the amplitude and phase of the transform; such a filter has complex transmittance. In some cases, certain frequencies from the spectrum are filtered out, which is achieved by a blocking filter. Experiments of Abbé and also of Porter for the verification of Abbé's theory of microscope imaging are beautiful examples of optical filtering. Figure 3.3 shows a schematic view of an optical filtering set-up. This is the well-known 4*f* processor. The filter, either as a mask or a transparency, is placed at the filter plane, where the Fourier transform of the input is displayed. A further Fourier transformation yields a filtered image. The technique is used to process images blurred by motion or corrupted by aberrations of the imaging system. One of the most common applications of optical filtering is to clean a laser beam using a tiny pinhole at the focus of a microscope objective. Such an arrangement of a microscope objective with a tiny pinhole at its focal plane is known as a spatial filter.

Phase-contrast microscopy exploits optical filtering in which the zeroth order is phase-advanced or phase-retarded by $\pi/2$, thereby converting phase variation in a phase object into intensity variations. The schlieren technique used in aerodynamics and combustion science to visualize a refractive index gradient is another example of optical filtering. The Foucault knife-edge test for examining the figure of a mirror is yet another example of Fourier filtering. One of the most commonly used techniques to clean a laser beam is spatial filtering, which filters out all Fourier components that

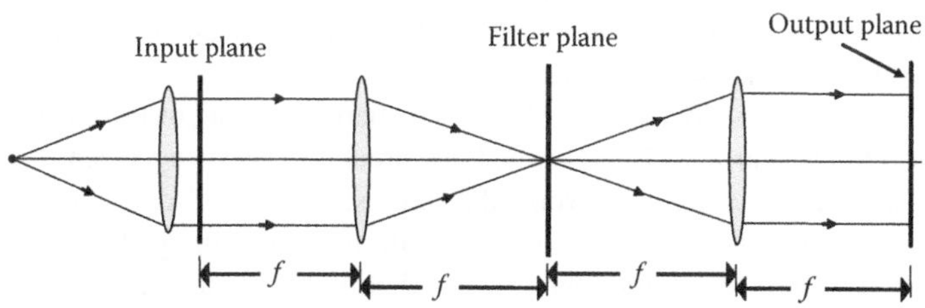

FIGURE 3.3 4*f* arrangement for optical filtering.

make the beam dirty. A special mask can be used at the aperture stop in an imaging lens to alter its transfer function, and hence selectively filter the band of frequencies for image formation. Optical filtering is used in data processing, image processing, and pattern recognition. Theta modulation and frequency modulation are used to encode an object so that certain features can be extracted by optical filtering.

3.6 OPTICAL COMPONENTS IN OPTICAL METROLOGY

Optical components can be classified as reflective, refractive, or diffractive.

3.6.1 Reflective Optical Components

Plane mirrors, spherical mirrors, parabolic mirrors, and ellipsoidal mirrors all fall into the reflective category. Mirrors are used for imaging. A plane mirror, although often used simply for changing the direction of light beam, can also be considered an imaging element that always forms a virtual image free from aberrations. On the other hand, a spherical mirror forms both a real and a virtual image: a convex mirror always forms a reduced virtual image, while a concave mirror can form both real and virtual images of an object, depending on the object's position and the focal length of the mirror. Mirror imaging is free from chromatic aberrations, but suffers from achromatic aberrations. A parabolic mirror is usually used as the primary in a telescope, while ellipsoidal mirrors are used in laser cavities.

Mirrors are usually metallic-coated. Silver, gold, chromium, and aluminum are among the metals used as reflective coatings. The reflectivity can be altered by depositing a dielectric layer on a metallic coating; an example is enhanced aluminum coating. Alternatively, a mirror can be coated with multiple layers of dielectric materials to provide the desired reflectivity.

3.6.2 Refractive Optical Components

Refractive optics includes both imaging and nonimaging optics. Lenses are used for imaging: a concave lens always forms a virtual image, while a convex lens can form both real and virtual images, depending on the object's position and the focal length of the lens. Lens imaging suffers from both achromatic and chromatic aberrations. A shape factor is used to reduce the achromatic aberrations of a single lens. A lens designer performs the task of designing a lens with optimized aberrations, bearing in mind both cost and ease of production. These lenses have several elements, including diffractive elements in special lenses.

In theory, we assume that a lens is free from aberrations and of infinite size so that diffraction effects can also be neglected. In practice, assuming a lens to be well designed and theoretically free from aberrations, it still suffers from diffraction because of its finite size. A point object at infinity is not imaged as a point even by an aberration-free lens, but rather as a distribution known as the Airy distribution.

The intensity distribution in the image of point source at infinity is given by $[2J_1(x)/x]^2$, where the argument x of the Bessel function $J_1(x)$ depends on the

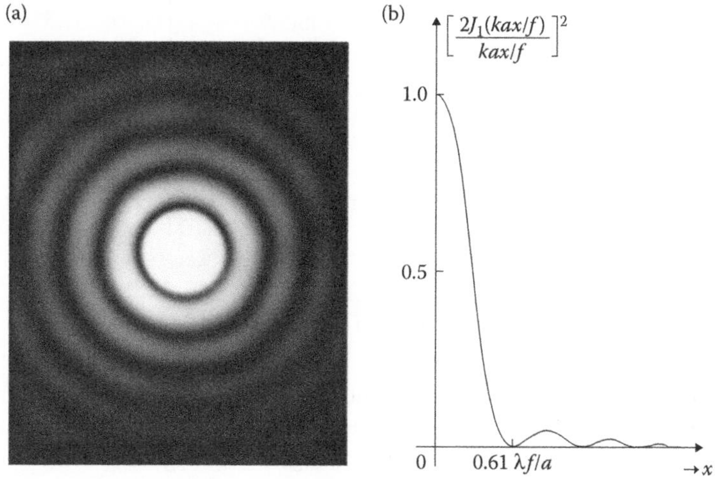

FIGURE 3.4 Diffraction at a lens of circular aperture. (a) Photograph of the intensity distribution at the focus when a plane wave is incident on it. (b) Intensity distribution.

f-number of the lens, the magnification, and the wavelength of light. This distribution is plotted in Figure 3.4b. Figure 3.4a shows a photograph of an Airy pattern. It can be seen that the image of a point source is not a point but a distribution, which consists of a central disk, called the Airy disk, within which most of the intensity is contained, surrounded by circular rings of decreasing intensity. The radius of the Airy disk is $1.22\lambda f/2a$, where $2a$ is the lens aperture and f is the focal length of the lens. The Airy disk defines the size of the image of a point source. It should be kept in mind that if the beam incident on the lens has a diameter smaller than the lens aperture, then $2a$ in the expression for the disk radius must be replaced by the beam diameter. The intensity distribution in the diffraction pattern outside the Airy disk is governed by the shape and size of the aperture.

If we carry out an analysis in the spatial frequency domain, the lens is considered to be a low-pass filter. The frequency response can be modified by apodization: the use of an appropriate aperture mask on the lens alters its frequency response dramatically. As an obvious example, the use of an annular aperture in a telescope enhances its angular resolution. Similar arguments apply when imaging by mirrors is considered.

Prisms function as both reflective and refractive components. When used as dispersive elements in spectrographs and monochromators, prisms function purely as refractive components. However, when used for bending and splitting of beam, they use both refraction and reflection. As an example, we consider a right-angle prism, which can be used to bend rays by 90° or by 180°, as shown in Figure 3.5.

When employed as in Figure 3.5b as a reflector, a prism can be used to introduce lateral shear. It may be observed that the beam bending is by total internal reflection. For deviating beams at other angles, special prisms are designed in which there can be metallic reflection. A Dove prism is used for image rotation. In interferometry, a corner cube is used as a reflector in place of a mirror owing to its insensitivity to

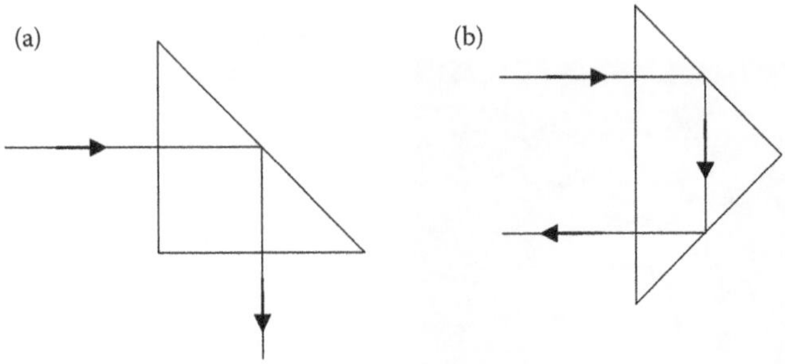

FIGURE 3.5 Right-angle prisms used to bend a beam by (a) 90° and (b) 180°.

angular misalignment. Two cemented right-angle prisms with dielectric or metallic coatings on the hypotenuse faces form a beam-splitter. The entrance and exit faces have antireflective coatings. A dielectric coating on the diagonal surface can be so designed that a cube beam-splitter also functions as a polarization beam-splitter. In many interferometers, plane-parallel plates are used both as beam-splitters and as beam-combiners. One of the surfaces has an antireflective coating. Plane-parallel plates with uncoated surfaces have been used as shear elements in shear interferometry. Sometimes, wedge plates are used as beam-splitters and beam-combiners, and also as shear elements.

Some optical elements are constructed from anisotropic materials such as calcite and quartz. Savart and half-shade plates are examples of such elements, as are Nichol, Wollaston, Rochon, and Senarmont prisms. These elements are used in polarization interferometers, polarimeters, and ellipsometers.

3.6.3 Diffractive Optical Components

A grating is an example of a diffractive optical element. In general, optical elements that utilize diffraction for their functions are classed as diffractive optical elements (DOEs). Computer-generated holograms (CGHs) and holographic optical elements (HOEs) also fall into this category. Their main advantages are flexibility in size, shape, layout, and choice of materials. They can carry out multiple functions simultaneously, for example bending and focusing of a beam. They are easy to integrate with the rest of the system. DOEs offer the advantages of reduction in system size, weight, and cost. Because of their many advantages, considerable effort has gone into designing and fabricating efficient DOEs. DOEs can be produced using a conventional holographic method. However, most such components are formed by either a laser or an electron-beam writing system on an appropriate substrate. Mass production is by hot embossing or ultraviolet embossing, or by injection molding. These elements can also be fabricated using diamond turning.

DOEs can be used as lenses, beam-splitters, polarization elements, or phase-shifters in optical systems. A zone plate is an excellent example of a diffractive lens that produces multiple foci. However, using multiple phase steps or a continuous phase profile, diffractive lenses with efficiency approaching 100% can be realized. Gratings can be designed that split the incident beam into two beams of equal intensity, functioning as a conventional beam-splitter. Alternatively, it can yield two beams of equal intensity in reflection as shown in Figure 3.6. These beams are then combined by another DOE, thereby making a very compact Mach–Zehnder interferometer.

A polarizer has been realized in a photoresist on a glass plate in which a deep binary grating is written. In fact, diffractive elements with submicrometer pitch exhibit strong polarization dependence. An interesting application of DOEs is in phase-shifting interferometry. For studying transient events, spatial phase-shifting is used such that three or four phase-shifted data are available from a single interferogram. This may be done by using multiple detectors or a spatial carrier with a single detector. A DOE has also been produced in such a way that when placed in a reference arm, it provides four 90° phase-shifted reference wavefronts. Thereby, four phase-shifted interferograms are obtained simultaneously, from which the phase at each pixel is obtained using a four-step algorithm.

Gratings (linear, circular, spiral, or concave) are dispersive elements, and hence are used mainly in spectrometers, spectrophotometers, and similar instrumements. When used with monochromatic radiation, gratings are useful elements in metrology. A low-frequency grating is used in moiré metrology. High-accuracy measurements are performed with gratings of fine pitch. A grating is usually defined as a periodic arrangement of slits (opaque and transparent regions). A collimated beam incident on a grating is diffracted into a number of orders. When a grating of pitch d is illuminated normally by a monochromatic beam of wavelength λ, the mth diffraction order is in the direction θ_m such that

$$d \sin \theta_m = m\lambda \quad \text{for } m = 0, \pm 1, \pm 2, \pm 3, \ldots \tag{3.18}$$

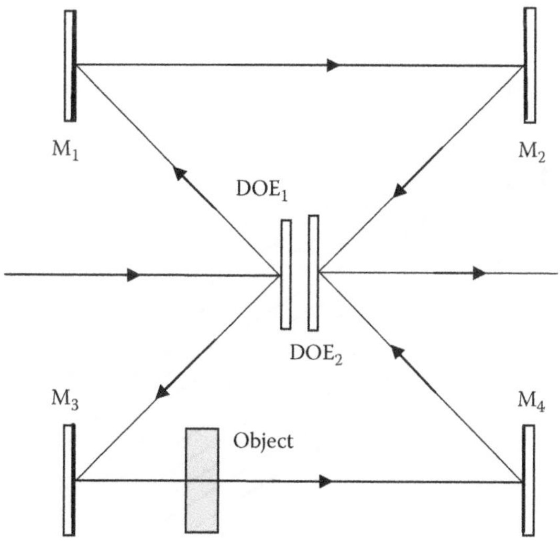

FIGURE 3.6 Mach–Zehnder interferometer using diffractive optical elements.

Such a grating will generate a large number of orders ($m \leq d/\lambda$, since $\theta_m \leq \pi/2$). Assuming small-angle diffraction (i.e., diffraction from a coarse grating) and diffraction in the first order, we obtain

$$\theta = \frac{\lambda}{d} \quad \text{or} \quad \frac{1}{d} = \frac{\theta}{\lambda} \tag{3.19}$$

Suppose a lens of focal length f is placed behind the grating, then this order will be focused to a spot such that $x = \lambda f/d$, or $1/d = x/\lambda f$, where x is the distance of the spot from the optical axis. The grating is assumed to be oriented such that its grating elements are perpendicular to the x axis. We define the spatial frequency μ of the grating as the inverse of the period d:

$$\mu = \frac{1}{d} = \frac{x}{\lambda f} \tag{3.20}$$

It is expressed as lines per millimeter (lines/mm). Thus, the spatial frequency μ and off-axis distance x are linearly related for a coarse grating. In fact, the spatial frequency is defined as $\mu = (\sin\theta)/\lambda$. It is thus obvious that a sinusoidal grating of frequency μ will generate only three orders: $m = 0, 1, -1$.

A real grating can be represented as a summation of sinusoidal gratings whose frequencies are integral multiples of the fundamental. When such a grating is illuminated normally by a collimated beam propagating in the z direction, a large number of spots are formed on either side of the z axis: symmetric pairs correspond to a particular frequency in the grating.

If a grating of pitch p is inclined with respect to the x axis as shown in Figure 3.7, then its spatial frequencies μ and ν along the x and y directions are $\mu = 1/d_x$ and

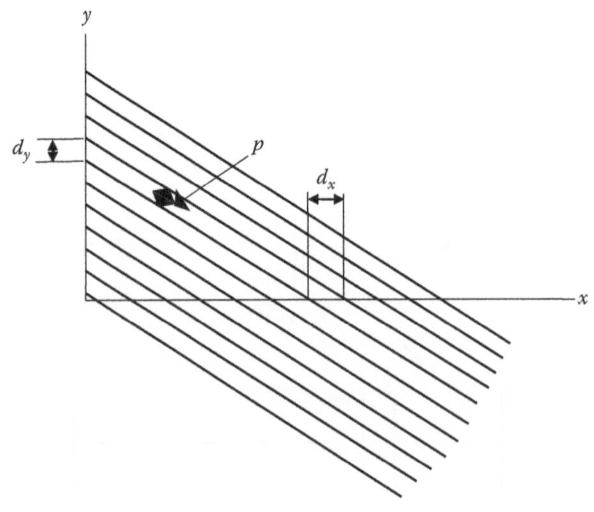

FIGURE 3.7 An inclined linear grating.

$\nu = 1/d_y$, such that

$$\rho^2 = \frac{1}{p^2} = \frac{1}{d_x^2} + \frac{1}{d_y^2} \tag{3.21}$$

where ρ is the spatial frequency of the grating. That is, $\rho^2 = \mu^2 + \nu^2$. Diffraction at the inclined sinusoidal grating again results in the formation of three diffraction orders, which will lie on neither the x nor the y axis. It can be shown that the diffraction spots always lie on a line parallel to the grating vector.

3.6.3.1 Sinusoidal Grating

Interference between two inclined plane waves of equal amplitude generates an interference pattern in which the intensity distribution is of the form

$$I(x) = 2I_0[1 + \cos(2\pi x/d_x)] \tag{3.22}$$

where d_x is the grating pitch and I_0 is the intensity of each beam. We associate a spatial frequency μ, which is the reciprocal of d_x; that is, $\mu = 1/d_x$. Thus, we can express the intensity distribution as

$$I(x) = 2I_0[1 + \cos(2\pi\mu x)] \tag{3.23}$$

Assuming an ideal recording material that maps the incident intensity distribution before exposure to the amplitude transmittance after exposure, the amplitude transmittance of the grating record is given by

$$t(x) = 0.5[1 + \cos(2\pi\mu x)] \tag{3.24}$$

The factor 0.5 appears owing to the fact that the transmittance varies between 0 and 1. Let a unit-amplitude plane wave be incident normally on the grating. The amplitude just behind the grating is

$$a(x) = 0.5(1 + \cos 2\pi\mu x) = 0.5(1 + 0.5e^{2\pi i\mu x} + 0.5e^{-2\pi i\mu x}) \tag{3.25}$$

This represents a combination of three plane waves: one wave propagating along the axis and other two propagating inclined to the axis. A lens placed behind the grating will focus these plane waves to three diffraction spots, each spot corresponding to a plane wave. These spots lie on the x axis in this particular case. If the transmittance function of the grating is other than sinusoidal, a large number of diffraction spots are formed.

Let us now recall the grating equation when a plane wave is incident normally; that is,

$$d_x \sin\theta_m = m\lambda$$

In the first order $m = 1$, and assuming small-angle diffraction, so that $\sin\theta \approx \theta$, we obtain

$$\frac{1}{d_x} = \mu = \frac{\theta}{\lambda}$$

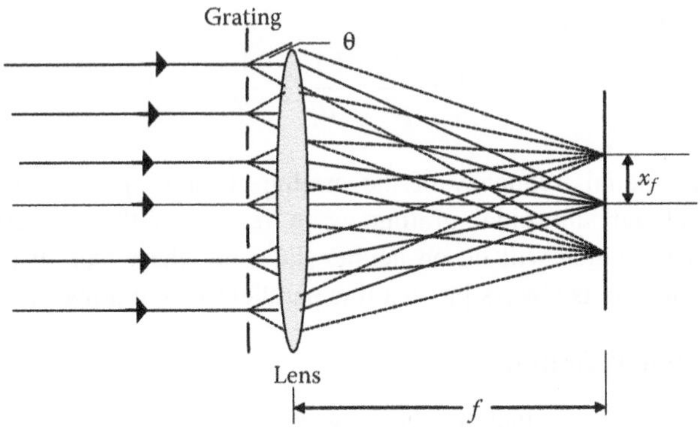

FIGURE 3.8 Diffraction at a sinusoidal grating.

From Figure 3.8, $\theta = x_f/f$, and hence $\mu = x_f/\lambda f$. Thus, the position of the spot is directly proportional to the frequency of the grating. This implies that diffraction at a sinusoidal grating generates three diffraction orders: the zeroth order lies on the optical axis and the first orders lie off-axis, with their displacements being proportional to the spatial frequency of the grating. These orders lie on a line parallel to the grating vector $(2\pi/d_x)$; the grating vector is oriented perpendicular to the grating elements. If the grating is rotated in its plane, these orders also rotate by the same angle.

A real grating generates many spots. These can easily be observed by shining an unexpanded laser beam onto the grating and observing the diffracted orders on a screen placed some distance away. This is due to the fact that the grating transmittance differs significantly from a sinusoidal profile. Since a grating is defined as a periodic structure of transparent and opaque parts, it gives rise to an infinite number of orders, subject to the condition that the diffraction angle $\theta \leq \pi/2$. It should be remembered that the linear relationship between spatial frequency and distance is valid only for low-angle diffraction or for coarse gratings.

3.6.4 Phase Grating

A phase grating has a periodic phase variation—it does not attenuate the beam. It can be realized either by thickness changes or refractive index changes or by a combination. One of the simplest methods is to record an interference pattern on a photographic emulsion and then bleach it. Assuming that the phase variation is linearly related to the intensity distribution, the transmittance of a sinusoidal phase grating can be expressed as

$$t(x) = \exp(i\phi_0)\exp[i\phi_m \cos(2\pi\mu_0 x)] \quad (3.26)$$

where ϕ_0 and ϕ_m are the constant phase and phase modulation, and μ_0 is the spatial frequency of the grating. This transmittance can be expressed as a Fourier–Bessel series. This implies that when such a grating is illuminated by a collimated beam, an infinite number of orders are produced, unlike the three orders from an amplitude sinusoidal grating. The phase grating, therefore, is intrinsically nonlinear. The intensity distribution in the various orders can be controlled by groove shape and modulation, and also by the frequency.

3.6.5 Diffraction Efficiency

Gratings can be either amplitude or phase gratings. The diffraction efficiency of an amplitude grating having a sinusoidal profile of the form $t(x) = 0.5 + 0.5\sin(2\pi\mu x)$ has a maximum value of 6.25%. However, if the profile is binary, the diffraction efficiency in first order is 10.2%. It will also produce a large number of orders. If the widths of the opaque and transparent parts are equal, then the even orders are missing. For a sinusoidal phase grating, the maximum diffraction efficiency is 33.9%; for a binary phase grating, it is 40.6%. If the grating is blazed, it may have 100% diffraction efficiency in the desired order. Thick phase gratings have high diffraction efficiencies, approaching 100%.

3.7 RESOLVING POWER OF OPTICAL SYSTEMS

Assume a point source emitting monochromatic spherical waves. An image of the point source can be obtained by following the procedure outlined in Section 3.4. The intensity distribution in the image of a point source is proportional to the square of the Fourier transform of the aperture function of an aberration-free imaging lens. As already mentioned, for a lens with circular aperture, the intensity distribution of the image of a point source is an Airy distribution, proportional to $[2J_1(x)/x]^2$. The image of a point source is not a point but a distribution, and hence two closely separated point sources can be seen distinctly in the image if they are adequately separated. The ability to see them distinctly depends on the resolving power of the imaging lens.

Resolving power is important for both imaging and dispersive instruments. In one case, one wants to know how close two objects could be seen as distinct; in the other, one wants to know how close two spectral lines could be seen as distinct. In both cases, the ability to see distinctly is governed by diffraction. In case of imaging, an imaging system intercepts only a part of the spherical waves emanating from a point source, thereby losing a certain amount of information. This results in an Airy distribution, with a central maximum surrounded by circular rings with decreasing intensity. For objects of the same intensity, it is easier to find a criterion of resolution. According to Rayleigh, two objects are just resolved when the intensity maximum of one coincides with the first intensity minimum of the other. The resulting intensity distribution shows two humps with a central minimum, as shown in Figure 3.9. The intensity of the central minimum in the case of a square aperture is 81% of the maximum. In practice, several resolution criteria have been proposed, but the most commonly used criterion is that due to Rayleigh.

However, in the case of two objects with one very much brighter than the other, it becomes extremely difficult to apply the Rayleigh criterion. Further, there are situations when the intensity distribution does not exhibit any secondary maxima. In such cases, the Rayleigh criterion states that two objects are just resolved when the intensity of the central minimum is 81% of the maximum.

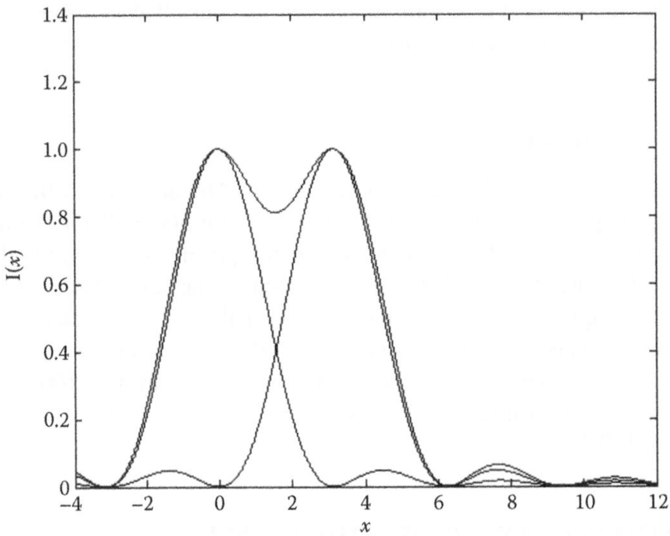

FIGURE 3.9 Rayleigh resolution criterion: maximum of first point image falls over the minimum of the second point image.

The Airy intensity distribution may be called an "instrumental function" of the imaging devices. Similarly, in the case of a grating used for spectral analysis, the intensity distribution is governed by $\sin^2\left(\frac{1}{2}N\delta\right)/\sin^2\left(\frac{1}{2}\delta\right)$, where δ is the phase difference between two consecutive rays. This is the instrumental function, which governs the resolution. An interferometer may therefore be described in terms of an instrumental function; a common situation where this arises is in Fourier-transform spectroscopy.

BIBLIOGRAPHY

1. A. Sommerfeld, *Optics*, Academic Press, New York, 1964.
2. W. T. Cathey, *Optical Information Processing and Holography*, Wiley, New York, 1974.
3. M. Born and E. Wolf, *Principles of Optics*, 7th edn, Cambridge University Press, Cambridge, 1999.
4. J. D. Gaskill, *Linear Systems, Fourier Transforms and Optics*, Wiley, New York, 1978.
5. D. Casasent (Ed.), *Optical Data Processing*, Springer-Verlag, Berlin, 1978.
6. S. H. Lee (Ed.), *Optical Data Processing—Fundamentals*, Springer-Verlag, Berlin, 1981.
7. R. S. Sirohi, *Wave Optics and Applications*, Orient Longmans, Hyderabad, 1993.
8. J. W. Goodman, *An Introduction to Fourier Optics*, McGraw-Hill, New York, 1996.
9. E. Hecht and A. R. Ganesan, *Optics*, 4th edn, Dorling Kindersley (India), New Delhi, 2008.

4 Phase-Evaluation Methods

Phase distribution is encoded in an intensity distribution as a result of interference phenomena, and is displayed in the form of an interference pattern. Phase distribution should therefore be retrievable from the interference pattern. Phase difference is related to parameters of interest such as the figure of a surface, height variations, and refractive index variations, which can be extracted from measurements of phase difference.

We employ interference phenomena in classical interferometry, in holographic interferometry, and in speckle interferometry to convert the phase of a wave of interest into an intensity distribution. Even in moiré methods, the desired information can be cast into the form of an intensity distribution. The fringes in an interference pattern are loci of constant phase difference. Earlier methods of extracting phase information, and consequently the related parameters of interest, were very laborious and time-consuming and suffered from the inherent inaccuracies of evaluation procedures. With the availability of desktop computers with enormous computational and processing power and CCD array detectors, many automatic fringe evaluation procedures have been developed.

There are many methods of phase evaluation. A comparison of these methods with respect to several parameters is presented in Table 4.1.

It should be noted that complexity and cost increase with increasing resolution. We will now discuss these methods in detail. During the course of this discussion, the significance of some of the parameters and characteristics of the various methods will become obvious.

4.1 INTERFERENCE EQUATION

Interference between two waves of the same frequency results in an intensity distribution of the form

$$I(x,y) = I_0(x,y)[1 + V(x,y)\cos\delta(x,y)] \tag{4.1}$$

where $I_0(x,y)$ is the low-frequency background, often called the DC background, $V(x,y)$ is the fringe visibility (or modulation), and $\delta(x,y)$ is the phase difference between the two waves, which is to be determined. Here (x,y) are the coordinates of a point, or of a pixel on an observation plane. The intensity distribution may be degraded

TABLE 4.1
Comparison of Some Important Methods of Phase Evaluation in Terms of Selected Parameters

Parameter	Method: Fringe Skeletonization	Phase-Stepping and Phase-Shifting	Fourier Transform	Temporal Heterodyning
Number of interferograms to be reconstructed	1	Minimum 3	1 (2)	1 per detection point
Resolution λ	1–1/10	1/10–1/100	1/10–1/30	1/100–1/1000
Evaluation between intensity extrema	No	Yes	Yes	Yes
Inherent noise suppression	Partially	Yes	No (Yes)	Partially
Automatic sign detection	No	Yes	No (Yes)	Yes
Necessary experimental manipulation	No	Phase-shift	No (Phase-shift)	Frequency
Experimental effort	Low	High	Low	Extremely high
Sensitivity to external influences	Low	Moderate	Low	Extremely high
Interaction by the operator	Possible	Not possible	Possible	Not possible
Speed of evaluation	Low	High	Low	Extremely low
Cost	Moderate	High	Moderate	Very high

by several factors, such as speckle noise, spurious fringe noise, and electronic noise. However, when all of these factors are included, the intensity distribution may still be expressed in the form

$$I(x,y) = a(x,y) + b(x,y)\cos\delta(x,y) \tag{4.2}$$

This equation has three unknowns: $a(x,y), b(x,y)$, and $\delta(x,y)$. We are mostly interested in obtaining $\delta(x,y)$, and hence will discuss the methods of phase evaluation only.

4.2 FRINGE SKELETONIZATION

The fringe skeletonization methods are computerized forms of the former manual fringe counting methods. The fringe pattern is received on a CCD detector. It is assumed that a local extremum of the intensity distribution corresponds to a maximum or minimum of the cosine function. (However, this requirement is not met by fringe patterns obtained in speckle photography.) The interference phase at pixels where an intensity maximum or minimum is detected is an even or odd integer multiple of π. The fringe skeletonization method seeks maxima or minima of an intensity distribution. The positions of maxima or minima in the fringe pattern are determined within several pixels. The method is thus similar to the manual method of locating the fringe maxima or minima, but is performed much more rapidly and with relatively high accuracy. Sign ambiguity is still not resolved.

Methods of fringe skeletonization can be divided into those based on fringe tracking and those related to segmentation. Both techniques require careful preprocessing to minimize speckle noise through low-pass filtering and to correct for uneven brightness distribution in the background illumination.

4.3 TEMPORAL HETERODYNING

In temporal heterodyning, a small frequency difference $\Delta\omega/2\pi = (\omega_1 - \omega_2)/2\pi$, which is less than 100 kHz, is introduced between the two interfering waves. The local intensity of the interference pattern then varies sinusoidally at the beat frequency $\Delta\omega/2\pi$. The intensity distribution can be expressed as

$$I(x, y; t) = a(x, y) + b(x, y)\cos[\Delta\omega t + \delta(x, y)] \tag{4.3}$$

There is no stationary stable interference pattern of the type familiar in interferometry; instead, the intensity at each point (x, y) varies sinusoidally at the beat frequency, and the interference phase at that point is transformed into the phase of the beat frequency signal. There is no way to measure the phase of this beat signal. However, as is obvious, the intensity at all points varies with the beat frequency, but the phases are different, being equal to the values of the interference phases at these points. As the beat frequency is sufficiently low, the phase difference between two points can be measured with very high accuracy independently of $a(x, y)$ and $b(x, y)$ by using two photodetectors and an electronic phase-meter. In this way, both the interpolation problem and the sign ambiguity of classical interferometry are solved. The method requires special equipment such as acousto-optic modulators for frequency shift and a phase-meter. The image (interference pattern) is scanned mechanically by photodetectors to measure the phase difference: one detector is fixed and the other is scanned, thereby measuring the interference phase with respect to the fixed photodetector. The measurement speed is low (typically 1 s per point), but the accuracy (typically $\lambda/1000$) and spatial resolution ($>10^6$ resolvable points) are extremely high.

In double-exposure holographic interferometry, the two interfering waves are released simultaneously from the hologram, and hence temporal heterodyning cannot be applied for phase evaluation. On the other hand, if a two-reference-wave set-up is used such that each state of the object is recorded with one of the waves as a reference wave, this technique can be implemented (see Chapter 6). In practice, two acousto-optic modulators in cascade are placed in the path of one of the reference waves. These modulators introduce frequency shifts of opposite signs. During recording, both modulators are driven at the same frequency, say, at 40 MHz, and hence the net frequency shift is zero. The initial and final states of the object are recorded sequentially with one reference wave at a time. The reconstruction, however, is performed with both reference waves simultaneously, with one modulator driven at 40 MHz and the other at 40.1 MHz. Therefore, the reference waves are of different frequencies, with a frequency difference of 100 kHz, resulting in an interference pattern that oscillates at that frequency.

One way to evaluate the interference pattern is to keep one detector fixed at the reference point while the other is scanned over the image. In this way, the phase difference is measured with respect to the phase of the reference point. The phase differences are stored, arranged, and displayed. Instead of two detectors, one could use an array of three, four, or five detectors to scan the whole image. In this way, the phase differences $\delta\delta_x$ and $\delta\delta_y$ between adjacent points along the x and y directions with known separations are measured. Numerical integration of recorded and stored phase differences gives the interference phase distribution.

4.4 PHASE-SAMPLING EVALUATION: QUASI-HETERODYNING

According to Equation 4.2, the intensity distribution in an interference pattern has three unknowns: $a(x,y), b(x,y)$, and $\delta(x,y)$. Therefore, if three intensity values at each point are available, we could set up three equations and solve them for the unknowns. These three intensity values are obtained by changing the phase of the reference wave. For the sake of generality, we set up n equations.

Two different approaches are available for quasi-heterodyning phase measurement: the phase-step method and the phase shift method. In the phase-step method, the local intensity $I_n(x,y)$ in the interference pattern is sampled at fixed phases α_n of the reference wave; that is,

$$I_n(x,y) = a(x,y) + b(x,y)\cos[\delta(x,y) + \alpha_n] \quad \text{for } n = 1,2,3,\ldots,N(n > 3) \tag{4.4}$$

As mentioned earlier, at least three intensity measurements, I_1, I_2, I_3, must be carried out to determine all the three unknowns.

In the phase-shifting or integrating bucket method, which is intended primarily for use with CCD detectors on which optical power is integrated by the detector, the phase of the reference wave is varied linearly and the sampled intensity is integrated over the phase interval $\Delta\alpha$ from $\alpha_n - \Delta\alpha/2$ to $\alpha_n + \Delta\alpha/2$. The intensity $I_n(x,y)$ sampled at a pixel at (x,y) is given by

$$\begin{aligned} I_n(x,y) &= \frac{1}{\Delta\alpha}\int_{\alpha_n-\Delta\alpha/2}^{\alpha_n+\Delta\alpha/2} \left\{a(x,y) + b(x,y)\cos\left[\delta(x,y) + \alpha(t)\right]\right\} d\alpha(t) \\ &= a(x,y) + \operatorname{sinc}(\Delta\alpha/2)\, b(x,y)\cos\left[\delta(x,y) + \alpha_n\right] \end{aligned} \tag{4.5}$$

This expression is equivalent to that of the phase-shifting method, the only difference being that the modulation $b(x,y)$ is multiplied by $\operatorname{sinc}(\Delta\alpha/2)$; there is a reduction in the contrast of fringes by $\operatorname{sinc}(\Delta\alpha/2)$. In this sense, the phase-shifting method is equivalent to the phase-stepping method, and the names are used synonymously. Furthermore, for data processing, both methods are handled in an identical manner.

4.5 PHASE-SHIFTING METHOD

For the solution of the nonlinear system of equations, we use the Gauss least-square approach and rewrite the intensity distribution as

$$I_n(x,y) = a(x,y) + b(x,y)\cos[\delta(x,y) + \alpha_n] = a + u\cos\alpha_n + v\sin\alpha_n \tag{4.6}$$

where $u(x,y) = b(x,y)\cos\delta(x,y)$ and $v(x,y) = -b(x,y)\sin\delta(x,y)$. We obtain a, u, and v by minimizing the errors; that is, the sum of the quadratic errors,

$$\sum_{n=1}^{N}\left[I_n(x,y) - (a + u\cos\alpha_n + v\sin\alpha_n)\right]^2$$

is minimized. Taking partial derivatives of this function with respect to a, u, and v and then equating these derivatives to zero gives a linear system of three equations:

$$\begin{pmatrix} N & \sum\cos\alpha_n & \sum\sin\alpha_n \\ \sum\cos\alpha_n & \sum\cos^2\alpha_n & \sum\sin\alpha_n\cos\alpha_n \\ \sum\sin\alpha_n & \sum\cos\alpha_n\sin\alpha_n & \sum\sin^2\alpha_n \end{pmatrix}\begin{pmatrix} a \\ u \\ v \end{pmatrix} = \begin{pmatrix} \sum I_n \\ \sum I_n\cos\alpha_n \\ \sum I_n\sin\alpha_n \end{pmatrix} \tag{4.7}$$

This system is to be solved pointwise. We thus obtain

$$a(x,y) = \frac{1}{N}\sum_{n=1}^{N} I_n \tag{4.8a}$$

$$b(x,y) = \frac{\sqrt{\left(\sum I_n\cos\alpha_n\right)^2 + \left(\sum I_n\sin\alpha_n\right)^2}}{N} \tag{4.8b}$$

$$\tan\delta(x,y) = -\frac{v}{u} = -\frac{\sum I_n\sin\alpha_n}{\sum I_n\cos\alpha_n} \tag{4.8c}$$

The interference phase is computed modulo 2π. Many algorithms have been mentioned in the literature. Some of these algorithms, derived from the interference equation, using three, four, up to eight steps, are presented in Table 4.2.

4.6 PHASE-SHIFTING WITH UNKNOWN BUT CONSTANT PHASE-STEP

There is a method known as the Carré method, which utilizes four phase-shifted intensity distributions; the phase-step need not be known, but must remain constant during phase-shifting. The four intensity distributions are expressed as

$$I_1(x,y) = a(x,y) + b(x,y)\cos\left[\delta(x,y) - 3\alpha\right] \tag{4.9a}$$

$$I_2(x,y) = a(x,y) + b(x,y)\cos\left[\delta(x,y) - \alpha\right] \tag{4.9b}$$

$$I_3(x,y) = a(x,y) + b(x,y)\cos\left[\delta(x,y) + \alpha\right] \tag{4.9c}$$

$$I_4(x,y) = a(x,y) + b(x,y)\cos\left[\delta(x,y) + 3\alpha\right] \tag{4.9d}$$

TABLE 4.2
Some Algorithms Derived from the Interference Equation

N	Phase-Step α_n	Expression for $\tan\delta(x, y)$
3	60° (0°, 60°, 120°)	$\dfrac{2I_1 - 3I_2 + I_3}{\sqrt{3}\,(I_2 - I_3)}$
3	90° (0°, 90°, 180°)	$\dfrac{I_1 - 2I_2 + I_3}{I_1 - I_3}$
3	120° (0°, 120°, 240°)	$\dfrac{\sqrt{3}\,(I_3 - I_2)}{2I_1 - I_2 - I_3}$
4	90° (0°, 90°, 180°, 270°)	$\dfrac{I_4 - I_2}{I_1 - I_3}$
4	60° (0°, 60°, 120°, 180°)	$\dfrac{5\,(I_1 - I_2 - I_3 + I_4)}{\sqrt{3}\,(2I_1 + I_2 - I_3 - 2I_4)}$
5	90° (0°, 90°, 180°, 270°, 360°)	$\dfrac{7\,(I_4 - I_2)}{4I_1 - I_2 - 6I_3 - I_4 + 4I_5}$
6	90° (0°, 90°, 180°, 270°, 360°, 450°)	$\dfrac{I_1 - I_2 - 6I_3 + 6I_4 + I_5 - I_6}{4(I_2 - I_3 - I_4 + I_5)}$
7	90° (0°, 90°, 180°, 270°, 360°, 450°, 540°)	$\dfrac{-I_1 + 7I_3 - 7I_5 + I_7}{-4I_2 + 8I_4 - 4I_6}$
8	90° (0°, 90°, 180°, 270°, 360°, 450°, 540°, 630°)	$\dfrac{-I_1 - 5I_2 + 11I_3 + 15I_4 - 15I_5 - 11I_6 + 5I_7 + I_8}{I_1 - 5I_2 - 11I_3 + 15I_4 + 15I_5 - 11I_6 - 5I_7 + I_8}$

For the sake of simplicity, the phase step is taken to be 2α. There are four unknowns, namely $a(x, y), b(x, y), \delta(x, y)$, and 2α, and hence it should be possible to obtain the values of all four from the four equations. However, we present the expressions only for the phase-step and interference phase:

$$\tan^2\alpha = \frac{3(I_2 - I_3) - (I_1 - I_4)}{(I_2 - I_3) + (I_1 - I_4)} \tag{4.10a}$$

$$\tan\delta(x, y) = \frac{(I_2 - I_3) + (I_1 - I_4)}{(I_2 + I_3) - (I_1 + I_4)}\tan\alpha \tag{4.10b}$$

These two equations are combined to yield the interference phase from the measured intensity values as follows:

$$\tan\delta(x, y) = \frac{\sqrt{\left[3(I_2 - I_3) - (I_1 - I_4)\right]\left[(I_2 - I_3) + (I_1 - I_4)\right]}}{(I_2 + I_3) - (I_1 + I_4)} \tag{4.11}$$

An improvement over the Carré method is realized by calculating the value of the phase-step over the whole pixel format, and then taking the average value for calculation. Constancy of the phase-step is also easily checked. The phase-step 2α is obtained from the expression

$$\cos 2\alpha = \frac{(I_1 - I_2) + (I_3 - I_4)}{2(I_2 - I_3)} \tag{4.12}$$

By assumption, the phase-step 2α must be constant over all the pixels. However, we take an average value, which smoothes out any fluctuations. The interference phase is then obtained as

$$\tan\delta(x,y) = \frac{(I_3 - I_2) + (I_1 - I_3)\cos\bar{\alpha}_1 + (I_2 - I_1)\cos 2\bar{\alpha}_1}{(I_1 - I_3)\sin\bar{\alpha}_1 + (I_2 - I_1)\sin 2\bar{\alpha}_1} \tag{4.13a}$$

$$\tan\delta(x,y) = \frac{(I_4 - I_3) + (I_2 - I_4)\cos\bar{\alpha}_1 + (I_3 - I_2)\cos 2\bar{\alpha}_1}{(I_2 - I_4)\sin\bar{\alpha}_1 + (I_3 - I_2)\sin 2\bar{\alpha}_1} \tag{4.13b}$$

where $2\bar{\alpha}_1$ is the average value of the phase-step 2α, averaged over all the pixels. The 2π steps of the interference phase distribution modulo 2π occur at different points in the image. This information can be used in the subsequent demodulation by considering the continuous phase variation of the two terms at each pixel.

There is another interesting algorithm, which also utilizes four intensity distributions but can be used for the evaluation of a very noisy interferogram. We consider an intensity distribution in the image of an object being studied using speckle techniques. The intensity distribution is expressed as

$$I(x,y) = a(x,y) + b(x,y)\cos(\phi_o - \phi_R), \tag{4.14}$$

where ϕ_o and ϕ_R are the phases of the speckled object wave and the reference wave; the phase difference $\phi = \phi_o - \phi_R$ is random. We capture four frames, the intensity distributions being expressed as

$$I_1(x,y) = a(x,y) + b(x,y)\cos(\phi - \pi/2) \tag{4.15a}$$

$$I_2(x,y) = a(x,y) + b(x,y)\cos\phi \tag{4.15b}$$

$$I_3(x,y) = a(x,y) + b(x,y)\cos\left[\phi + \delta(x,y)\right] \tag{4.15c}$$

$$I_4(x,y) = a(x,y) + b(x,y)\cos\left[\phi + \delta(x,y) + \pi/2\right] \tag{4.15d}$$

Here $\delta(x,y)$ is the phase introduced by deformation and is to be determined. From these equations, we obtain

$$I_3 - I_2 = -2b\sin(\phi + \delta/2)\sin(\delta/2) \tag{4.16a}$$

$$I_4 - I_1 = -2b\sin(\phi + \delta/2)\cos(\delta/2) \tag{4.16b}$$

From Equations 4.16a and 4.16b, the deformation phase is obtained easily as

$$\tan\frac{\delta(x,y)}{2} = \frac{I_3 - I_2}{I_4 - I_1} \tag{4.17}$$

The phase is obtained modulo 2π, as in other algorithms. However, the phase shift has to be exactly $\pi/2$.

All of the algorithms listed in Table 4.2 and utilizing a 90° phase shift require the phase-step to be exactly 90°; otherwise, errors are introduced. Simulation studies have indicated that the magnitude of errors continues to decrease with increasing number of phase-steps. An eight-step phase shift algorithm is practically immune to phase shift errors.

Phase-shifting methods are commonly used in all kinds of interferometers. The phase shift is usually introduced by calibrated PZT shifters attached to a mirror. This mechanical means of shifting the phase is quite common and convenient to use. There are various other methods, which rely on polarization and refractive index changes of the medium by an electrical field, and so on.

Advantages of the phase-shifting technique are (i) almost real-time processing of interferograms, (ii) removal of sign ambiguity, (iii) insensitivity to source intensity fluctuations, (iv) good accuracy even in areas of very low fringe visibility, (v) high measurement accuracy and repeatability, and (vi) applicability to holographic interferometry, speckle interferometry, moiré methods, and so on. Further, we can calculate the visibility and background intensity from the measured intensity values.

4.7 SPATIAL PHASE-SHIFTING

Temporal phase-shifting as described above requires stability of the interference pattern over the period of phase-shifting. However, in situations where a transient event is being studied or where the environment is highly turbulent, the results from temporal phase-shifting could be erroneous. To overcome this problem, the phase-shifted interferograms can be captured simultaneously using multiple CCD array detectors and appropriate splitting schemes. Recently, a pixelated array interferometer has also been developed that captures four phase-shifted interferograms simultaneously. All of these schemes, however, add to the cost and complexity of the system. Another solution is to use spatial phase-shifting. This requires an appropriate carrier. The intensity distribution in the interferogram can be expressed as

$$I(x, y) = a(x, y) + b(x, y)\cos\left[2\pi f_0 x + \phi(x, y)\right] \tag{4.18}$$

where f_0 is the spatial carrier along the x direction. The spatial carrier is chosen such that it has either three or four pixels on the adjacent fringes to apply three-step or four-step algorithms. There are several ways to generate the carrier frequency. The simplest is to tilt the reference beam appropriately. Polarization coding and gratings are also frequently used. We describe a procedure that is used in speckle metrology (Chapter 7).

We consider the experimental set-up shown in Figure 4.1. This arrangement is employed for the measurement of the in-plane displacement component.

The aperture separation is chosen such that the fringe width is equal to the width of three pixels of the CCD camera for the 120° phase shift algorithm, and to the width of four pixels for the 90° phase shift algorithm, as shown in Figure 4.2. The average speckle size is nearly equal to the fringe width. The intensity distribution $I(x_n, y)$ at

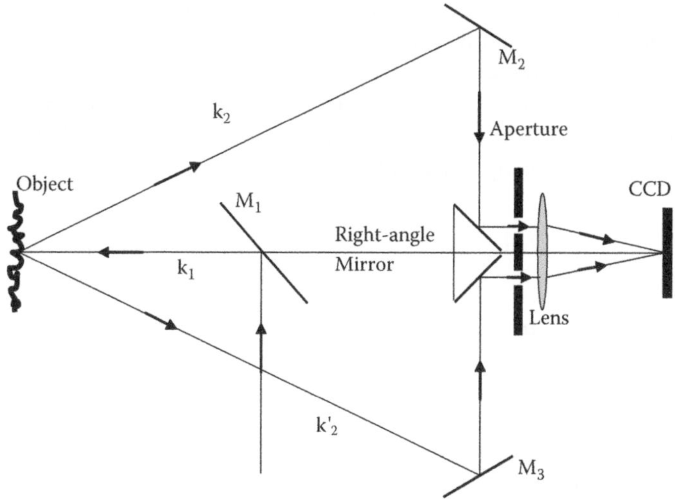

FIGURE 4.1 Measurement of in-plane component using spatial phase-shifting.

the nth pixel along the x direction is given by

$$I(x_n, y) = a(x_n, y) + b(x_n, y)\mathrm{sinc}\left(\frac{\Phi}{2}\right)\cos\left[\phi(x_n, y) + n\beta + C\right], \quad n = 1, 2, 3, \ldots, N \tag{4.19}$$

where $\phi(x_n, y)$ is the phase to be measured,

$$\beta = \frac{2\pi}{\lambda}\frac{d}{v}d_\mathrm{p}$$

is the phase change from one pixel to the next (which is separated by d_p along the x direction),

$$\Phi = \frac{2\pi}{\lambda}\frac{d}{v}d_\mathrm{t}$$

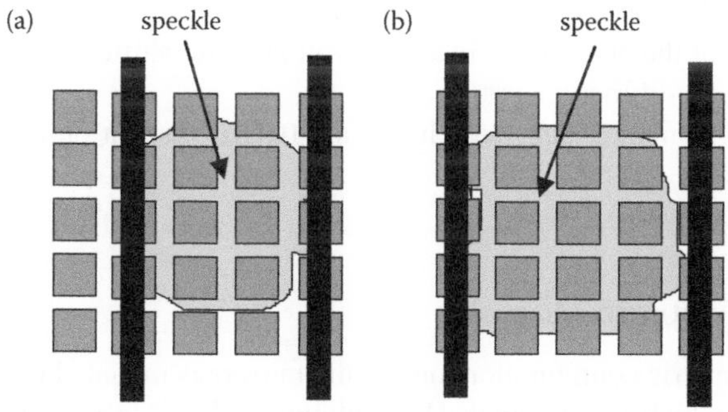

FIGURE 4.2 (a) 120° spatial phase-shifting. (b) 90° spatial phase-shifting.

is the phase shift angle over which the pixel of width d_t integrates the intensity, C is a constant phase offset, and N is the number of pixels in a row. Owing to object deformation, the initial phase $\phi_1(x_n, y)$ changes to $\phi_2(x_n, y) = \phi_1(x_n, y) + \delta(x_n, y)$. If $\phi_1(x_n, y)$ and $\phi_2(x_n, y)$ are measured, the phase difference $\delta(x_n, y)$ can be obtained. In order to calculate $\phi_n = \phi(x_n, y)$, we need at least three adjacent values of $I(x_n, y)$. If these values are picked up at 120° phase intervals, then the phase ϕ_n is given by

$$\phi_n = \tan^{-1}\left(\sqrt{3}\frac{I_{n-1} - I_{n+1}}{2I_n - I_{n-1} - I_{n+1}}\right) \quad \text{mod } \pi \tag{4.20}$$

$$I_{n+m} = a + b\,\text{sinc}\left(\frac{\Phi}{2}\right)\cos\left[\phi_n + \frac{2\pi}{\lambda}\frac{d}{v}d_p(n+m)\right],$$
$$n = 2, 3, \ldots, N-1; \quad m = -1, 0, +1 \tag{4.21}$$

The constant C has been dropped from Equation 4.21. It may be seen that in spatial phase-shifting, the modified phase

$$\psi_n = \phi_n + \frac{2\pi}{\lambda}\frac{d}{v}d_p n \quad \text{mod } \pi$$

rather than the speckle phase ϕ_n mod π, is reconstructed. Therefore, the phase offset

$$\frac{2\pi}{\lambda}\frac{d}{v}d_p n$$

must be subtracted in a later step.

4.8 METHODS OF PHASE-SHIFTING

A number of methods for phase shifting have been proposed in the literature. They usually introduce phase shifts sequentially. Some methods have also been adopted for simultaneous phase-shifts, and thus are useful for studying dynamical events. These methods include

- PZT-mounted mirror
- Tilt of a glass plate between exposures
- Rotation of the phase-plate in a polarization phase-shifter
- Motion of a diffraction grating
- Use of a computer-generated hologram (CGH) written on a spatial light modulator
- Special methods

4.8.1 PZT-MOUNTED MIRROR

In an interferometric configuration, one of the mirrors is mounted on lead zirconate titanate (PZT), which can be actuated by applying a voltage. In a Michelson interferometer, a shift of the mirror by $\lambda/8$ will introduce a phase shift of $\pi/2$ ($\lambda/4$ in path

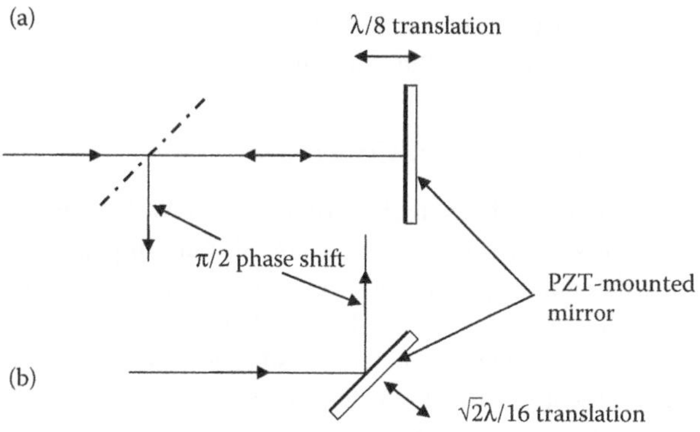

FIGURE 4.3 Phase-shifting by a PZT-driven mirror: (a) Michelson interferometer configuration; (b) Mach–Zehnder interferometer configuration.

change). Of course, any amount of phase shift can be introduced by a PZT-mounted mirror. In a Mach–Zehnder interferometer, the mirrors are inclined at 45°, and hence the mirror has to be shifted by $\sqrt{2}\lambda/16$ to introduce a phase shift of $\pi/2$. Figure 4.3 shows schematic representations of phase-shifting in Michelson and Mach–Zehnder interferometers. PZT-mounted mirrors can be used in any interferometric set-up, but the magnitude of the shift of mirror has to be calculated for each configuration. Care should be taken to avoid overshoot and hysteresis.

4.8.2 Tilt of Glass Plate

A glass plate of constant thickness and refractive index is introduced in the reference arm of the interferometer. The plate is tilted by an appropriate angle to introduce a required phase shift, as shown in Figure 4.4. It may be noted that the tilt also results in a lateral shift of the beam, although this is negligibly small. When a plate of refractive index μ, and uniform thickness t is placed normally to the beam, it introduces a path change $(\mu - 1)t$. When it is inclined such that the incident beam strikes it at an angle i, the plate introduces a path difference $\mu t \cos r - t \cos i$, where r is the angle of refraction in the plate. Tilt of the plate by an angle i thus introduces a net path difference $\mu t(1 - \cos r) - t(1 - \cos i)$. The plate is therefore tilted by an

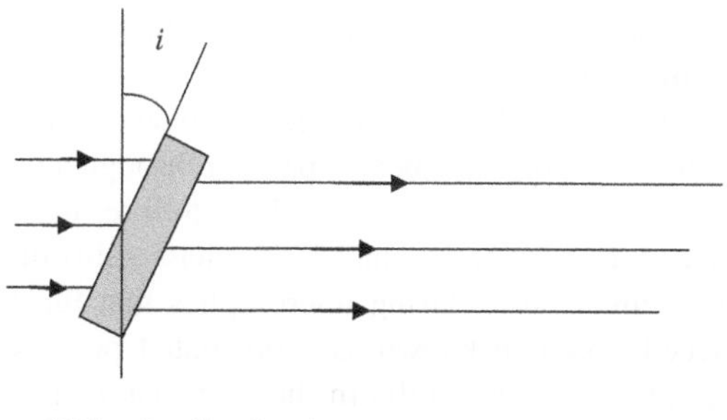

FIGURE 4.4 Phase-shifting by tilt of a plane parallel plate.

appropriate angle between exposures to introduce the desired phase step. This method is applicable only with collimated illumination.

4.8.3 Rotation of Polarization Component

Polarization components such as half-wave plates (HWPs), quarter-wave plates (QWPs), and polarizers have been used as phase-shifters in phase-shifting interferometry by rotating them in the path of the interfering beams. Different configurations of polarization-component phase-shifters have been used for phase-shifting, depending on their location in the interferometer. Possible locations are the input end, the output end, or one of the arms of the interferometer. Figure 4.5 shows configurations of phase-shifters at different locations in an interferometer.

Figure 4.5a shows a phase-shifter for the input end. It consists of a rotating half-wave plate followed by a quarter-wave plate fixed at an angle of 45°. The way in which polarization phase-shifters work can be understood with the help of the Jones calculus. The input to the interferometer is linearly polarized at 0°. The polarized light after the quarter-wave plate (QWP) Q is split by a polarization beam-splitter (PBS) into two orthogonal linearly polarized beams. The QWPs Q_1 and Q_2 serve to rotate the plane of polarization of the incoming beam by 90°, so that both the beams proceed towards the polarizer P oriented at 45°, which takes a component from each beam to produce an interference pattern. A rotation of the half-wave plate (HWP) H shifts the phase of the interference pattern. The shift of the phase is four times the angle of rotation of the HWP.

A phase-shifter for the output end is shown in Figure 4.5b. It consists of a quarter-wave plate at 45° followed by a rotating polarizer. Figure 4.5c shows a phase-shifter for use in one of the arms of an interferometer. The phase-shifter consists of a fixed quarter-wave plate and a rotating quarter-wave plate. A rotating polarizer P (Figure 4.5b) and a rotating QWP Q_3 (Figure 4.5c) will produce phase modulation at twice the rotation angle.

The intensity distribution is measured at different orientations of the rotating polarization component. A high degree of accuracy can be achieved by mounting the rotating component on a precision-divided circle with incremental or coded position information suitable for electronic processing.

Polarization interferometers for phase-shifting as discussed above can also be used with liquid crystals, which can provide variable retardation. Likewise, an electro-optic effect can also be used to produce a variable phase shift in a polarization interferometer. Polarization components have also been employed successfully in white-light phase-shifting interferometry.

In a variant of the method in which the polarizer is rotated, with the right circularly polarized and left circularly polarized beams passing through a polarizer inclined at an angle α, a phase change of 2α is introduced. Therefore, using a CGH together with polarizers oriented at 0°, 45°, 90°, and 135°, phase shifts of $\pi/2, \pi, 3\pi/2$ and 2π are introduced simultaneously. Using micro-optics that function as polarizers, WYCO has introduced a system known as a pixellated phase-shifter. This introduces $\pi/2, \pi, 3\pi/2$, and 2π phase shifts in the four quadrants, thereby achieving simultaneous phase-shifting.

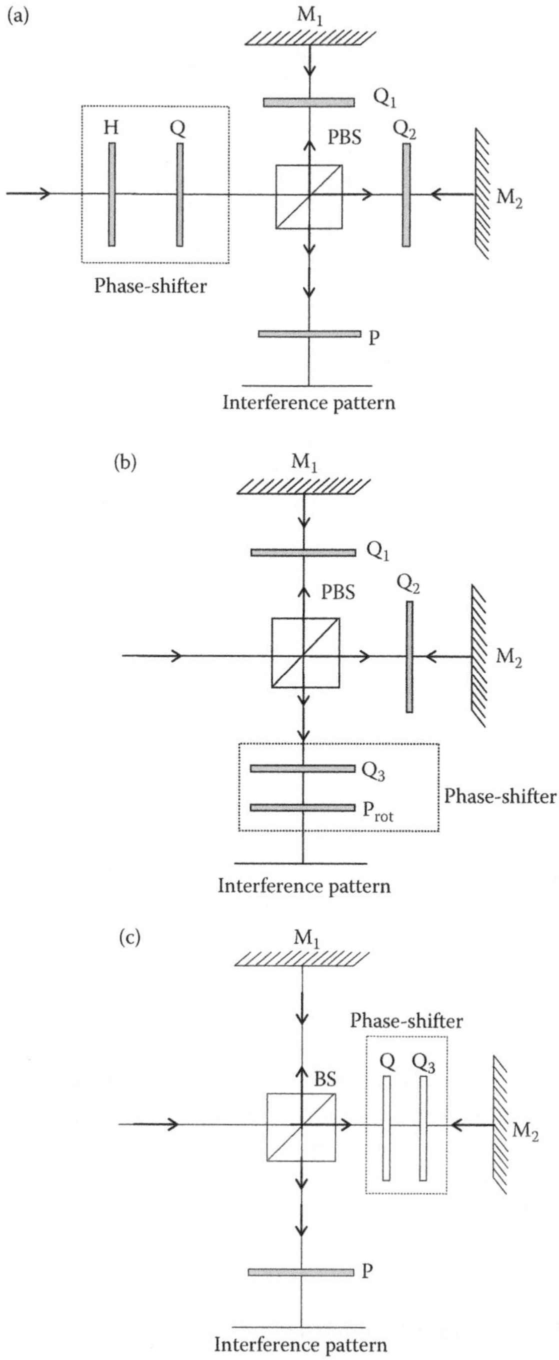

FIGURE 4.5 Schematic diagrams of polarization phase-shifters for use (a) at the input, (b) at the output, and (c) in the reference arm of an interferometer. PBS, polarizing beam-splitter; M_1, M_2, mirrors; P, fixed polarizer, Q, Q_1, Q_2, fixed quarter-wave plates; H, rotating half-wave plate; Q_3, rotating quarter-wave plate; P_{rot}, rotating polarizer.

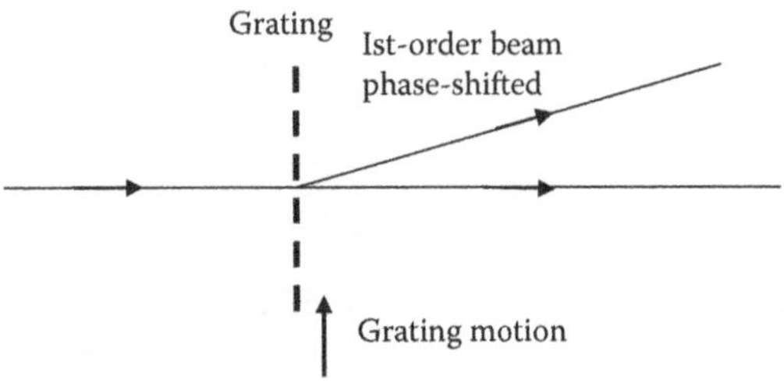

FIGURE 4.6 Phase-shifting by translation of a grating.

4.8.4 Motion of a Diffraction Grating

A diffraction grating produces several orders; the first-order beam is used for phase-shifting. Translation of the grating in its plane by its pitch p introduces a phase change of 2π in the diffracted first-order beam. Therefore, the grating is translated by $p/4$ and the frame is captured (Figure 4.6). Successive frames are captured by translating the grating by $p/4$ steps. Alternatively the grating is moved at a constant velocity v, and the frame is captured at instants $Np/4v$, where $N = 1, 2, 3$, and 4 for a four-step algorithm with $\pi/2$ step.

4.8.5 Use of a CGH Written on a Spatial Light Modulator

A phase shift of $\pi/2$ can be introduced using a CGH written on a spatial light modulator. A CGH of a known wavefront, usually a plane wavefront, is written using a phase-detour method. If each cell of the CGH is divided into four elements, and these elements are filled in according to the phase at that cell, then their shift by one element will introduce a global shift of $\pi/2$. Thus, the filled elements over the whole CGH are shifted by one element each after successive frames, to introduce sequential phase shifts of 0, $\pi/2$, π, and $3\pi/2$.

4.8.6 Special Methods

Common-path interferometers are used for optical testing, because to their insensitivity to vibrations and refractive index changes over the optical path. Since both test and reference beams traverse the same path, it is difficult to separate reference and test beams for phase-shifting. A clever arrangement has been suggested using a birefringent scatter-plate. There have also been some modifications of the Smartt point-diffraction interferometer.

4.9 FOURIER TRANSFORM METHOD

The Fourier transform method is used in two ways: (i) without a spatial frequency carrier and (ii) with a spatial frequency carrier added to it (spatial heterodyning). We will first describe the Fourier transform method without a spatial carrier. We again express the intensity distribution in the interference pattern as

$$I(x, y) = a(x, y) + b(x, y) \cos \delta(x, y)$$

With the cosine function written in complex exponentials, that is,

$$2 \cos \delta(x, y) = e^{i\delta(x,y)} + e^{-i\delta(x,y)}$$

the intensity distribution can be expressed as

$$\begin{aligned} I(x, y) &= a(x, y) + 2b(x, y)e^{i\delta(x,y)} + 2b(x, y)e^{-i\delta(x,y)} \\ &= a(x, y) + c(x, y) + c^*(x, y) \end{aligned} \tag{4.22}$$

where $c(x, y)$ and $c^*(x, y)$ are now complex functions. We take the Fourier transform of this distribution, yielding

$$I(\mu, \nu) = A(\mu, \nu) + C(\mu, \nu) + C^*(-\mu, -\nu) \tag{4.23}$$

with μ and ν being the spatial frequencies. Since the intensity distribution $I(x, y)$ is a real-valued function in the spatial domain, its Fourier transform is Hermitian in the frequency domain; that is,

$$I(\mu, \nu) = I^*(-\mu, -\nu) \tag{4.24}$$

The real part of $I(\mu, \nu)$ is even and the imaginary part is odd. The amplitude distribution $|I(\mu, \nu)|^{1/2}$ is point-symmetric with respect to the DC term $I(0, 0)$. Now it can be seen that $A(\mu, \nu)$ contains the zero peak $I(0, 0)$ and the low-frequency content due to slow variation of the background. $C(\mu, \nu)$ and $C^*(-\mu, -\nu)$ carry the same information, but with sign ambiguity. By bandpass filtering in the spatial frequency domain, $A(\mu, \nu)$ and one of the terms $C(\mu, \nu)$ or $C^*(-\mu, -\nu)$ is filtered out. The remaining spectrum $C^*(-\mu, -\nu)$ or $C(\mu, \nu)$ is not Hermitian, so the inverse Fourier transform gives complex $c(x, y)$ with nonvanishing real and imaginary parts. The interference phase $\delta(x, y)$ is obtained as follows:

$$\delta(x, y) = \tan^{-1} \frac{\text{Im}\{c(x, y)\}}{\text{Re}\{c(x, y)\}} \tag{4.25}$$

where Re{ } and Im{ } represent real and imaginary parts, respectively.

4.10 SPATIAL HETERODYNING

In spatial heterodyning, a carrier frequency is added to the interference pattern. The spatial frequency is chosen higher than the maximum frequency content in $a(x, y)$, $b(x, y)$, and $\delta(x, y)$. The intensity distribution in the interference pattern can then be expressed as

$$I(x, y) = a(x, y) + b(x, y) \cos[\delta(x, y) + 2\pi f_0 x]$$

where f_0 is the spatial carrier frequency along the x direction. In interferometry, the carrier is usually added by tilting the reference mirror. There are other ways of introducing the carrier in holographic interferometry, speckle interferometry, and electronic speckle pattern interferometry. The Fourier transform of this expression can be expressed as

$$I(\mu,\nu) = A(\mu,\nu) + C(\mu - f_0,\nu) + C^*(-\mu - f_0,-\nu) \tag{4.26}$$

Since the spatial carrier frequency is chosen higher than the maximum frequency content in $a(x,y), b(x,y)$, and $\delta(x,y)$, the spectra $A(\mu,\nu), C(\mu - f_0,\nu)$, and $C^*(-\mu - f_0,-\nu)$ are separated. The spectrum $A(\mu,\nu)$ is centered on $\mu = 0, \nu = 0$ and carries the information about the background. The spectra $C(\mu - f_0,\nu)$ and $C^*(-\mu - f_0,-\nu)$ are placed at $\mu = f_0, \nu = 0$, and $\mu = -f_0, \nu = 0$, that is, symmetrically about the DC term. If, by means of an appropriate bandpass filter, $A(\mu,\nu)$ and $C^*(-\mu - f_0,-\nu)$ are eliminated and subsequently $C(\mu - f_0,\nu)$ is shifted by f_0 towards the origin, thereby removing the carrier, we can obtain $c(x,y)$ by inverse Fourier transformation. The interference phase is then obtained from the real and imaginary parts of $c(x,y)$.

If, instead of $C(\mu,\nu)$, the inverse transform of $C^*(-\mu,-\nu)$ is taken, this results in $-\delta(x,y)$. The sign ambiguity is always present when a single interferogram is evaluated. Information about the sign of the phase is obtained when an additional phase-shifted interferogram is available. Let us now write down the expressions for the intensity distributions in the two interferograms, with one of them phase-stepped by α:

$$I_1(x,y) = a(x,y) + b(x,y)\cos\delta(x,y)$$

$$I_2(x,y) = a(x,y) + b(x,y)\cos[\delta(x,y) + \alpha]$$

Theoretically, the value of α must be in the range $0 < \alpha < \pi$; in practice, a value in the range $\pi/3 < \alpha < 2\pi/3$ is recommended. If this condition is fulfilled, the exact value of α need not be known. Again, we can express these intensity distributions in terms of complex exponentials and then take the Fourier transforms as described earlier. After bandpass filtering and taking the inverse Fourier transforms of the spectra belonging to each intensity distribution, we obtain

$$c_1(x,y) = 2b(x,y)e^{i\delta(x,y)}$$

$$c_2(x,y) = 2b(x,y)e^{-i[\delta(x,y)+\alpha]}$$

The phase step α is calculated pointwise from the expression

$$\alpha(x,y) = \tan^{-1}\frac{\mathrm{Re}\{c_1(x,y)\}\mathrm{Im}\{c_2(x,y)\} - \mathrm{Im}\{c_1(x,y)\}\mathrm{Re}\{c_2(x,y)\}}{\mathrm{Re}\{c_1(x,y)\}\mathrm{Re}\{c_2(x,y)\} + \mathrm{Im}\{c_1(x,y)\}\mathrm{Im}\{c_2(x,y)\}} \tag{4.27}$$

The knowledge of $\alpha(x,y)$ is used for determination of the sign-corrected interference phase distribution $\delta(x,y)$ from the expression

$$\delta(x,y) = \mathrm{sign}\{\alpha(x,y)\}\tan^{-1}\frac{\mathrm{Im}\{c_1(x,y)\}}{\mathrm{Re}\{c_1(x,y)\}} \tag{4.28}$$

The phase is unwrapped as usual.

BIBLIOGRAPHY

1. A. Gerrard and J. M. Burch, *Introduction to Matrix Methods in Optics,* Wiley, London, 1975.
2. E. Wolf (Ed.), *Progress in Optics*, Vol. 26, Elsevier Science, Amsterdam, 1988.
3. E. Wolf (Ed.), *Progress in Optics*, Vol. 28, Elsevier Science, Amsterdam, 1990.
4. W. Osten, *Digital Processing and Evaluation of Interference Images*, Akademie Verlag, Berlin, 1991.
5. D. W. Robinson and G. T. Reid (Eds.), *Interferogram Analysis–Digital Fringe Measurement Techniques*, Institute of Physics Publishing, Bristol, 1993.
6. T. Kreis, *Holographic Interferometry, Principles and Methods*, Akademie Verlag Series in Optical Metrology Vol. 1, Akademie Verlag, Berlin, 1996.
7. P. K. Rastogi (Ed.), *Optical Measurement Techniques and Applications*, Artech House, Boston, 1997.
8. D. Malacara, M. Servin, and Z. Malacara, *Optical Testing: Analysis of Interferograms*, Marcel Dekker, New York, 1998.
9. C. Ghiglia and M. D. Pritt, *Two Dimensional Phase Unwrapping: Theory, Algorithms and Software*, Wiley, New York, 1998.
10. D. Malacara, M. Servín, and Z. Malacara, *Interferogram Analysis for Optical Testing*, 2nd edn, CRC Press, Boca Raton, 2005.

ADDITIONAL READING

1. R. C. Jones, A new calculus for the treatment of optical systems. VII. Properties of *N* matrices, *J. Opt. Soc. Am.*, 38, 671–685, 1948.
2. D. A. Tichenor and V. P. Madsen, Computer analysis of holographic interferograms for non-destructive testing, *Opt. Eng.*, 8, 469–473, 1979.
3. T. M. Kreis and H. Kreitlow, Quantitative evaluation of holographic interferograms under image processing aspects, *Proc. SPIE*, 210, 196–202, 1979.
4. M. Takeda, H. Ina, and S. Kobayashi, Fourier-transform method of fringe pattern analysis for computer based topography and interferometry, *J. Opt. Soc. Am.*, 72, 156–160, 1982.
5. T. Yatagai and M. Idesawa, Automatic fringe analysis for moiré topography, *Opt. Lasers Eng.*, 3, 73–83, 1982.
6. K. Itoh, Analysis of phase unwrapping algorithms, *Appl. Opt.*, 21, 2470–2470, 1982.
7. P. Hariharan, B. F. Oreb, and N. Brown, A digital phase-measurement system for real-time holographic interferometry, *Opt. Commun.*, 41, 393–396, 1982.
8. H. E. Cline, W. E. Lorensen, and A. S. Holik, Automatic moiré contouring, *Appl. Opt.*, 23, 1454–1459, 1984.
9. M. P. Kothiyal and C. Delisle, Optical frequency shifter for heterodyne interferometry using counter-rotating wave plates, *Opt. Lett.*, 9, 319–321, 1984.
10. V. Srinivasan, H. C. Liu, and M. Halioua, Automated phase measuring profilometry of 3-D diffuse objects, *Appl. Opt.*, 23, 3105–3108, 1984.
11. K. Creath, Phase-shifting speckle interferometry, *Appl. Opt.*, 24, 3053–3058, 1985.
12. K. A. Stetson and W. R. Brohinsky, Electrooptic holography and its applications to hologram interferometry, *Appl. Opt.*, 24, 3631–3637, 1985.
13. K. A. Nugent, Interferogram analysis using an accurate fully automatic algorithm, *Appl. Opt.*, 24, 3101–3105, 1985.
14. K. Andresen, The phase shift method applied to moiré image processing, *Optik*, 72, 115–119, 1986.

15. G. T. Reid, R. C. Rixon, S. J. Marshall, and H. Stewart, Automatic on-line measurements of 3-D shape by shadow casting moiré topography, *Wear*, 109, 297–304, 1986.
16. K. Andresen and D. Klassen, The phase shift method applied to cross grating moiré measurement, *Opt. Lasers Eng.*, 7, 101–114, 1987.
17. C. Roddier and F. Roddier, Interferogram analysis using Fourier transform techniques, *Appl. Opt.*, 26, 1668–1673, 1987.
18. M. Owner-Petersen and P. Damgaard Jensen, Computer-aided electronic speckle pattern interferometry (ESPI): Deformation analysis by fringe manipulation, *NDT International (UK)*, 21, 422–426, 1988.
19. K. Creath, Phase-measurement interferometry techniques, In *Progress in Optics* (ed. E. Wolf), Vol. 26, 349–393, North-Holland, Amsterdam, 1988.
20. Y. Morimoto, Y. Seguchi, and T. Higashi, Application of moiré analysis of strain using Fourier transform, *Opt. Eng.*, 27, 650–656, 1988.
21. M. Kujawińska and D. W. Robinson, Multichannel phase-stepped holographic interferometry, *Appl. Opt.*, 27, 312–320, 1988.
22. J. J. J. Dirckx and W. F. Decraemer, Phase shift moiré apparatus for automatic 3D surface measurement, *Rev. Sci. Instrum.*, 60, 3698–3701, 1989.
23. J. M. Huntley, Noise-immune phase unwrapping algorithm, *Appl. Opt.*, 28, 3268–3270, 1989.
24. M. Takeda, Spatial-carrier fringe pattern analysis and its applications to precision interferometry and profilometry: An overview, *Ind. Metrol.*, 1, 79–99, 1990.
25. M. Kujawińska, L. Salbut, and K. Patorski, 3-channel phase-stepped system for moiré interferometry, *Appl. Opt.*, 30, 1633–1637, 1991.
26. J. Kato, I. Yamaguchi, and S. Kuwashima, Real-time fringe analysis based on electronic moiré and its applications, In *Fringe'93* (ed. W. Jüptner and W. Osten), 66–71, Akademie Verlag, Berlin, 1993.
27. J. M. Huntley and H. Saldner, Temporal phase-unwrapping algorithm for automated interferogram analysis, *Appl. Opt.*, 32, 3047–3052, 1993.
28. C. Joenathan, Phase-measuring interferometry: New methods and error analysis, *Appl. Opt.*, 33, 4147–4155, 1994.
29. G. Jin, N. Bao, and P. S. Chung, Applications of a novel phase shift method using a computer-controlled mechanism, *Opt. Eng.*, 33, 2733–2737, 1994.
30. S. Yoshida, R. W. Suprapedi, E. T. Astuti, and A. Kusnowo, Phase evaluation for electronic speckle-pattern interferometry deformation analyses, *Opt. Lett.*, 20, 755–757, 1995.
31. S. Suja Helen, M. P. Kothiyal, and R. S. Sirohi, Achromatic phase-shifting by a rotating polarizer, *Opt. Commun.*, 154, 249–254, 1998.
32. B. V. Dorrío and J. L. Fernández, Phase-evaluation methods in wholefield optical measurement techniques, *Meas. Sci. Technol.*, 10, R33–R55, 1999.
33. S. Suja Helen, M. P. Kothiyal, and R. S. Sirohi, Phase shifting by a rotating polarizer in white light interferometry for surface profiling, *J. Mod. Opt.*, 46, 993–1001, 1999.
34. H. Zhang, M. J. Lalor, and D. R. Burton, Robust accurate seven-sample phase-shifting algorithm insensitive to nonlinear phase shift error and second-harmonic distortion: A comparative study, *Opt. Eng.*, 38, 1524–1533, 1999.
35. S. Suja Helen, M. P. Kothiyal, and R. S. Sirohi, White light interferometry with polarization phase-shifter at the input of the interferometer, *J. Mod. Opt.*, 47, 1137–1145, 2000.
36. J. H. Massig and J. Heppner, Fringe-Pattern Analysis with High Accuracy by Use of the Fourier-Transform Method: Theory and Experimental Tests, *Appl. Opt.*, 40, 2081–2088, 2001.

37. M. Afifi, A. Fassi-Fihri, M. Marjane, K. Nassim, M. Sidki, and S. Rachafi, Paul wavelet-based algorithm for optical phase distribution evaluation, *Opt. Commun.*, 211, 47–51, 2002.
38. M. B. North-Morris, J. VanDelden and J.C. Wyant, Phase-shifting birefringent scatter-plate interferometer, *Appl. Opt.*, 41, 668–677, 2002.
39. K. Kadooka, K. Kunoo, N. Uda, K. Ono, and T. Nagayasu, Strain analysis for moiré interferometry using the two-dimensional continuous wavelet transform, *Exp. Mech.*, 43, 45–51, 2003.
40. C.-S. Guo, Z.-Y. Rong, H.-T. Wang, Y. Wang, and L. Z. Cai, Phase-shifting with computer-generated holograms written on a spatial light modulator, *Appl. Opt.*, 42, 6875–6879, 2003.
41. J. Novak, Five-step phase-shifting algorithms with unknown values of phase shift, *Optik,* 114, 63–68, 2003.
42. Y. Fu, C. J. Tay, C. Quan, and H. Miao, Wavelet analysis of speckle patterns with a temporal carrier, *Appl. Opt.*, 44, 959–965, 2005.
43. J. Millerd, N. Brock, J. Hayes, B. Kimbrough, M. Novak, M. North-Morris, and J. C. Wyant, Modern approaches in phase measuring metrology, *Proc. SPIE*, 5856, 1–22, 2005.
44. A. Jesacher, S. Fürhapter, S. Bernet, and M. Ritsch-Marte, Spiral interferogram analysis, *J. Opt. Soc. Am. A,* 23, 1400–1409, 2006.
45. B. Bhaduri, N. K. Mohan, and M. P. Kothiyal, Cyclic-path digital speckle shear pattern interferometer: Use of polarization phase-shifting method, *Opt. Eng.*, 45, 105604, 2006.
46. K. Patorski and A. Styk, Interferogram intensity modulation calculations using temporal phase shifting: Error analysis, *Opt. Eng.*, 45, 085602, 2006.
47. S. K. Debnath and M. P. Kothiyal, Experimental study of the phase shift miscalibration error in phase-shifting interferometry: Use of a spectrally resolved white-light interferometer, *Appl. Opt.*, 46, 5103–5109, 2007.
48. Y. Fu, G. Pedrini, and W. Osten, Vibration measurement by temporal Fourier analyses of a digital hologram sequence, *Appl. Opt.*, 46, 5719–5727, 2007.
49. L. R. Watkins, Phase recovery from fringe patterns using the continuous wavelet transform, *Opt. Lasers Eng.*, 45, 298–303, 2007.
50. Y. H. Huang, S. P. Ng, L. Liu, Y. S. Chen, and M. Y. Y. Hung, Shearographic phase retrieval using one single specklegram: A clustering approach, *Opt. Eng.*, 47, 054301, 2008.

5 Detectors and Recording Materials

Detection in optical measurement is a very important step in the measurement process. A detector converts incident optical energy into electrical energy, which is then measured. Many physical effects have been used to detect optical radiation, for example, photo-excitation and photo-emission, the photoresistive effect, and the photothermal effect. Detectors based on the photo-excitation or photo-emission of electrons are by far the most sensitive and provide the highest performance. These detectors are, however, small-area detectors, measuring/sampling optical energy over a very small area. In the techniques and methods described in this book, large-area or image detectors are employed except in heterodyne holographic interferometry. These include photographic plates, photochromics, thermoplastics, photorefractive crystals, and charge-coupled device/complementary metal-oxide semiconductor (CCD)/(CMOS) arrays, among others. CCD arrays, which are now used frequently, consist of an array of small-area (pixel) detectors based on photo-excitation. We therefore first discuss the use of semiconductor devices as detectors, then photomultiplier tubes (PMTs), and finally image detectors.

A variety of detectors are used to detect or measure optical signals. Photodetectors fall into two general categories: thermal detectors and quantum detectors. In thermal detectors, the incident thermal energy is converted to heat, which in turn may cause a rise in temperature and a corresponding measurable change in resistance, electromotive force (emf), and so on. These detectors are not used in the techniques described in this book, and hence will not be discussed here. Examples of quantum detectors are PMTs, semiconductor photodiodes, photoconductors, and phototransistors. In such detectors, the incident optical energy, that is, the photons, cause the emission of electrons in PMTs or the generation of electron-hole pairs in diodes. A resistance change is directly produced in photoconductors by absorption of photons.

5.1 DETECTOR CHARACTERISTICS

Detectors are compared by the use of several defined parameters. These are (i) responsivity R, (ii) detectivity D or D^*, (iii) noise equivalent power (NEP), (iv) noise, (v) spectral response, and (vi) frequency response. The responsivity is the ratio of the output signal (in amperes or volts) to the incident intensity, and is measured in μA/(mW/cm^2). For photodiodes, the quantum efficiency, which is the number of electron-hole pairs per incident photon, is sometimes given. Responsivity is proportional to quantum efficiency. The detectivity is the ratio of the responsivity to the noise current (voltage) produced by the detector. This, therefore, is the signal-to-noise ratio (SNR) divided by intensity. A parameter that is often used is D^*, which includes the dependence on noise frequency bandwidth Δf and detector area A, and is connected to the detectivity through the relation

$$D^* = D(A\Delta f)^{1/2} \tag{5.1}$$

NEP is the reciprocal of D. It is thus the light intensity required to produce an SNR of 1. There are several sources of noise, including Johnson (thermal) noise, shot noise, and generation–recombination noise. In all of these cases, the noise current is proportional to $(\Delta f)^{1/2}$. Spectral response refers to the variation of responsivity as a function of wavelength. Finally, the frequency response of a detector refers to its ability to respond to a chopped or modulated beam. Most solid state detectors behave like low-pass filters, with the responsivity being given by

$$R = R_0 \frac{1}{\left(1 + \omega^2\tau^2\right)^{1/2}} \tag{5.2}$$

where R_0 is the responsivity at zero frequency, $f = \omega/2\pi$ is the frequency of modulation, and τ is the time constant. The cut-off frequency $f_c = (2\pi\tau)^{-1}$, where $R = R_0/\sqrt{2}$.

5.2 DETECTORS

The incident photon creates an electron-hole pair in semiconductor devices and causes the release of an electron in a PMT. There is thus a frequency below which the detector does not respond to optical energy. The frequency ν of the radiation must be greater than the threshold frequency ν_{th}, where $h\nu_{th} = E_g, E_g$ being the band gap in a semiconductor or the work function of the cathode material in a PMT, and h is Planck's constant.

5.2.1 PHOTOCONDUCTORS

Photoconductor detectors are of two types: intrinsic and extrinsic. In an intrinsic semiconductor, an incident photon, on absorption, excites an electron from the valence band to the conduction band. This requires that the photon energy $h\nu \gg E_g$, where E_g is the band gap. The cut-off wavelength λ_c is given by

$$\lambda_c(\mu\text{m}) = 1.24/E_g(\text{eV}) \tag{5.3}$$

By choosing an appropriate E_g, the spectral response of the detector can be tailored. In fact, absorption of photons results in the generation of both excess electrons and holes, which are free to conduct under an applied electric field. The excess conductivity is called the intrinsic photoconductivity.

When the incident photons excite the electrons from donor levels to the conduction band or holes from acceptor levels to the valence band, the excess conductivity is called the extrinsic photoconductivity. In this case, excitation creates only one type of excess carriers—electrons or holes—by ionizing a donor or an acceptor, respectively. The spectral response is determined by the donor or the acceptor energies. For donor excitation, the cut-off wavelength is given by

$$\lambda_c = \frac{1.24}{E_c - E_d} \tag{5.4}$$

where E_c and E_d are the energies of the conduction band and the donor level, respectively. Since $E_c - E_d$ or $E_a - E_v$ (where E_a and E_v are the energies of the acceptor level and the valence band, respectively) is much smaller than E_g, these detectors are sensitive in the long-wavelength infrared region.

When a beam of light of intensity I and frequency ν is incident on a photoconductor detector, the rate of generation of carriers can be expressed as follows:

$$G = \eta \frac{I}{h\nu} \tag{5.5}$$

where η is the quantum efficiency. The photocurrent i generated in the external circuit is

$$i = \eta \frac{e}{h\nu}\left(\frac{\tau_0}{\tau_d}\right) I \tag{5.6}$$

where τ_0 is the lifetime of the carriers, τ_d is the transit time (the time required by the carriers to traverse the device), and the factor τ_0/τ_d is interpreted as photoconductive gain. It can be seen that the photocurrent is proportional to the light intensity.

Photoconductors exhibit a memory effect; that is, at a given illumination, the photoconductor may have several resistance values, depending on the previous history of illumination. These devices are not stable, and hence are not suited for precise and accurate measurements. These are, however, inexpensive, simple, bulky, and durable devices.

5.2.2 Photodiodes

The commonly known photodiodes include the simple p–n junction diode, the p–i–n diode, and the avalanche diode. The fabrication and operation of photodiodes are based on p–n technology. A p–n junction is formed by bringing p-type and n-type materials together. The dominant charge carriers in n-type and p-type semiconductors are electrons and holes, respectively. At the instant when the materials are joined together, there is an almost infinite concentration gradient across the junction for both the electrons and the holes (Figure 5.1a). Consequently, the electrons and holes diffuse in opposite directions. The diffusion process, however, does not continue indefinitely, since a potential barrier is developed that opposes the diffusion. For every free electron leaving an n-type region, an immobile positive charge is left behind. The amount of positive charge increases as the number of departing electrons increases. Similarly,

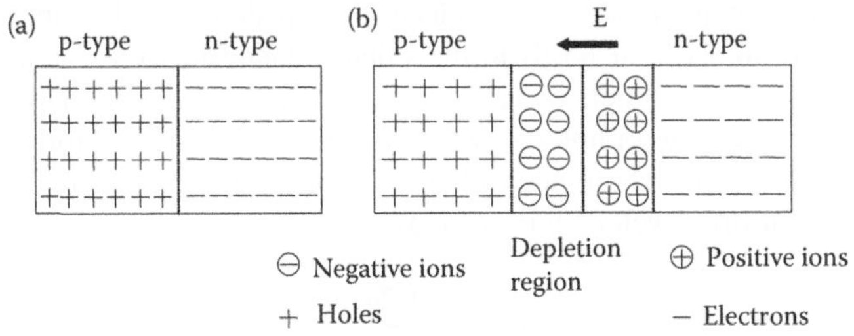

FIGURE 5.1 p–n junctions: (a) just formed; (b) at equilibrium.

as the holes depart from the p-type region, immobile negative charges build up in the p-type region. The mobile electrons and holes combine and neutralize each other, forming a region on both sides of the junction where there are no free charge carriers. This region is called the depletion region and acts as an insulator owing to the absence of free carriers. The immobile charges of opposite polarity generate a barrier voltage, which opposes further diffusion (Figure 5.1b). The barrier voltage depends on the semiconductor material. Its value for silicon is 0.7 V. The junction appears like a capacitor, that is, with a nonconductive material separating the two conductors.

When forward bias is applied to the junction, it opposes the barrier voltage, thereby reducing the thickness of the depletion region and increasing the junction capacitance. When the bias reaches the barrier voltage, the depletion region is eliminated and the junction becomes conductive. When the reverse bias is applied, the depletion region is widened, the junction capacitance is reduced, and the junction stays nonconductive. The reverse-bias junction, however, can conduct current when free carriers are introduced into it. These free carriers can be introduced by incident radiation. When the radiation falls in the depletion region or within the diffusion length around it, electron-hole pairs are created. The electrons (as minority carriers in the p-type region) will drift toward the depletion region, cross it, and hence contribute to the external current. Similarly, holes created by photo-absorption in the n-type region will drift in the opposite direction, and will contribute to the current. However, within the depletion region, the electron-hole pairs will be separated, and electrons and holes will drift from each other in opposite directions; thus, both will contribute to the current flow.

If we consider only the photons absorbed within the diffusion length to the depletion region and neglect the recombination of the generated electron-hole pairs, the photocurrent I is given by

$$i = e\eta \frac{I}{h\nu} \tag{5.7}$$

In a p–n junction diode, most of the applied bias voltage appears across the depletion region. Thus, only those pairs formed within the region or capable of diffusion into this region can be influenced by the externally applied field and contribute to the external current.

Figure 5.2 shows a schematic of a simple p–n junction diode. A heavily doped p-type material and lightly doped n-type material form the p–n junction. Because of different doping levels, the depletion region extends deeper into the n-type material.

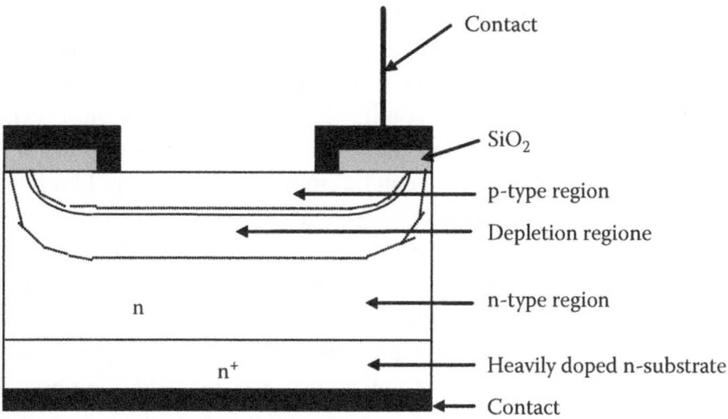

FIGURE 5.2 Schematic of a p–n junction diode.

At the bottom of the diode there is usually a region of n^+-type material, which is the substrate. The substrate terminates at the bottom electrical contact. The top electrical contact is fused to the p-type semiconductor. An insulating layer of silicon dioxide is provided, as shown in Figure 5.2.

The energy gap of most semiconductors is of the order of 1 eV, which corresponds to an optical wavelength of about 1 μm. Thus, photodiodes can respond to light in the spectral range from ultraviolet to near-infrared. However, the penetration of the photons through the various regions depends on the frequency of the incident radiation. Ultraviolet radiation is strongly absorbed at the surface, while infrared radiation can penetrate deep into the structure. In a Schottky photodiode, the p–n junction is replaced by a metal semiconductor junction. A thin layer (<10 nm) of gold is deposited on the n-type semiconductor. This layer is almost transparent and forms the metal–semiconductor junction. A schematic of a Schottky diode is shown in Figure 5.3. Since the depletion region commences right at the semiconductor surface, Schottky photodiodes have superior ultraviolet response.

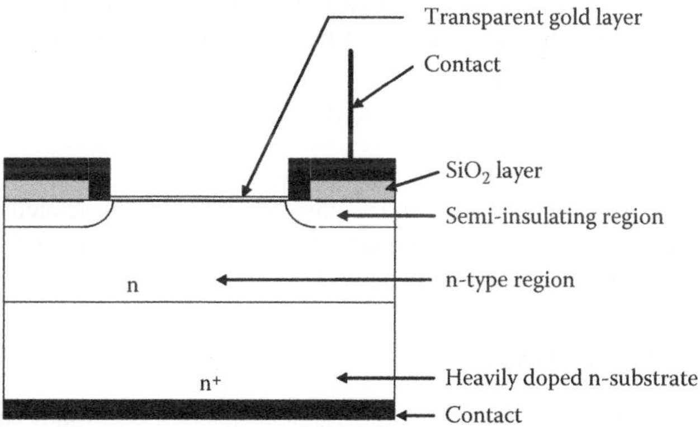

FIGURE 5.3 Schematic of a Schottky diode.

Since the minority carriers may have to travel some distance—up to the diffusion length before being transported across the p–n junction—the response of a p–n junction diode is relatively slow. This is a particularly serious drawback in some semiconducting materials such as silicon in which the depletion region is small compared to the diffusion length. This drawback is overcome by utilizing a p–i–n structure. The thickness of the depletion region can be controlled by doping. In a p–i–n diode, an intrinsic region, which has a very high resistivity, is sandwiched between the p- and n-regions. Figure 5.4a shows the microstructure of a p–i–n diode. The diode is usually operated with a reverse-bias voltage. This expands the depletion region. Furthermore, the voltage drop occurs mostly across the intrinsic layer (Figure 5.4b). When a photon whose energy exceeds the threshold value enters the intrinsic region, electron-hole pairs are created. Under the action of the applied electric field, the photogenerated electrons and holes are swept swiftly toward n- and p-type regions, respectively, and create a signal current. Figure 5.4c shows a cross-section of a typical p–i–n diode. The avalanche photodiode is a junction diode with an internal gain mechanism.

5.2.3 Photomultiplier Tube

The photomultiplier tube (PMT) is one of the most sensitive optical detectors: a photon flux as low as one photon per second can be detected. Many PMT designs exist. Figure 5.5 shows a cross-section of a commonly used PMT. It consists of a cathode, a series of dynodes, and an anode in a squirrel cage. A photon of frequency ν incident on the cathode ejects an electron, provided that $h\nu > \phi$, where ϕ is the work function of the cathode material. A variety of photocathodes cover the spectral range from 0.1 to 11 μm. Figure 5.6 shows the responsivities and quantum efficiencies of some photocathode materials.

The series of dynodes constitutes a low-noise amplifier. These dynodes are at progressively higher positive potential. A photoelectron emitted from the cathode is accelerated toward the first dynode. Secondary electrons are ejected as a result of the collision of the photoelectron with the dynode surface. This is the first stage of amplification. These secondary electrons are accelerated toward the more positive dynode, and the process is repeated. The amplified electron beam is collected by the anode. The multiplication of electrons at each dynode or the electron gain A depends on the dynode material and the potential difference between the dynodes. The total gain G of the PMT is given by

$$G = A^n \tag{5.8}$$

where n is the number of dynodes. The typical gain is around 10^6. PMTs are extremely fast and sensitive, but are expensive and require sophisticated associated electronics. These tubes are influenced by stray magnetic fields. However, there are PMT designs that are not affected by magnetic fields.

5.3 IMAGE DETECTORS

The detectors discussed so far are single-element detectors. In many applications, such as electronic speckle pattern interferometry (ESPI) and robotic vision, spatially varying information—say an image—is to be recorded. This is achieved by area

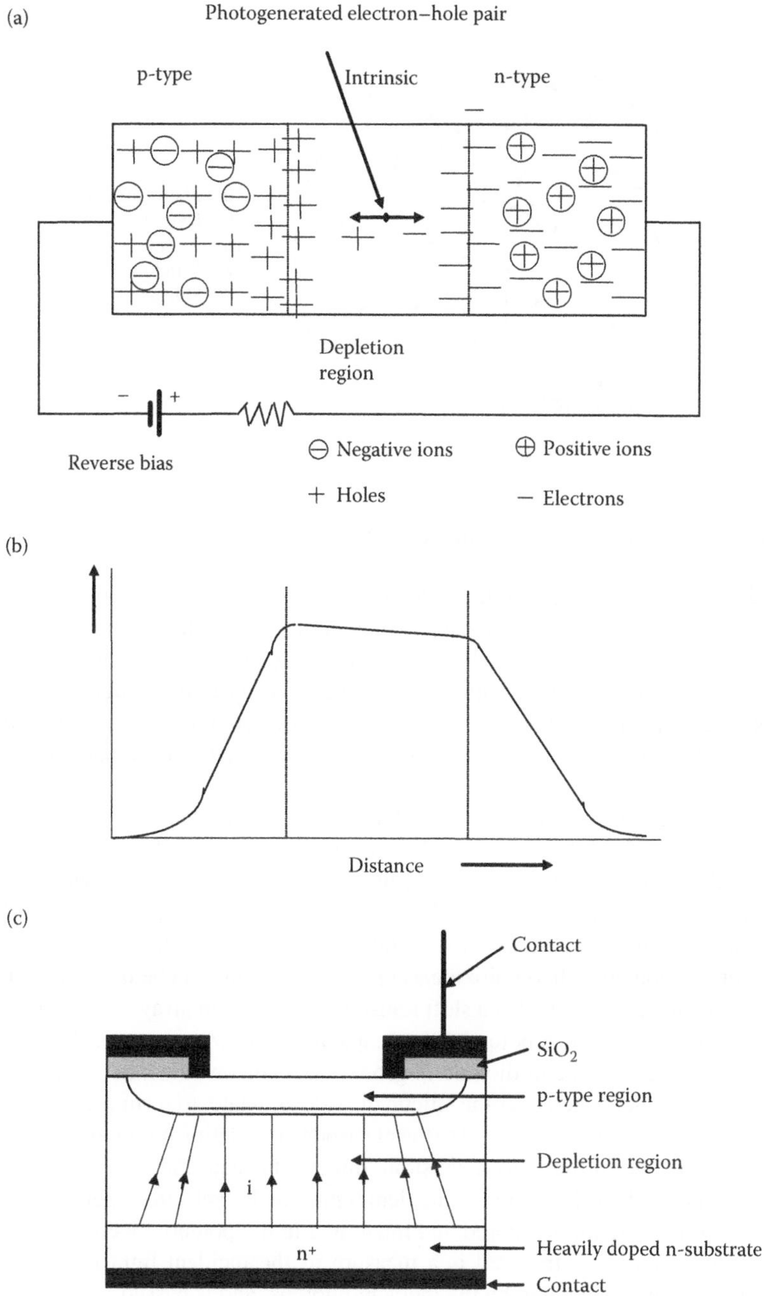

FIGURE 5.4 (a) Microstructure of a p–i–n diode. (b) Variation of electric field with distance. (c) Schematic of a p–i–n photodiode.

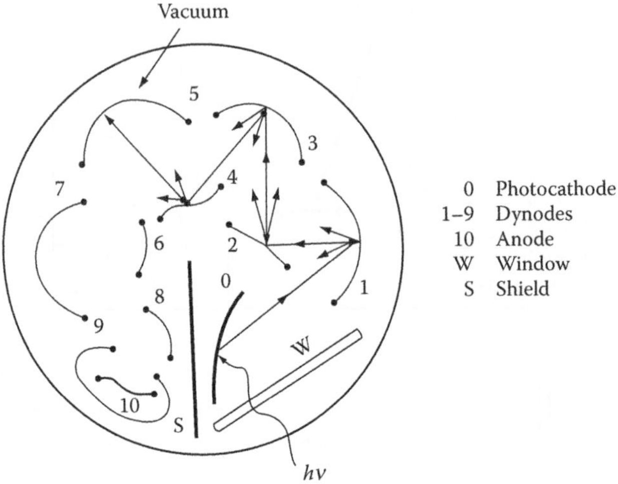

FIGURE 5.5 Cross-section of a squirrel-cage PMT.

array detectors. It is also possible to record an image with a linear array detector by scanning. There are three mechanisms at play in imaging with detector arrays. First, the image is intercepted by the elements of the array and converted to an electrical charge distribution, which is proportional to the intensity in the image. Second, the charges are read out as elements of an image while retaining correlation with the position on the array where each charge was generated. Finally, the image information is displayed or stored.

Area or focal plane arrays could be based on a single detector, but this would require many wires and processing electronics. The concept of a CCD makes the retrieval of detector signals easy and eliminates the need for a maze of wires. A CCD in its simplest form is a closely spaced array of metal-insulator–semiconductor (MIS) capacitors. The most important is the metal-oxide–semiconductor (MOS) capacitor, made from silicon and silicon dioxide as the insulator. This can be made monolithic. The basic structure of a CCD is a shift register formed by an array of closely spaced potential-well capacitors. A potential-well capacitor is shown schematically in Figure 5.7a. A thin layer of silicon dioxide is grown on a silicon substrate. A transparent electrode is then deposited over the silicon dioxide as a gate, to form a tiny capacitor. When a positive potential is applied to the electrode, a depletion region or an electrical potential is created in the silicon substrate directly beneath the gate. Electron-hole pairs are generated on absorption of incident light. The free electrons generated in the vicinity of the capacitor are stored and integrated in the potential well. The number of electrons (charge) in the well is a measure of the incident light intensity. In a CCD, the charge is generated by the incident photons, and is passed between spatial locations and detected at the edge of the CCD. The charge position in the MOS array of capacitors is controlled electrostatically by voltage levels. With appropriate application of these voltage levels and their relative phases, the capacitor can be used to store and transfer the charge packet across the semiconductor substrate in a controlled manner.

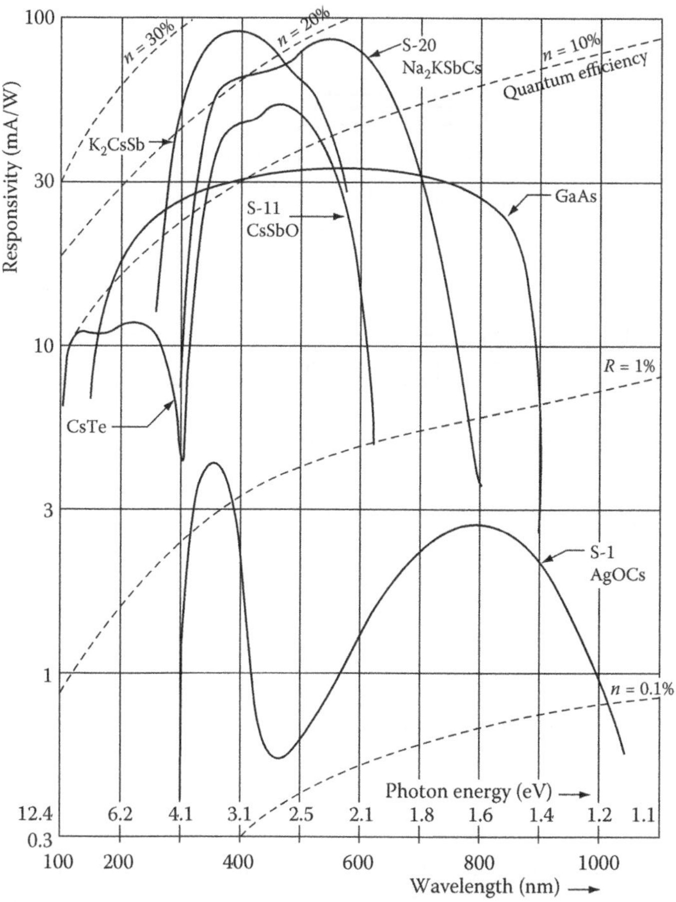

FIGURE 5.6 Responsivity and quantum efficiency of some common cathode materials. (From Uiga, E., *Optoelectronics*, Prentice Hall, Englewood Cliffs, NJ, 1995. With permission.)

Figure 5.7b illustrates the principle of charge transfer through the propagation of potential wells in a three-phase clocking layout. In phase ϕ_1, gates G_1 and G_4 are turned on, while all other gates are turned off. Hence, electrons are collected in wells W_1 and W_4. In phase ϕ_2, gates G_1, G_2, G_4, and G_5 are turned on. Therefore, wells W_1 and W_2, and W_4 and W_5 merge into wider wells. In phase ϕ_3, gates G_1 and G_4 are turned off, while G_2 and G_5 are left on. The electrons stored earlier in W_1 are now shifted to W_2. Similarly, the electrons stored in W_4 are now shifted to W_5. By repeating this process, all charge packets will be transferred to the edge of the CCD, where they are read by external electronics. Area detector arrays with pixels in numbers of 256×256 to 4096×4096 are available. The center-to-center distances of the pixels range from 10 to 40 μm, but most commonly used devices have a pixel separation of 12.7 μm. A dynamic range of 1000 to 1 is quite common. The sensitivity of video-rate CCD cameras is of the order of 10^{-8} W/cm^2, which

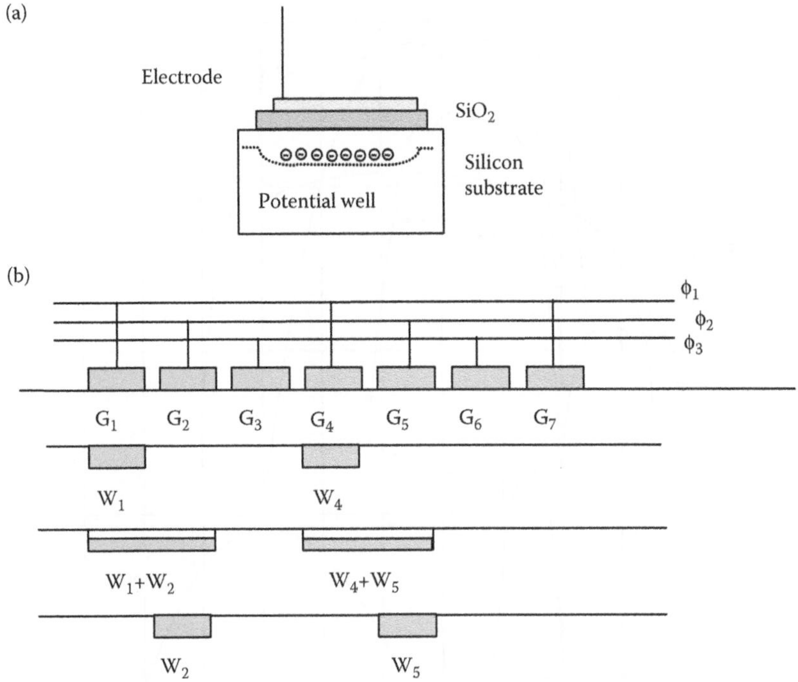

FIGURE 5.7 (a) Structure of a CCD pixel. (b) Mechanism of charge transfer in a three-phase clocking layout.

corresponds to about 0.05 lux. CCD cameras are available with exposure times as small as 1/10,000 second, achieved by electronic shuttering, and an SNR of 50 dB.

CMOS sensors are interesting alternatives to CCD sensors in optical metrology. CMOS sensors are relatively cheap and have lower power consumption. Other physical characteristics include random access, which allows fast readout of a small area of interest, and physical layout, which enables active electronic components to be located on each pixel and prevents blooming. CMOS sensors also have disadvantages, including lower sensitivity due to a smaller fill factor, higher temporal noise, higher pattern noise, higher dark current, and a nonlinear characteristic curve.

As a rule, the pixels in CCD sensors are built from MOS capacitors in which the electrons generated by photon absorption during the exposure are stored. The maximum number of electrons that can be stored in a pixel is the full-well capacity. In interline transfer (IT) and frame transfer (FT) sensors, the electrons are shifted into separate storage cells at the end of each exposure time. After this shifting, the next image can be exposed. During this exposure, the charge in the storage cells is shifted pixel by pixel into the sense node (readout node), where it is converted into the output voltage.

In CMOS sensors, the single pixels are built from photodiodes. Electronic components such as storage cells, transistors for addressing, and amplifiers can be assigned to every pixel. This is why such sensors are called active pixel sensors (APS). There are two types: integrating and nonintegrating sensors. Nonintegrating sensors provide a pixel signal that depends on the instantaneous current in the photodiode (direct readout sensor). Owing to their nonlinear current-to-voltage conversion, these sensors usually have a logarithmic characteristic curve. In integrating sensors, the depletion-layer capacity of the photodiode is usually used for charge storage. The characteristic curve of such sensors is slightly nonlinear.

For integrating sensors with a rolling shutter, the exposure and the readout of the single lines occur sequentially. In integrating sensors with a global shutter, each pixel has its own storage cell. All pixels are exposed at the same time, and at the end of the exposure time the charges in all pixels are shifted simultaneously into storage cells. Afterwards, the storage cells are read out sequentially. Owing to the random access to individual pixels, it is possible to read out a region of interest (ROI) of the whole sensor. High frame rates can thereby be achieved for small ROIs.

The recording of a dynamic process, such as the measurement of time-dependent deformations with ESPI, requires a camera with a suitable frame rate and the possibility of simultaneous exposure of all pixels. Therefore, only IT and FT sensors are suitable for CCD cameras, and only integrating sensors with a global shutter are suitable in the case of CMOS cameras.

5.3.1 Time-Delay and Integration Mode of Operation

Conventional CCD cameras are restricted to working with stationary objects. Object motion during exposure blurs the image. Time-delay and integration (TDI) is a special mode of CCD cameras, which provides a solution to the blurring problem. In the TDI mode, the charges collected from each row of detectors are shifted to the neighboring sites at a fixed time interval. As an example, consider four detectors operating in TDI mode as shown in Figure 5.8. The object is in motion, and at time t_1, its image is formed on the first detector, which creates a charge packet. At time t_2, the image moves to the second detector. Simultaneously, the pixel clock moves the charge packet to the well under the second detector. Here, the image creates an additional charge, which is added to the charge shifted from the first detector. Similarly, at time t_3, the image moves to the third detector and creates a charge. Simultaneously, the charge from the second detector is moved to the well of the third detector. The charge thus increases linearly with the number N of detectors in TDI. Therefore, the signal increases with N. The noise also increases, but as $\sqrt{N}$, and hence the SNR increases as $\sqrt{N}$. The well capacity limits the maximum number of TDI elements that can be used. As is obvious, the charge packet must always be in synchronism with the image for the camera to work in TDI mode. The mismatch between the image scan rate and clock rate can adversely smear the output.

5.4 RECORDING MATERIALS

We will now discuss the materials that record the image, that is, the intensity distribution. The recording is required either to keep a permanent record or to provide an input for measurement and processing. Photographic emulsions are by far the most

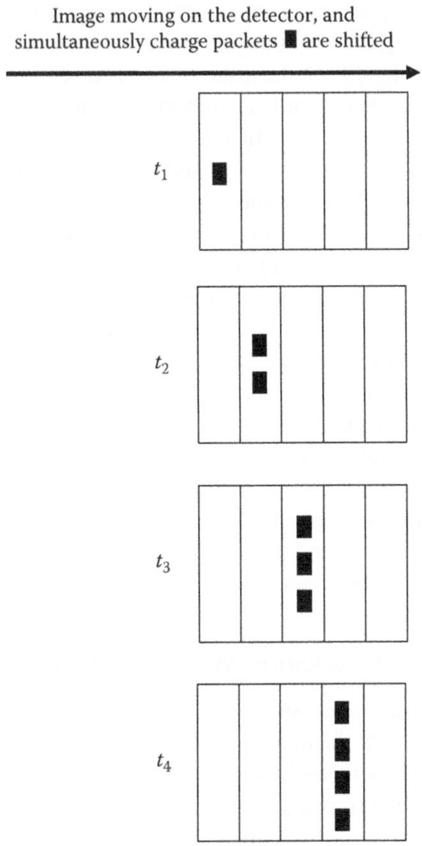

FIGURE 5.8 TDI mode of operation.

sensitive and widely used recording medium both for photography and for holography. Several other recording media have been developed; these are listed in Table 5.1. We will discuss some characteristics of these media.

5.4.1 Photographic Films and Plates

A photographic film/plate consists of silver halide grains distributed uniformly in a gelatin matrix deposited in a thin layer on a transparent substrate: either a glass plate or an acetate film. When the photographic emulsion is exposed to light, the silver halide grains absorb optical energy and undergo a complex physical change; that is, a latent image is formed. The exposed film is then developed. The development converts the halide grains that have absorbed sufficient optical energy into metallic silver. The film is then fixed, which removes the unexposed silver halide grains while leaving the metallic silver. The silver grains are largely opaque at the optical frequency. Therefore, the processed film will exhibit spatial variation of opacity depending on the density of the silver grains in each region of the transparency.

TABLE 5.1
Recording Media for Photography and Holography

S. No.	Class of Material	Spectral Range (nm)	Recording Process	Spatial Frequencies (lines/mm)	Types of Grating	Processing	Readout Process	Maximum Diffraction Efficiency (%)
1	Photographic materials	400–700 (<1300)	Reduction to Ag metal grains	>3000	Plane/volume amplitude	Wet chemical	Density change	5
			Bleached to silver salts		Plane/volume phase	Wet chemical	Refractive index change	20–50
2	Dichromated gelatin	250–520 and 633	Photo-crosslinking	>3000	Plane phase, volume phase	Wet chemical followed by heat	Refractive index change	30 > 90
3	Photoresists	UV-500	Photo-crosslinking or photopolymerization	<3000	Surface relief/ phase-blazed reflection	Wet chemical	Surface relief	30–90
4	Photopolymers	UV-500	Photopolymerization	~200–1500 bandpass	Volume phase	None or post-exposure and post- heating	Refractive index change or surface relief	10–85
5	Photoplastics/ photoconductor thermoplastics	Nearly panchromatic for PVK TNK photoconductor	Formation of an electrostatic latent image with electric-field-produced deformation of heated plastic	400–1000 bandpass	Plane phase	Corona charge and heat	Surface relief	6–15
6	Photochromics	300–450	Generally photo-induced new absorption bands	>3000	Volume absorption	None	Density change	1–2
7	Ferroelectric crystals	400–650	Electro-optic effect/ photorefractive effect	>3000	Volume phase	None	Refractive index change	90

Photographic emulsions (silver halides) are widely used for photography (incoherent recording) and holography (coherent recording). They are sensitive over a wide spectral range and also offer a very wide resolution range and good dynamic response. An apparent drawback of photographic materials is that they require wet development and fixing processing, followed by drying. However, the development process provides a gain of the order of a million, which amplifies the latent image formed during the exposure.

It was mentioned earlier that the spatial variation of light intensity incident on a photo-emulsion is converted into the variation of density of metallic silver grains; consequently, its transmission becomes spatially variable. The transmission function τ is related to the photographic density D, the density of the metallic silver grains per unit area, by

$$D = -\log \tau \tag{5.9}$$

The transmission function τ is defined by

$$\tau(x, y) = \frac{I_t(x, y)}{I_i(x, y)} \tag{5.10}$$

where $I_t(x, y)$ and $I_i(x, y)$ are the transmitted and incident intensities, respectively. The reflection losses at the interfaces are ignored. The transmission is averaged over a very tiny area around the point (x, y).

One of the most commonly used descriptions of the photosensitivity of a photographic film is the Hurter–Driffield (H&D) curve. This is a plot of the density D versus the logarithm of the exposure E. The exposure E is defined as the energy per unit area incident on the film and is given by $E = IT$, where I is the incident intensity and T is the exposure time. Figure 5.9 illustrates a typical H&D curve for a photographic negative. If the exposure is below a certain level, the density is independent of exposure. This minimum density is usually referred to as gross fog. In the "toe" of the curve, the density begins to increase with $\log E$. There follows a region where the density increases linearly with the logarithm of the exposure; this is the linear region of the curve, and the slope of this linear region is referred to as the film gamma γ. Finally, the curve saturates in a region called the "shoulder." There is no change in density with the logarithm of the exposure after the "shoulder." However, with very large exposures, solarization takes place.

The linear region of the H&D curve is generally used in conventional photography. A film with a large value of γ is called a high-contrast film, while a film with low γ is a low-contrast film. The γ of the film also depends on the development time and the developer. It is thus possible to obtain a prescribed value of γ by a judicious choice of film, developer, and development time.

Since the emulsion is usually used in the linear portion of the H&D curve, the density D when the film is given an exposure E can be expressed as

$$D = \gamma_n \log E - D_0 = \gamma_n \log(\mathrm{IT}) - D_0 \tag{5.11}$$

where the subscript n means that a negative film is being used and D_0 is the intercept. Equation 5.11 can be written in terms of the transmission function τ_n of the negative

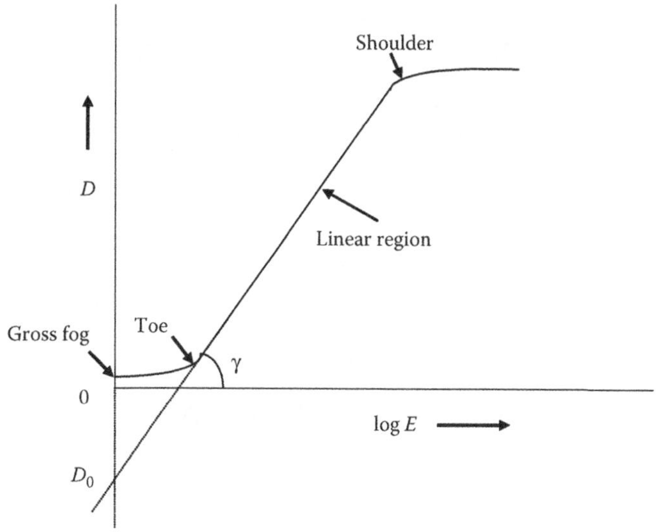

FIGURE 5.9 The Hurter–Driffield curve.

film as

$$\tau_n = K_n I^{-\gamma_n} \tag{5.12}$$

where $K_n = 10^{D_0} T^{-\gamma_n}$ is a positive constant. Equation 5.12 relates the incident intensity to the transmission function of the film after development. It can be seen that the transmission is a highly nonlinear function of the incident intensity for any positive value of γ_n. In many applications, it is required to have either a linear or a power-law relationship. This, however, requires a two-step process. In the first step, a negative transparency is made in the usual fashion, which will have a gamma γ_{n1}. The transmission of this negative transparency is given by

$$\tau_{n1} = K_{n1} I^{-\gamma_{n1}} \tag{5.13}$$

In the second step, the negative transparency is illuminated by a uniform intensity I_0 and the light transmitted is used to expose a second film, which will have a gamma γ_{n2}. This results in a positive transparency with transmission τ_p, given by

$$\tau_p = K_{n2}(I_0 \tau_{n1})^{-\gamma_{n2}} = K_{n2} I_o^{-\gamma_{n2}} K_{n1}^{-\gamma_{n2}} I^{\gamma_{n1}\gamma_{n2}} = K_p I^{\gamma_p} \tag{5.14}$$

where K_p is another positive constant and $\gamma_p = \gamma_{n1}\gamma_{n2}$ is the overall gamma of the two-step process. Evidently, a positive transparency does provide a linear mapping of intensity when the overall gamma is unity.

When photographic emulsions are used for holography or, in general, for coherent optical systems, the H&D curve is never used. Instead, the plot of amplitude transmittance $\mathbf{t}(x, y)$ versus exposure E is used. The amplitude transmittance is usually complex, since film introduces both amplitude and phase variations to the incident

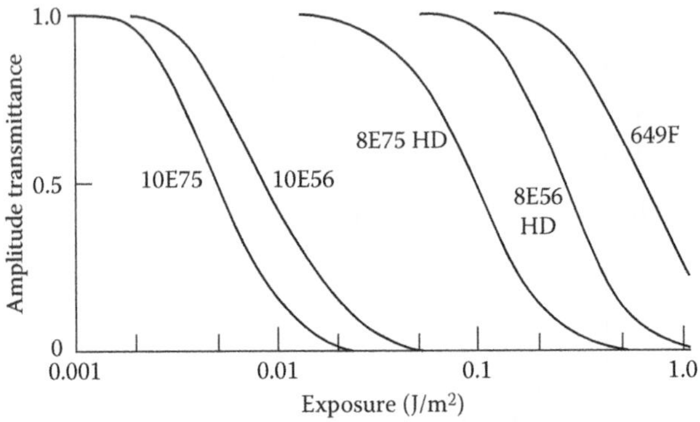

FIGURE 5.10 Plots of amplitude transmittance |**t**| versus exposure E for several holographic emulsions.

plane wave. However, if the film is used in a liquid gate, the phase variations can be eliminated. The amplitude transmittance then is given by the square root of the transmission function; that is, $|\mathbf{t}(x, y)| = \sqrt[+]{\tau(x, y)}$. Typical plots of $|\mathbf{t}(x, y)|$ versus E for several holographic emulsions are shown in Figure 5.10. Holographic recording is generally carried out in the linear region of the $|\mathbf{t}(x, y)|$ versus E curve.

Since formation of the latent image does not cause any changes in the optical properties during exposure, it is possible to record several holograms in the same photographic emulsion without any interaction between them. The information can be recorded in the form of either transmittance variations or phase variations. For recording in the form of phase variations, the amplitude holograms are bleached. Bleaching converts the metallic silver grains back to transparent silver halide crystals. Also, holographic information can be recorded in the volume, provided that the emulsion is sufficiently thick and has enough resolution.

5.4.2 Dichromated Gelatin

Dichromated gelatin is, in some respects, an ideal recording material for volume phase holograms, since it has large refractive index modulation capability, high resolution, and low absorption and scattering. The gelatin layer can be deposited on a glass plate and sensitized. Alternatively, the photographic plates can be fixed, rinsed, and sensitized. Ammonium, sodium, and potassium dichromates have been used as sensitizers. Most often, ammonium dichromate is used for sensitization. Sensitized gelatin thus obtained is called dichromated gelatin. Dichromated gelatin exhibits sensitivity in the wavelength range 250–520 nm. The sensitivity at 514 nm is about a fifth of that at 448 nm. Gelatin can also be sensitized at the red wavelength of a He–Ne laser by the addition of methyl blue dye.

The exposure causes crosslinking between gelatin chains and alters swelling properties and solubility. Treatment with warm water dissolves the unexposed gelatin and thereby forms a surface relief pattern. However, much better holograms can be obtained if the gelatin film is processed to obtain a modulation of the refractive index. During processing, rapid dehydration of the gelatin film is carried out in an isopropanol bath at an elevated temperature. This creates a very large number of small vacuoles in the gelatin layer, and hence modulates the refractive index. It is also suggested that the formation of complexes of a chromium(III) compound, gelatin, and isopropanol in the hardened areas is also partly responsible for the refractive index modulation. Phase holograms in gelatin are very efficient, directing more than 90% of the incident light into the useful image.

5.4.3 Photoresists

Photoresists are organic materials that are sensitive in the ultraviolet and blue regions. They are all relatively slow, and thus require very long exposure. Usually, a thin layer (about 1 μm) is obtained either by spin coating or spray coating on the substrate. This layer is then baked at around 75°C. On exposure, one of three processes takes place: formation of an organic acid, photo-crosslinking, or photopolymerization.

There are two types of photoresists: negative and positive. In negative photoresists, the unexposed regions are removed during development, while in positive photoresists, the exposed regions are removed during development. A surface relief recording is thus obtained. A grating recorded on a photoresist can be blazed by ion bombardment. Blazed grating can also be recorded by optical means. Relief recording offers the advantage of replication using thermoplastics.

When exposure is made in a negative photoresist from the air–film side, the layer close to the substrate is the last to photolyze. This layer will be simply dissolved away during development, since it has not photolyzed fully. This nonadhesion of negative photoresists in holographic recording is a serious problem. To overcome this problem, the photoresist is exposed from the substrate–film side so as to photolyze the resist better at the substrate–film interface. On the other hand, positive photoresists do not have this problem, and hence are preferred. One of the most widely used positive photoresists is the Shipley AZ-1350. The recording can be done at the 458 nm wavelength of an Ar^+ laser or at the 442 nm of a He–Cd laser.

5.4.4 Photopolymers

Photopolymers are also organic materials. Photopolymers for use in holography can be in the form of either a liquid layer enclosed between glass plates or a dry layer. Exposure causes photopolymerization or crosslinking of the monomer, resulting in refractive index modulation, which may or may not be accompanied by surface relief. Photopolymers are more sensitive than photoresists, and hence require moderate exposure. They also possess the advantage of dry and rapid processing. Thick polymer materials such as poly-methyl methacrylate (PMMA) and cellulose acetate butyrate (CAB) are excellent candidates for volume holography, since the refractive index change on exposure can be as large as 10^{-3}. Two photopolymers are commercially available: the Polaroid DMP 128 and the Du Pont OmniDex. The Polaroid DMP 128 uses dye-sensitized photopolymerization of a vinyl monomer incorporated in a polymer matrix, which is coated on a glass or plastic substrate. Coated plates or films can be exposed with blue, green, and red light. Du Pont OmniDex film consists of a polyester base coated with a photopolymer, and is used for contact copying of master holograms with UV radiation.

The response of photopolymers is band-limited because of the limitation imposed by the diffusion length of the monomer at the lower end of the response curve and the length of the polymer at the higher end.

5.4.5 Thermoplastics

A thermoplastic is a multilayer structure having a substrate coated with a conducting layer, a photoconductor layer, and a thermoplastic layer. A photoconductor that works well is the polymer poly-*N*-vinylcarbazole (PVK) to which is added a small amount of the electron donor 2,4,7-trinitro-9-fluorenone (TNF). The thermoplastic is a natural tree resin, Staybelite. The thermoplastics for holographic recording combine the advantages of high sensitivity and resolution, dry and nearly instantaneous *in situ* development, erasability, and high readout efficiency.

The recording process involves a number of steps. First, a uniform electrostatic charge is established on the surface of the thermoplastic in the dark by means of a corona discharge assembly. The charge is capacitively divided between the photoconductor and the thermoplastic layers. In the second step, the thermoplastic is exposed; the exposure causes the photoconductor to discharge its voltage at the illuminated regions. This does not cause any variation in the charge distribution on the thermoplastic layer; the electric field in the thermoplastic layer remains unchanged. In the third step, the surface is charged again by the corona discharge assembly. In this process, the charge is added at the exposed regions. Therefore, an electric field distribution, which forms a latent image, is now established. In the fourth step, the thermoplastic is heated to its softening point, thereby developing the latent image. The thermoplastic layer undergoes local deformation as a result of the varying electric field across it, becoming thinner wherever the field is higher (the illuminated regions) and thicker in the unexposed areas. Rapid cooling to room temperature freezes the deformation; the recording is now in the form of a surface relief. The recording is stable at room temperature, but can be erased by heating the thermoplastic to a temperature higher than that used for development. At the elevated temperature, the surface tension evens out the thickness variations and hence erases the recording. This is the fifth step—the erasure step. Figure 5.11 shows the whole recording process. The thermoplastic can be reused several hundred times. The response of these devices is band-limited, depending on the thickness of the thermoplastic and other factors.

5.4.6 Photochromics

Materials that undergo a reversible color change on exposure are called photochromic materials. Photochroism occurs in a variety of materials, both organic and inorganic. Organic photochromics have a limited life and are prone to fatigue. However, organic films of spiropyran derivatives have been used for hologram recording in darkening mode at 633 nm. Inorganic photochromics are either crystals or glasses doped with selected impurities: photochroism is due to a reversible charge transfer between two species of electron traps. Recording in silver halide photochromic glasses has been

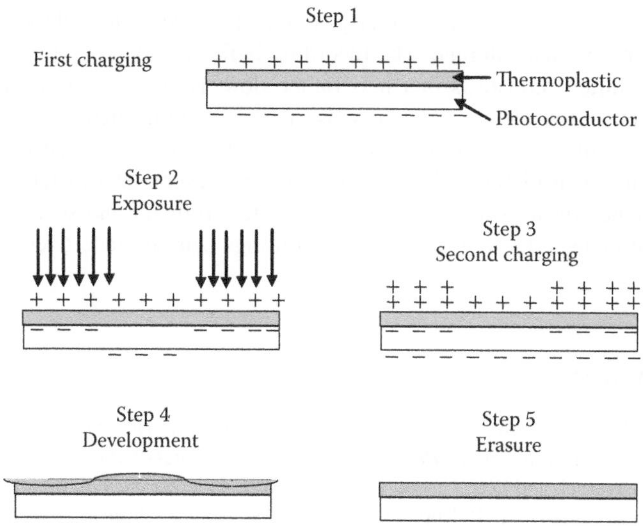

FIGURE 5.11 The record–erase cycle for a thermoplastic material.

done in darkening mode at 488 nm and in bleaching mode at 633 nm. Doped crystals of CaF_2 and SrO_2 have been used in bleaching mode at 633 nm. The sensitivity of photochromics is very low, because the reaction occurs at a molecular level. For the same reason, they are essentially grain-free and have resolution in excess of 3000 lines/mm. Inorganic photochromics have large thicknesses, and hence a number of holograms can be recorded in them. They do not require any processing, and can be reused almost indefinitely. In spite of all these advantages, these materials have limited applications, owing to their low diffraction efficiency (<0.02) and low sensitivity.

5.4.7 Ferroelectric Crystals

Certain ferroelectric crystals, such as lithium niobate ($LiNbO_3$), lithium tantalate ($LiTaO_3$), barium titanate ($BaTiO_3$), and strontium barium niobate (SBN), exhibit small changes in refractive index when exposed to intense light. The photo-induced refractive index change can be reversed by an application of heat and light. The mechanism of recording in these crystals is as follows: exposure to light frees trapped electrons, which then migrate through the crystal lattice and are again trapped in adjacent unexposed or low-intensity regions. The migration usually occurs through diffusion or an internal photovoltaic effect. This produces a spatially varying net space-charge distribution and a corresponding electric field distribution. The electric field modulates the refractive index through the electro-optic effect and creates a volume phase grating. These are real-time recording materials; the records are stable, since the charges are bound to the localized traps. The recording can, however, be erased by illuminating it with a light beam of wavelength that can release the trapped electrons.

Lithium niobate crystals—particularly Fe-doped—have been used for holographic interferometry and data storage. The recording is fixed by temperature. The disadvantage of lithium niobate is that it is rather slow. Higher sensitivity is obtained with photoconductive electro-optic crystals such as bismuth silicon oxide (BSO) and bismuth germanium oxide (BGO) by the application of an external electric field. These crystals are available in the form of thin slices several centimeters in diameter. Barium titanate crystal is used in holographic interferometry and speckle photography because of its very slow response. Recordings can be made over a very wide spectral range.

BIBLIOGRAPHY

1. J. W. Goodman, *Introduction to Fourier Optics*, McGraw-Hill, New York, 1968, 1996.
2. H. M. Smith, *Holographic Recording Materials*, Springer-Verlag, Berlin, 1977.
3. E. L. Dereniak and D.G. Crowe, *Optical Radiation Detectors*, Wiley, New York, 1984.
4. R. S. Sirohi and M. P. Kothiyal, *Optical Components, Systems, and Measurement Techniques*, Marcel Dekker, New York, 1991.
5. M. P. Petrov, S. I. Stepanov, and A. V. Khomenko, *Photorefractive Crystals in Coherent Optical Systems*, Springer-Verlag, Berlin, 1991.
6. J. T. Luxon and D. E. Parker, *Industrial Lasers and Their Applications*, Prentice Hall, Englewood Cliffs, NJ, 1992.
7. H. I. Bjelkhagen, *Silver-Halide Recording Materials: For Holography and their Processing*, Springer-Verlag, Berlin, 1993.
8. E. Uiga, *Optoelectronics*, Prentice Hall, Englewood Cliffs, NJ, 1995.
9. S. S. Jha (Ed.), *Perspectives in Optoelectronics*, World Scientific, Singapore, 1995.
10. J. E. Stewart, *Optical Principles and Technology for Engineers*, Marcel Dekker, New York, 1996.
11. P. Hariharan, *Optical Holography*, Cambridge University Press, Cambridge, 1996.
12. H. I. Bjelkhagen (Ed.), *Holographic Recording Materials*, SPIE Milestone MS 130, SPIE Optical Engineering Press, Bellingham, WA, 1996.
13. G. C. Holst, *CCD Arrays, Cameras, and Displays*, JCD Publishing/SPIE Optical Engineering Press, Bellingham, WA, 1996.
14. F. T. S. Yu and X.Yang, *Introduction to Optical Engineering*, Cambridge University Press, Cambridge, 1997.
15. G. C. Holst and T. S. Lomheim, *CMOS/CCD Sensors and Camera Systems*, JCD Publishing/SPIE Optical Engineering Press, Bellingham, WA, 2007.

ADDITIONAL READING

1. J. C. Urbach and R. W. Meier, Thermoplastic xerographic holography, *Appl. Opt.*, 5, 666–667, 1966.
2. T. A. Shankoff, Phase holograms in dichromated gelatin, *Appl. Opt.*, 7, 2101–2105, 1968.
3. M. J. Beesley and T. G. Castledine, The use of photoresist as a holographic recording medium, *Appl. Opt.*, 9, 2720–2724, 1970.
4. W. S. Colburn and K. A. Haines, Volume hologram formation in photopolymer materials, *Appl. Opt.*, 10, 1636–1641, 1971.

5. B. Smolinska, Relief hologram formation and replication in hardened dichromated PVA films, *Acta Phys. Pol.*, A40, 327–332, 1971.
6. E. G. Ramberg, Holographic information storage, *RCA Rev.*, 33, 5–53, 1972.
7. D. Meyerhofer, Phase-holograms in dichromated gelatin, *RCA Rev.*, 33, 110–130, 1972.
8. T. L. Credelle and F. W. Spong, Thermoplastic media for holographic recording, *RCA Rev.*, 33, 206–226, 1972.
9. R. A. Bartolini, Characteristics of relief holograms recorded on photoresists, *Appl. Opt.*, 13, 129–139, 1974.
10. A. Graube, Advances in bleaching methods for photographically recorded holograms, *Appl. Opt.*, 13, 2942–2946, 1974.
11. K. Biedermann, Information storage materials for holography and optical data storage, *Opt. Acta*, 22, 103–124, 1975.
12. B. L. Booth, Photopolymer materials for holography, *Appl. Opt.*, 14, 593–601, 1975.
13. S. L. Norman and M. P. Singh, Spectral sensitivity and linearity of Shipley AZ 1350 J photoresist, *Appl. Opt.*, 14, 818–820, 1975.
14. S. K. Case and R. Alferness, Index modulation and spatial harmonic generation in dichromated gelatin films, *Appl. Phys.*, 10, 41–51, 1976.
15. R. A. Bartolini, H. A. Weakliem, and B. F. Williams, Review and analysis of optical recording media, *Ferroelectrics*, 11, 393–396, 1976.
16. R. A. Bartolini, Optical recording media review, *Proc. SPIE*, 123, 2–9, 1977.
17. B. L. Booth, Photopolymer laser recording materials, *J. Appl. Photographic Eng.*, 3, 24–30, 1977.
18. D. Casasent and F. Caimi, Photodichroic crystals for coherent optical data processing, *Opt. Laser Technol.*, 9, 63–68, 1977.
19. T. C. Lee, J. W. Lin, and O. N. Tufte, Thermoplastic photoconductor for optical recording and storage, *Proc. SPIE*, 123, 74–77, 1977.
20. B. J. Chang, Dichromated gelatin as holographic storage medium, *Proc. SPIE*, 177, 71–81, 1979.
21. K. Blotekjaer, Limitations on holographic storage capacity of photochromic and photorefractive media, *Appl. Opt.*, 18, 57–67, 1979.
22. B. J. Chang and C. D. Leonard, Dichromated gelatin for the fabrication of holographic elements, *Appl. Opt.*, 18, 2407–2417, 1979.
23. T. Kubota and T. Ose, Methods of increasing the sensitivity of methylene-blue sensitized dichromated gelatin, *Appl. Opt.*, 18, 2538–2539, 1979.
24. P. Hariharan, Holographic recording materials: Recent development, *Opt. Eng.*, 19, 636–641, 1980.
25. P. Hariharan, Silver halide sensitised gelatin holograms: Mechanism of hologram formation, *Appl. Opt.*, 25, 2040–2042, 1986.
26. S. Calixto, Dry polymer for holographic recording materials, *Appl. Opt.*, 26, 3904–3910, 1987.
27. R. T. Ingwall, M. Troll, and W. T. Vetterling, Properties of reflection holograms recorded on Polaroid's DMP-128 photopolymer, *Proc. SPIE*, 747, 67–73, 1987.
28. N. V. Kukhtarev and V. V. Mauravev, Dynamic holographic interferometry in photorefractive crystals, *Opt. Spectrosc. (USSR)*, 64, 656–659, 1988.
29. R. Changkakoti and S. V. Pappu, Methylene blue sensitised DCG holograms: A study of their storage and reprocessibility, *Appl. Opt.*, 28, 340–344, 1989.
30. R. A. Lessard and J. J. Couture, Holographic recordings in dye/polymer systems for engineering applications, *Proc. SPIE*, 1183, 75–89, 1989.

31. R. T. Ingwall and M. Troll, Mechanism of hologram formation in DMP-128 photopolymer, *Opt. Eng.*, 28, 586–591, 1989.
32. S. Redfield and L. Hesselink, Data storage in photorefractives revisited, *Opt. Comput.*, 63, 35–45, 1989.
33. W. K. Smothers, B. M. Monroe, A. M. Weber, and D. E. Keys, Photopolymers for holography, *Proc. SPIE*, 1212, 20–29, 1990.
34. F. Ledoyen, P. Bouchard, D. Hennequin, and M. Cormier, Physical model of a liquid thin film: Application to infrared holographic recording, *Phys. Rev. A*, 41, 4895–4902, 1990.
35. G. D. Savant and J. L. Jannson, Optical recording materials, *Proc. SPIE*, 1461, 79–90, 1991.
36. S. A. Zager and A. M. Weber, Display holograms in Du Pont's Omnidex™ films, *Proc. SPIE*, 1461, 58–67, 1991.
37. F. P. Shvartsman, Dry photopolymer embossing: Novel photo-replication technology for surface relief holographic optical elements, *Proc. SPIE*, 1507, 383–391, 1991.
38. E. Bruzzone and F. Mangili, Calibration of a CCD camera on a hybrid coordinate measuring machine for industrial metrology, *Proc. SPIE*, 1526, 96–112, 1991.
39. J. L. Salter and M. F. Loeffler, Comparison of dichromated gelatin and Du Pont HRF-700 photopolymer as media for holographic notch filters, *Proc. SPIE*, 1555, 268–278, 1991.
40. R. D. Rallison, Control of DCG and non silver halide emulsions for Lippmann photography and holography, *Proc. SPIE*, 1600, 26–37, 1991.
41. T. J. Cvetkovich, Holography in photoresist materials, *Proc. SPIE*, 1600, 60–70, 1991.
42. K. Kurtis and D. Psaltis, Recording of multiple holograms in photopolymer films, *Appl. Opt.*, 31, 7425–7428, 1992.
43. R. A. Lessard, C. Malouin, R. Changkakoti, and G. Mannivannan, Dye-doped polyvinyl alcohol recording materials for holography and non-linear optics, *Opt. Eng.*, 32, 665–670, 1993.
44. U. S. Rhee, H. J. Caulfield, J. Shamir, C. S. Vikram, and M. M. Mirsalehi, Characteristics of the Du Pont photopolymer for angularly multiplexed page-oriented holographic memories, *Opt. Eng.*, 32, 1839–1847, 1993.
45. D. Dirksen and G. von Bally, Holographic double-exposure interferometry in near real time with photorefractive crystals, *J. Opt. Soc. Am. B*, 11, 1858–1863, 1994.
46. K. Meerholz, B. L. Volodin, Sandalphon, B. Kippelen, and N. Peyghambarlan, A photorefractive polymer with high optical gain and diffraction efficiency near 100%, *Nature*, 371, 497–500, 1994.
47. Jean-Pierre Fouassler and F. Morlet-Savary, Photopolymers for laser imaging and holographic recording: Design and reactivity of photosensitizers, *Opt. Eng.*, 35, 304–312, 1996.
48. V. V. Vlad, D. Malacara-Hernandez, and A. Petris, Real-time holographic interferometry using optical phase conjugation in photorefractive materials and direct spatial phase reconstruction, *Opt. Eng.*, 35, 1383–1388, 1996.
49. D. Litwiller, CMOS vs. CCD: Facts and Fiction, *Photonic Spectra*, 35, 154–158, 2001.
50. H. Helmers and M. Schellenberg, CMOS vs. CCD sensors in speckle interferometry, *Opt. Laser Technol.*, 35, 587–595, 2003.
51. S. R. Guntaka, V. Toal, and S. Martin, Holographic and electronic speckle pattern interferometry using a photopolymer recording material, *Strain 40*, 79–81, 2004.
52. D. Litwiller, CMOS vs. CCD: Maturing technologies, maturing markets, *Photonic Spectra*, 39(8), 54–61, 2005.
53. A. C. Sullivan, M. W. Grabowski, and R. R. McLeod, Three-dimensional direct-write lithography into photopolymer, *Appl. Opt.*, 46, 295–301, 2007.

6 Holographic Interferometry

6.1 INTRODUCTION

Holography was born out of the very challenging technological problem of improving the resolution of the electron microscope, which was limited by the spherical aberration of the electron lenses. Gabor therefore invented a two-step process, the first step involving recording without the lenses and the second step being reconstruction. The technique was demonstrated with microscopic objects and with spatially and temporally filtered radiation from a mercury lamp. This was necessary, since high-resolution recording materials and coherent sources were not available at that time.

After its invention in 1948, holography remained practically dormant until the arrival of the laser, since a long-coherence-length source was needed to record a hologram of an object. Earlier recordings of three-dimensional (3D) objects were made on Kodak 649F plates, with very impressive results. Holography, therefore, came to be known as 3D photography. It, however, is more than ordinary 3D photography, since it provides 3D views with changing perspectives.

Holography records the complex amplitude of a wave coming from an object (the object wave), rather than the intensity distribution in the image, as is the case in photography. Holography literally means "total recording," that is, recording of both the phase and the amplitude of a wave. The detectors in the optical regime respond to the intensity (energy) of the wave, and hence phase information is to be converted into intensity variations. This is accomplished by interferometry. A reference wave is added to the object wave at the recording plane. The recording is done on a variety of media, including photographic emulsions, photopolymers, and thermoplastics. The record is called a hologram. The hologram is like a window with a memory. Different perspectives of the scene are seen through different portions of the hologram. In addition to recording holograms of 3D objects, several new applications of holography have emerged, holographic interferometry being one of these.

Holographic interferometry (HI) has emerged as a technique of unparalleled applications, since it provides interferometric comparison of real objects or events separated in time and space. Various kinds of HI have been developed: real-time, double-exposure, time-average, etc. Further, it can be performed with one reference wave, two reference waves, and so on. These reference waves can be of the same or different wavelengths. The reference wave can come from the same side of the hologram as the object wave or from the other side. HI can be performed with a continuous-wave laser or a pulsed laser. The record can be made on a photographic emulsion, a thermoplastic, a photopolymer, a charge-coupled device (CCD), or other media. Digital holography provides for comparison of objects situated at different locations. Small objects can be studied for their responses to external agency. The possibilities are endless, and so are the applications. This chapter presents some of these applications, along with the relevant theoretical background.

6.2 HOLOGRAM RECORDING

An object is illuminated by a wave from a laser, and the diffracted field is received on a recording plate lying in the (x, y) plane. A reference wave is added to this field at the recording plane, as shown in Figure 6.1a. The diffracted field from the object constitutes the object wave, which is represented by $\mathbf{O}(x, y) = O_0(x, y)\exp[i\phi_o(x, y)]$, where $O_0(x, y)$ is the amplitude of the object wave and $\phi_o(x, y)$ is its phase. The complex amplitude $\mathbf{R}(x, y)$ of the reference wave at the recording plane is expressed

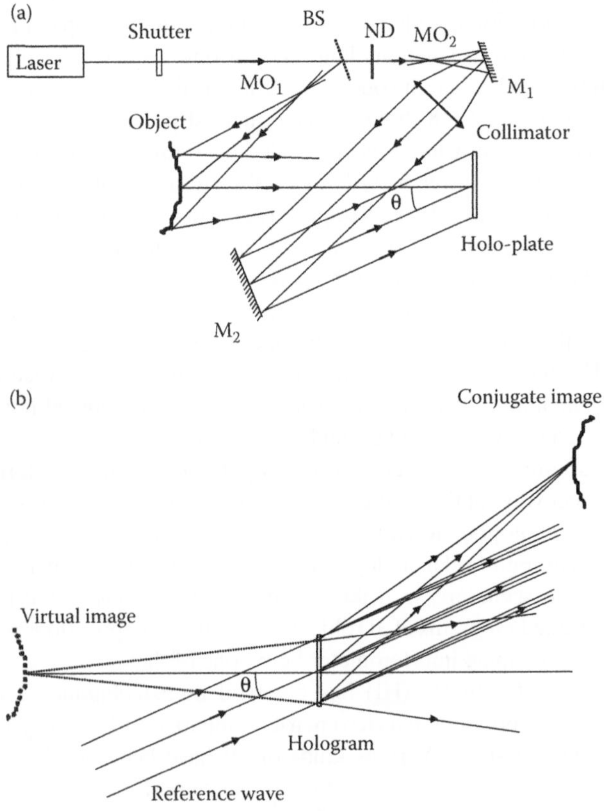

FIGURE 6.1 (a) Recording of a hologram. (b) Its reconstruction.

as $\mathbf{R}(x,y) = R\exp(2\pi i\nu_R y)$, where ν_R is the spatial frequency of the wave. We have taken a plane wave incident at an angle θ with the z axis as a reference wave; hence, $\nu_R = (\sin\theta)/\lambda$. These waves are derived from the same wave (source), and hence are coherent with each other. The total amplitude at the recording plane is

$$\mathbf{A}(x,y) = \mathbf{O}(x,y) + \mathbf{R}(x,y) \tag{6.1}$$

Therefore, the intensity distribution $I(x,y)$ on the recording plane is given by

$$I(x,y) = O_0^2(x,y) + R^2 + 2O_0(x,y)R\cos[2\pi\nu_R y - \phi_o(x,y)] \tag{6.2}$$

It can thus be seen that both the amplitude $O_0(x,y)$ variations and the phase $\phi_o(x,y)$ variations have been converted into intensity variations, to which the recording material responds. We assume that the recording material is a photographic emulsion, say, a holographic plate or a film. The intensity is recorded over an appropriate time period T, resulting in an exposure variation $E(x,y) = I(x,y)T$. After development, the plate/film is called a hologram. On the hologram, the exposure variation is converted into density variation or amplitude transmittance variation. The amplitude transmittance is complex, since the hologram introduces both amplitude and the phase variations—the latter as a result of thickness variations. The phase variations can be eliminated if the hologram is placed in a liquid gate. Further, if the phase variations are shared between two wavefronts, they are not seen when these wavefronts interfere. We assume here, however, that the amplitude transmittance of the hologram is proportional to the exposure incident during recording. Under this assumption, the amplitude transmittance $t(x,y)$ of the hologram is expressed as

$$t(x,y) = t_0 - \beta E(x,y) \tag{6.3}$$

where β is a constant dependent on processing parameters, exposure, and so on.

6.3 RECONSTRUCTION

This hologram is placed back in the same position held during recording, and is illuminated by the reference wave. The field just behind the hologram is given by

$$\begin{aligned} t(x,y)\mathbf{R}(x,y) = t_0'\mathbf{R}(x,y) - \beta T\mathbf{R}(x,y)\{O_0^2(x,y) \\ + 2O_o(x,y)R\cos[2\pi\nu_R y - \phi_o(x,y)]\} \end{aligned} \tag{6.4}$$

where $t_0' = t_0 - \beta TR^2$ is the modified DC transmittance. Equation 6.4 can also be written as

$$\begin{aligned} t(x,y)\mathbf{R}(x,y) = t_0'\mathbf{R}(x,y) - \beta T\mathbf{R}(x,y)[O_0^2(x,y) \\ + \mathbf{O}(x,y)\mathbf{R}^*(x,y) + \mathbf{O}^*(x,y)\mathbf{R}(x,y)] \end{aligned} \tag{6.5}$$

where * signifies the complex conjugate. It can be seen that there are four waves just behind the hologram, of which the wave $t_0'\mathbf{R}(x,y)$ is the uniformly attenuated reference wave. The second wave, $-\beta T\mathbf{R}(x,y)O_0^2(x,y)$, also propagates in the direction of the reference wave. Owing to the slow spatial variation of $O_0^2(x,y)$, there is a diffracted field around the direction of the reference wave. The third wave, $-\beta TR^2\mathbf{O}(x,y)$, is the original object wave multiplied by a constant, $-\beta TR^2$. This wave propagates in the direction of the object wave and has all the attributes of the object wave, except that its spatial dimensions are restricted by the size of the hologram. The negative sign signifies that a phase change of π has taken place owing to the wet development process. The fourth wave, $-\beta T\mathbf{R}^2(x,y)\mathbf{O}^*(x,y)$, represents a conjugate wave, which propagates in a different direction. This wave can also be written as $-\beta TR^2O(x,y)\exp[2\pi i 2\nu_R y - \phi_o(x,y)]$; the phase of the reference wave is modulated by that of the object wave, but it essentially travels in the direction ϕ, where $\sin\phi = 2\sin\theta$. Figure 6.1b shows that several waves are generated during the reconstruction step.

The recording and reconstruction geometries just described are the off-axis holography geometries due to Leith and Upatnieks. However, the reference beam can be added axially to the object beam. This is in-line holography or Gabor holography. In-line holography has applications in particle size measurements, etc.

6.4 CHOICE OF ANGLE OF REFERENCE WAVE

One of the shortcomings of Gabor holography is that all the diffracted waves propagate in the same direction. These waves are angularly separated if the reference wave is added at an angle during the recording of the hologram. The question is, how large should the angle be? In fact, the off-axis reference wave acts as a carrier of object information. It also results in very fine interference fringes, into which object information is coded. The fringe frequency is given by $(\sin\theta)/\lambda$, where θ is the mean angle between the reference and object waves. The recording medium should be capable of resolving this fringe frequency. Further, as has been described earlier, the various waves are angularly separated by nearly θ for small angles. However, a restriction is placed on θ by the simple condition that the spectra of the various waves should not overlap. If the bandwidth of the object wave is ν_o, then it is easily seen from Figure 6.2 that the spectra will be just separated when $3\nu_o = \nu_R$. This immediately gives the smallest angle θ_{min} as $\theta_{min} = \sin^{-1}(3\lambda\nu_o)$. For angles smaller than θ_{min}, the various diffracted beams will overlap.

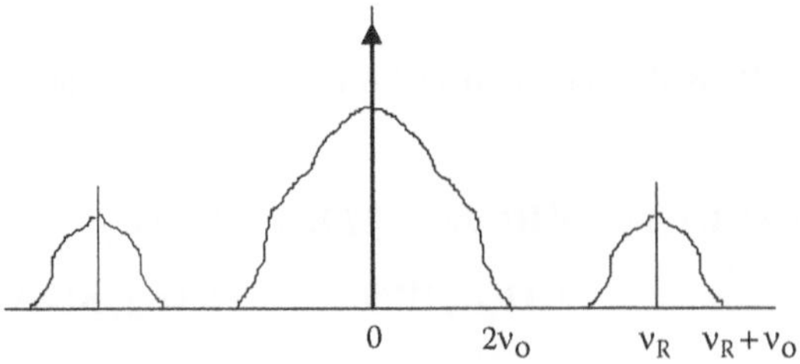

FIGURE 6.2 Fourier transform of the transmittance of a hologram.

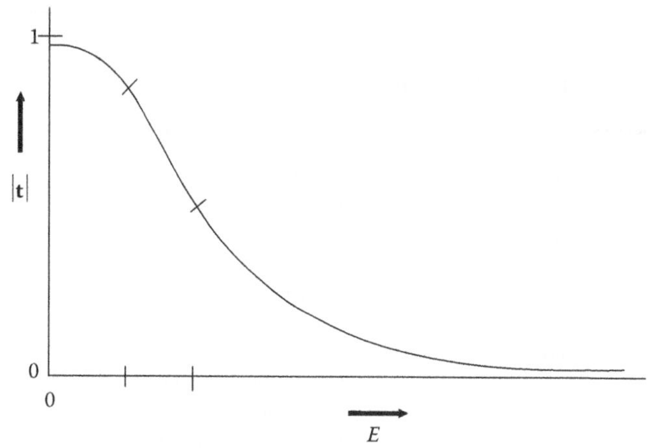

FIGURE 6.3 Amplitude transmittance |**t**| versus exposure E curve for a typical photographic emulsion.

6.5 CHOICE OF REFERENCE WAVE INTENSITY

It has been tacitly assumed above that the amplitude transmittance of the hologram is proportional to the exposure. This is valid only when one operates over a very small portion of the t–E curve, which is shown in Figure 6.3. In order to meet this condition, the reference wave should be 3–10 times stronger than the object wave. If this condition is not met, then nonlinearities will begin to play a role, and higher-order images will be produced.

6.6 TYPES OF HOLOGRAMS

Holograms are classified in a variety of ways. Table 6.1 summarizes these.

6.7 DIFFRACTION EFFICIENCY

Diffraction efficiency is a measure of the amount of light that goes into forming the useful image when the hologram is illuminated by the reference wave. It is calculated for the ideal situation of interference between plane waves. For a thin amplitude transmission hologram, the maximum diffraction efficiency is 6.25%. This value can be improved by bleaching; that is, the amplitude hologram is converted into a phase hologram. It can then reach up to 33.6% for a thin phase transmission hologram. The diffraction efficiency for a thick phase hologram can approach 100%: essentially, then, the entire incident light is utilized for image formation and there is only one diffraction order.

6.8 EXPERIMENTAL ARRANGEMENT

The experimental arrangement (Figure 6.1a) consists of a laser, a beam-splitter, beam-expanding optics, and the recording medium. We describe these in detail.

TABLE 6.1
Classification of Hologram Types

Properties	Hologram Types
Transmission function	Amplitude
	Phase
Region of diffraction	Fresnel
	Fraunhofer
	Fourier
	Image plane
Recording geometry	On-axis/Gabor/in-line
	Off-axis/Leith–Upatnieks/carrier-frequency
	Reflection/Denisyuk
Exposure	Single-exposure/real-time/live-fringe
	Double-exposure/lapsed-time/frozen-fringe
	Multiple-exposure
	Time-averaged
Emulsion thickness/Q-parameter	Thin
	Thick
Reconstruction	Monochromatic-light
	White-light
Recording configuration	Transmission
	Reflection
	Rainbow
	Digital
Object type	Transmission-object/phase-object
	Transmission-objects—diffuse illumination
	Opaque-object

6.8.1 Lasers

There are a large number of lasers from which to choose. Some of these are described below:

- *Ruby laser*. This is a pulsed laser with a pulse width of the order of 10 ns. Its coherence length can be greater than 2 m when it is used with an etalon, and it can deliver more than 1 J in a single pulse. Lasers for HI have dual-pulse facilities, with variable pulse separation for dynamical studies.
- *Argon–ion laser*. This is a continuous-wave laser with output exceeding 5 W. It gives a multiline output, and hence a dispersion prism is mounted to select a line. Further, an intercavity etalon increases the coherence length to a usable length for studying moderate-sized objects. It can be used with photoresisst and thermoplastics as recording media.
- *He–Ne laser*. This is the most commonly used laser for holography and HI, with a power output in the range 25–35 mW. The coherence length is greater than 20 cm.

- *Nd : YAG semiconductor pumped, first-harmonic green.* This laser is finding acceptance for holography. It offers a very long coherence length, with continuous-wave operation. The output ranges from 40 to 150 mW.
- *He–Cd laser.* This is a continuous-wave laser with output in the range 40–150 mW. It is suitable for recording on photoresists. Both blue and ultraviolet lines are usable for holography.
- *Semiconductor or diode lasers (LDs).* These can be run in both continuous-wave and pulsed modes. Continuous-wave lasers are preferred for holography, but are not very common. They require temperature and current stability. They can output up to 500 mW in continuous-wave mode.

6.8.2 Beam-Splitters

Usually, an unexpanded laser beam is split into two or more beams. For this, a glass plate (preferably a wedge plate) serves as a good beam-splitter. However, when an expanded collimated beam is to be split, a cube beam-splitter or a plane parallel plate splitter with one side antireflection-coated is recommended. Pellicle beam-splitters are also used. In some cases, polarization optics (Wollaston prisms) are employed as beam-splitters.

6.8.3 Beam-Expanders

With high-power lasers, a suitable concave or negative lens is used for beam expansion; otherwise, a microscope objective 5×, 10×, 20×, 40× (45×), or 60× (63×) can be used. As a result of dust particles on the optical surfaces, the beam is usually dirty; that is, it has circular rings, specks, etc. It can be cleaned by placing a pinhole at the focus of the microscope objective. The arrangement is known as spatial filtering. The pinhole size must be matched with the microscope objective. An achromatic lens is placed in the beam such that the pinhole is at its focus. This results in an expanded collimated beam—hence the name "beam-expander."

6.8.4 Object-Illumination Beam

For small objects, collimated-beam illumination is preferred as the propagation vector of the illumination beam is the same over the whole surface. For large objects, a spherical diverging wave is used. If the object is cylindrical in shape, a cylindrical diverging wave is preferred. Objects must be located in the coherence volume. Special recording geometries that relax the coherence requirement may be used.

6.8.5 Reference Beam

Either a diverging spherical wave or a plane wave can be used as a reference wave. However, if the hologram is to be reconstructed at a later time, or at a different location, the use of a plane wave is recommended. The beam is overexpanded to make it uniform over the hologram plane. The reference-beam intensity must be 3–10 times stronger than the object-beam intensity at the hologram plane for linear recording.

6.8.6 Angle between Object and Reference Beams

The object wave is a diffuse wave. Therefore, the angle between the object wave and the reference wave (even when the latter is a plane wave) will vary over the recording plane. However, we take the mean angle. The fringe frequency on the recording plane will be approximately $(\sin\theta)/\lambda$, where θ is the mean angle between the reference wave and the object wave incident nearly normally on the recording plane. The recording medium should be able to resolve this fringe structure. Too large an angle puts a higher demand on the resolution. A small angle, on reconstruction, may not result in separation of the beams. Therefore, the mean angle between the object wave and the reference wave has to be chosen judiciously.

6.9 HOLOGRAPHIC RECORDING MATERIALS

A variety of recording materials are available. They have been described in Chapter 5. The choice is made based on wavelength sensitivity, resolution, sensitivity, and other properties.

6.10 HOLOGRAPHIC INTERFEROMETRY

HI is used to compare two waves from real objects by the process of interference. These two waves are usually from an initial unstressed state and a final stressed state of an object. It is, however, assumed that the microstructure of the surface does not change as a result of loading: the two comparison waves differ owing to path-difference changes rather than microstructural changes. HI can be performed in a variety of ways. The most common methods are real-time, double-exposure, and time-averaged. Several novel configurations have been developed, depending on the application and also to exploit the strengths of HI. They are discussed at appropriate places below.

6.10.1 Real-Time HI

A hologram of the object is recorded, processed, and placed back in the experimental set-up at exactly the same location that it occupied during recording. Reconstruction of the hologram by the reference wave generates a replica of the original object wave, which propagates in the direction of the original wave. This wave is, however, phase-shifted by π owing to wet photographic development. Since the object wave is also present, it will undergo diffraction on passage through the hologram: a wave transmitted by the DC transmittance of the hologram will propagate in the original direction. Therefore, there are two waves: one released from the hologram by interaction of the reference wave and the other transmitted by the DC transmittance of the hologram. These waves are identical in all respects except for a phase change of π, and hence produce a dark field on interference. If the object is now loaded, the object wave carries the deformation phase, and hence an interference pattern is observed. This pattern changes in real time with the change in load. One can therefore monitor the response of the object to an external loading agency continuously until the fringes become too fine to be resolved. The technique is also known as single-exposure or live-fringe HI.

Mathematically, we can explain the procedure by writing—the transmittance of the single-exposure hologram as

$$t(x,y) = t_0' - \beta T\left\{O_0^2(x,y) + 2O_0(x,y)R\cos[2\pi\nu_R y - \phi_o(x,y)]\right\} \quad (6.6)$$

where $t_0' = t_0 - \beta TR^2$, with t_0' being a constant. Interrogation by the reference wave $\mathbf{R}(x,y)$ releases the object wave, which is now given by $-\beta TR^2\mathbf{O}(x,y)$, while the directly transmitted object wave is written as $[t_0' - \beta TO_0^2(x,y)]\mathbf{O}(x,y)e^{i\delta}$, where δ is the phase change introduced by the deformation. Interference between these two waves results in an intensity distribution that can be expressed as

$$I = \left(\beta TR^2\right)^2 O_0^2\left[1 + \frac{\left(t_0 - \beta TR^2 - \beta TO_0^2\right)^2}{\left(\beta TR^2\right)^2} - \frac{2\left(t_0 - \beta TR^2 - \beta TO_0^2\right)}{\beta TR^2}\cos\delta\right] \quad (6.7)$$

Assuming $\beta TR^2 \gg \beta TO_0^2$, which is usually true, Equation 6.7 can be rewritten as

$$I = I_0\left[1 + \frac{\left(t_0 - \beta TR^2\right)^2}{\left(\beta TR^2\right)^2} - \frac{2\left(t_0 - \beta TR^2\right)}{\beta TR^2}\cos\delta\right] \quad (6.8)$$

Dark fringes are formed wherever $\delta = 2m\pi$, m being an integer. The contrast η of the fringes is given by

$$\eta = \frac{2\beta TR^2(t_0 - \beta TR^2)}{(\beta TR^2)^2 + (t_0 - \beta TR^2)^2}$$

It can be seen that the contrast of the fringes can be controlled by the reference-wave intensity during reconstruction. In general, the contrast is less than unity, but, by appropriate increase of the reference-wave intensity, it is possible to achieve unit-contrast fringes. For example, if the bias transmittance t_0 is equal to $2\beta TR^2$, then a unit-contrast fringe pattern results.

6.10.2 Double-Exposure HI

Here, both exposures, belonging to the initial and final states of the object, are recorded sequentially on the same photographic plate. The total exposure recorded can be expressed as

$$\beta T[I_1(x,y) + I_2(x,y)] \quad (6.9)$$

The transmittance of the hologram is given by

$$t(x,y) = t_0'' - \beta T\{2O_0^2(x,y) + [\boldsymbol{O}(x,y) + \boldsymbol{O}(x,y)e^{i\delta}]\mathbf{R}^*(x,y) + [\mathbf{O}^*(x,y) + \mathbf{O}^*(x,y)e^{-i\delta}]\mathbf{R}(x,y)\} \quad (6.10)$$

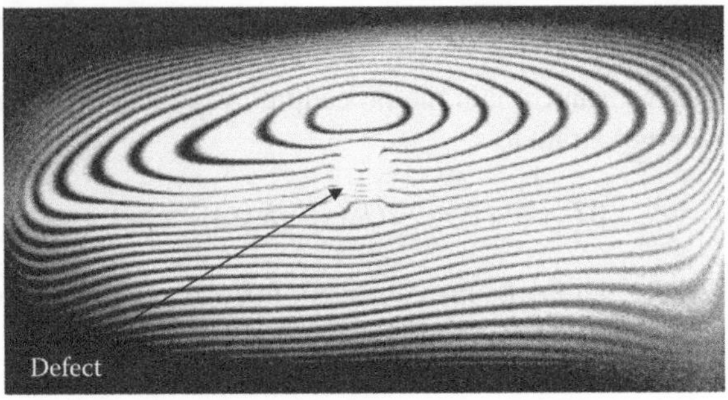

FIGURE 6.4 Double-exposure interferogram of a pipe with a defect.

where $t_0'' = t_0 - 2\beta TR^2$. When the double-exposure hologram is reconstructed, both object waves are released simultaneously, and interfere to produce a unit-contrast fringe pattern characteristic of the deformation. The intensity distribution in the fringe pattern is

$$I = 2(\beta TR^2)^2 O_0^2(1 + \cos\delta) = I_0(1 + \cos\delta) \tag{6.11}$$

Bright fringes are formed wherever $\delta = 2m\pi$, m being an integer. The technique is also known as frozen-fringe or lapsed-time HI. It may be noted that only two states of the object are being compared by this technique. Figure 6.4 shows a double-exposure interferogram of a pipe with a defect. The pipe was loaded by an application of hydraulic pressure between exposures. It can be seen that the defect is a region of thinner wall thickness. Table 6.2 gives a comparison between real-time and double-exposure HI.

6.10.3 Time-Average HI

Time-average HI is utilized for the study of vibrating bodies such as musical instruments. As the name suggests, the recording is carried out over a time period that is several times the period of vibration. Since an object vibrating sinusoidally spends most of the time at the locations of maximum displacement, a recording of such a vibrating object is equivalent to a double-exposure record. However, the intensity distribution in the reconstruction is modified considerably owing to the time of excursion between these extreme positions. To study this phenomenon, we can write for the instantaneous intensity on the recording plane,

$$I(x, y; t) = O_0^2(x, y) + R^2 + \mathbf{O}^*(x, y; t)\mathbf{R}(x, y) + \mathbf{O}(x, y; t)\mathbf{R}^*(x, y) \tag{6.12}$$

where the object wave $\mathbf{O}(x, y; t)$ is expressed as

$$\mathbf{O}(x, y; t) = O_0 e^{i\phi_o(x,y)} e^{i\delta(x,y;t)} \tag{6.13}$$

The phase $\delta(x, y; t)$ represents the phase change introduced by the vibration. If the body is vibrating with amplitude $A(x, y)$ at a frequency ω and is illuminated and

TABLE 6.2
Comparison of Single- and Double-Exposure HI

Single-Exposure HI	Double-Exposure HI
Monitors/compares different states of the object (due to loading) with the initial state over a range determined by usable fringe density and speckle noise.	Compares only the final state with the initial state.
In situ processing is recommended; otherwise the hologram has to be repositioned very accurately. Once the experimental set-up is disturbed, the hologram cannot be used for comparison.	Hologram can be reconstructed at leisure, even at a different location.
Fringe contrast is poor, but can be improved.	Unit-contrast fringe pattern is obtained, since the diffraction efficiency is equally shared by the two beams.
Bright fringes are formed when $\delta = (2m + 1)\pi$. A phase change of π takes place owing to wet photographic development.	Bright fringes are formed when $\delta = 2m\pi$.
Usually performed to obtain the correct parameters for double-exposure HI.	Usually used for quantitative evaluation.

observed along directions making angles θ_1 and θ_2 with the local normal, then the phase difference δ can be written as

$$\delta(x, y; t) = \frac{2\pi}{\lambda} A(x, y)\,(\cos\theta_1 + \cos\theta_2)\sin\omega t \tag{6.14}$$

The intensity distribution given by Equation 6.12 is recorded over a period T much longer than the period of vibration; that is, it is recorded over a large number of vibration cycles. The average intensity recorded over the period T is

$$I(x, y) = \frac{1}{T}\int_0^T I(x, y; t)\,dt \tag{6.15}$$

This record, on development, is called a time-average hologram, and the procedure is known as time-average HI. The hologram is reconstructed with the reference wave. Since, on reconstruction, the various waves generated are separable, we consider only the amplitude of the desired wave, that is,

$$\begin{aligned} a(x, y) &= -\beta T R^2 \frac{1}{T}\int_0^T \mathbf{O}(x, y; t)\,dt \\ &= -\beta T R^2 \mathbf{O}(x, y) \frac{1}{T}\int_0^T e^{(2\pi i/\lambda)A(x,y)(\cos\theta_1 + \cos\theta_2)\sin\omega t}\,dt \end{aligned} \tag{6.16}$$

The time integral $T^{-1}\int_0^T e^{(2\pi i/\lambda)A(x,y)(\cos\theta_1+\cos\theta_2)\sin\omega t}\,dt$ is called the characteristic function of the sinusoidal vibration, and is denoted by M_T. Thus, the intensity distribution in the reconstructed image is given by

$$I_{\text{rec}}(x,y) = a(x,y)a^*(x,y) = \beta^2T^2R^4O_0^2|M_T|^2 \quad (6.17)$$

The characteristic function has been generalized for various types of motion (Table 6.3). The characteristic function for sinusoidal motion with phase difference $\delta(x,y;t)$ as described by Equation 6.14 can be obtained analytically. It is given by

$$M_T = \frac{1}{T}\int_0^T \mathrm{e}^{(2\pi i/\lambda)A(x,y)(\cos\theta_1+\cos\theta_2)\sin\omega t}\,dt = J_0\left(\frac{2\pi}{\lambda}A(x,y)(\cos\theta_1+\cos\theta_2)\right) \quad (6.18)$$

where $J_0(x)$ is the Bessel function of zeroth order and first kind. The intensity distribution in the reconstructed image will be

$$I_{\text{rec}}(x,y) = I_0(x,y)\left[J_0\left(\frac{2\pi}{\lambda}A(x,y)(\cos\theta_1+\cos\theta_2)\right)\right]^2 \quad (6.19)$$

The variation of $[J_0(x)]^2$ with the argument x is shown in Figure 6.5. It can be seen that the reconstructed image is modulated by the function $[J_0(x)]^2$. Thus, the image is

TABLE 6.3
Characteristic Function $|M(t)|^2$

HI Type	Displacement	$\lvert M(t)\rvert^2$
Real-time	Static (**L**)	$1+c^2-2c\cos(\mathbf{k}\cdot\mathbf{L})$
	Harmonic of amplitude $\mathbf{A}(x,y)$	$1+c^2-2cJ_0(\mathbf{k}\cdot\mathbf{A})$
Real-time with reference fringes	Harmonic	$1+c^2-2c\cos(\mathbf{k}\cdot\mathbf{L})J_0(\mathbf{k}\cdot\mathbf{A})$
Real-time stroboscopic	Harmonic: pulses at $\omega t=\pi/2$ and $3\pi/2$	$1+c^2-2c\cos(2\mathbf{k}\cdot\mathbf{A})$
Double-exposure	Static (**L**)	$\cos^2(\mathbf{k}\cdot\mathbf{L}/2)$
Double-exposure stroboscopic	Harmonic: pulses at $\omega t=\pi/2$ and $3\pi/2$	$\cos^2(\mathbf{k}\cdot\mathbf{A})$
Time-average	Harmonic, of amplitude $\mathbf{A}(x,y)$	$[J_0(\mathbf{k}\cdot\mathbf{A})]^2$
	Constant velocity $\mathbf{L}_r=\mathbf{v}T$	$\mathrm{sinc}^2(\mathbf{k}\cdot\mathbf{L}_r/2)$
	Constant acceleration from rest	$\dfrac{C^2\left(\sqrt{\mathbf{k}\cdot\mathbf{A}}\right)+S^2\left(\sqrt{\mathbf{k}\cdot\mathbf{A}}\right)}{(2/\pi)(\mathbf{k}\cdot\mathbf{A})}$
Time-average	Irrationally related modes	$J_0(\mathbf{k}\cdot\mathbf{A}_1)J_0(\mathbf{k}\cdot\mathbf{A}_2)$
Temporally frequency-translated	Harmonic motion	$[J_m(\mathbf{k}\cdot\mathbf{A})]^2$
	Amplitude-modulated reference wave $f_r(t)=e^{i(\omega t-\Delta)}$	$[J_m(\mathbf{k}\cdot\mathbf{A})]^2\cos^2\Delta$
	Phase-modulated reference wave $f_r(t)=e^{iM_R\sin\omega t}$	$[J_m(\mathbf{k}\cdot\mathbf{A}-M_R)]^2$

Note: c is the contrast, and C and S are Fresnel cosine and sine integrals.

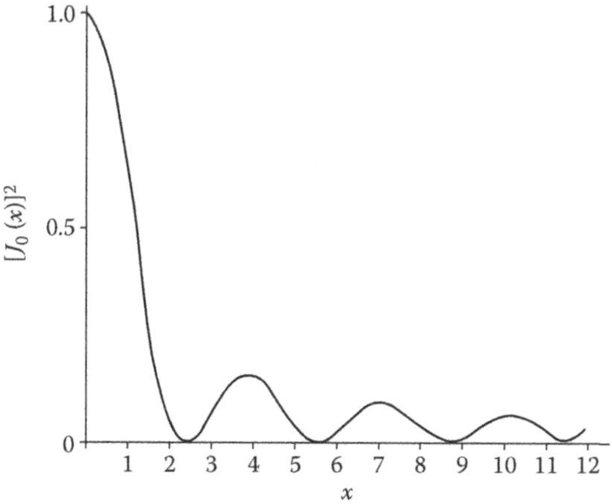

FIGURE 6.5 $[J_0(x)]^2$ distribution, which defines the intensity distribution in time-averaged HI fringes for a sinusoidally vibrating object.

covered by the fringes. Figure 6.6 shows the time-average interferogram of an edge-clamped diaphragm vibrating in the second-harmonic mode. The zero intensity in the image occurs at the zeros of the function $[J_0(x)]^2$. It can also be seen that the intensity is maximum at the stationary region ($A(x, y) = 0$), and decreases to zero when

$$\frac{2\pi}{\lambda} A(x, y)(\cos\theta_1 + \cos\theta_2) = 2.4048 \tag{6.20}$$

This gives the amplitude of vibration at the first zero of the Bessel function. Assuming illumination and observation directions along the normal to the surface of the object,

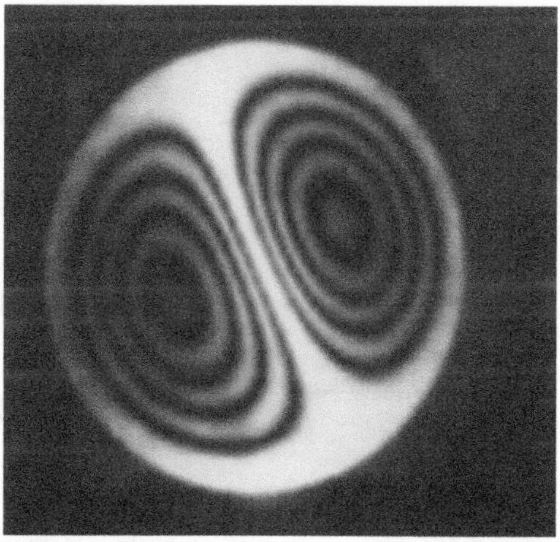

FIGURE 6.6 Time-averaged interferogram of an edge-clamped diaphragm vibrating in second-harmonic mode. (From J. W. Goodman, *Introduction to Fourier Optics*, McGraw-Hill, New York, 1968. With permission.)

the amplitudes at the first and successive zeros of the Bessel function are given by

$$A(x, y) = \frac{\lambda}{4\pi} \times 2.4048, \tag{6.21a}$$

$$= \frac{\lambda}{4\pi} \times 5.5200, \tag{6.21b}$$

$$= \frac{\lambda}{4\pi} \times 8.6537, \tag{6.21c}$$

$$= \frac{\lambda}{4\pi} \times 11.7915 \tag{6.21d}$$

Since the intensity falls off rapidly, high amplitudes of vibration are difficult to monitor by this technique. In brief, time-average HI gives the vibration map (phase of the vibration is lost) over a range of frequencies; that is, the amplitude of the vibration can be measured.

Let us consider a simple example of a cantilever as shown in Figure 6.7. It is illuminated normally and observed in the same direction. We assume illumination by the red radiation of a He–Ne laser. The amplitude of vibration can be expressed as

$$z(x, t) = z(x) \sin \omega t \tag{6.22}$$

The intensity distribution in the image is given by

$$I_{\text{rec}}(x, y) = I_0(x, y) \left[J_0\left(\frac{4\pi}{\lambda} A(x, y) \right) \right]^2 \tag{6.23}$$

The fringes run parallel to each other, and, for higher amplitudes, are almost equidistant. Dark fringes are located where the amplitudes of vibration are 0.12, 0.28, 0.44, 0.59 μm, ...

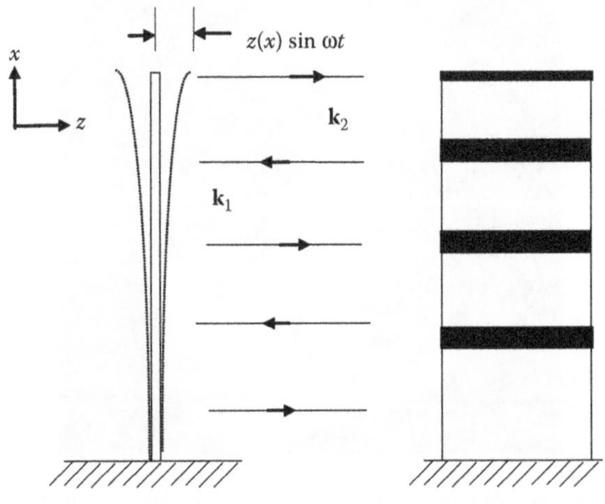

FIGURE 6.7 Vibrating cantilever and simulated fringe pattern as obtained by time-averaged HI.

6.10.4 Real-Time, Time-Average HI

In this technique, a single-exposure record of a stationary object is first made. After development, the hologram is precisely relocated. The object is now set in vibration and viewed through the hologram. Interference between the wave reconstructed from the hologram and the directed transmitted wave from the vibrating object will yield an intensity distribution that is proportional to $[1 - J_0(x)]^2$, where $x = (4\pi/\lambda)A(x, y)$. Thus, the nodal points on the object appear dark. The contrast of this fringe pattern is very low. While observing the real-time pattern, the laser beam can be chopped at the frequency of vibration. The method is then equivalent to the real-time HI for static objects. The fringes correspond to the displacement between the initial position and the position when the laser beam illuminates the object. By varying the time at which the light pulse illuminates the object during its vibration cycle, the displacement at different phases of vibration can be mapped.

6.11 FRINGE FORMATION AND MEASUREMENT OF DISPLACEMENT VECTOR

We have mentioned that object deformation causes a phase change between the two waves, at least one of which is derived from the hologram. This phase change is responsible for the formation of the fringe pattern. In this section, we address two issues:

- How is the phase change related to the deformation vector?
- Where are the fringes really localized?

Let us consider a point P on the object surface. Owing to loading, this point moves to a different location P′. The vector distance PP′ is the deformation vector **d**. We consider the geometry as shown in Figure 6.8 to calculate the phase difference introduced by the deformation. The phase difference δ at the observation point O is given by

$$\delta = \mathbf{k}_1 \cdot \mathbf{r}_1 + \mathbf{k}_2 \cdot \mathbf{r}_2 - (\mathbf{k}_1 + \Delta\mathbf{k}_1) \cdot \mathbf{r}_1' - (\mathbf{k}_2 + \Delta\mathbf{k}_2) \cdot \mathbf{r}_2' \tag{6.24}$$

We can express $\mathbf{r}_1'$ and $\mathbf{r}_2'$ as

$$\mathbf{r}_1 + \mathbf{d} = \mathbf{r}_1' \tag{6.25a}$$

$$\mathbf{r}_2' + \mathbf{d} = \mathbf{r}_2 \tag{6.25b}$$

Substituting for $\mathbf{r}_1'$ and $\mathbf{r}_2'$, we obtain

$$\begin{aligned}
\delta &= \mathbf{k}_1 \cdot \mathbf{r}_1 + \mathbf{k}_2 \cdot \mathbf{r}_2 - (\mathbf{k}_1 + \Delta\mathbf{k}_1) \cdot (\mathbf{r}_1 + \mathbf{d}) - (\mathbf{k}_2 + \Delta\mathbf{k}_2) \cdot (\mathbf{r}_2 - \mathbf{d}) \\
&= \mathbf{k}_1 \cdot \mathbf{r}_1 + \mathbf{k}_2 \cdot \mathbf{r}_2 - \mathbf{k}_1 \cdot \mathbf{r}_1 - \Delta\mathbf{k}_1 \cdot \mathbf{r}_1 - \mathbf{k}_1 \cdot \mathbf{d} \\
&\quad - \Delta\mathbf{k}_1 \cdot \mathbf{d} - \mathbf{k}_2 \cdot \mathbf{r}_2 + \mathbf{k}_2 \cdot \mathbf{d} + \Delta\mathbf{k}_2 \cdot \mathbf{d} - \Delta\mathbf{k}_2 \cdot \mathbf{r}_2 \\
&= (\mathbf{k}_2 - \mathbf{k}_1) \cdot \mathbf{d}
\end{aligned} \tag{6.26}$$

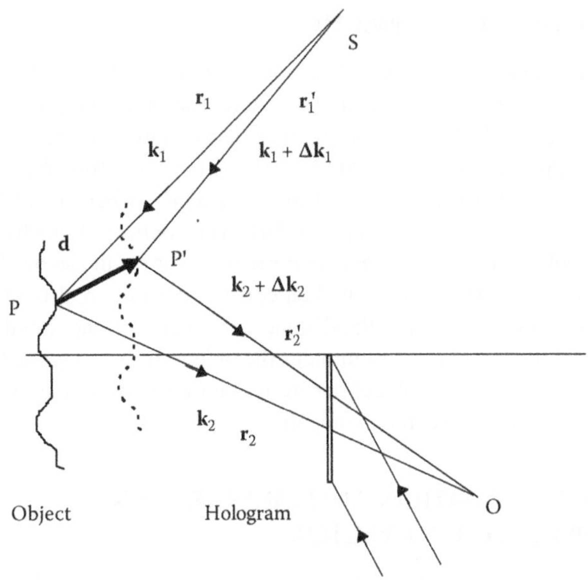

FIGURE 6.8 Calculation of the phase difference.

This expression has been obtained under the experimental conditions where the displacement vector is of the order of a few micrometers and the distances r_1 and r_2 involved could be from a few tens of centimetres to metres. When this is taken into account, the vectors $\mathbf{r}_1$ and $\mathbf{r}_2$ are perpendicular to $\mathbf{\Delta k}_1$ and $\mathbf{\Delta k}_2$, respectively, and hence their product terms are automatically zero, while the remaining two terms, $\mathbf{\Delta k}_1 \cdot d$ and $\mathbf{\Delta k}_2 \cdot d$, which represent second-order contributions, are also neglected.

This expression tells us when and how the fringe pattern is formed; it does not tell us where it actually is formed. The situation is quite different from that encountered in classical interferometry. In HI, the fringes are normally localized not on the object surface but on a surface in space. This surface is known as the surface of localization. Knowledge of the location of the surface of localization helps in determining the deformation vector. In fact, the surface of localization depends on several factors, including the deformation vector. Several theories have been presented in the literature. Needless to say, it is necessary to know about fringe localization in detail before any quantitative evaluation is carried out.

6.12 LOADING OF THE OBJECT

Since HI sees only the change in the state of the object, the object must be subjected to external agency to change its state. This can be performed by any one of the following methods:

- Mechanical loading
- Thermal loading
- Pressure/vacuum loading

- Vibratory or acoustic loading
- Impact loading

The most commonly used method is mechanical loading, in which the object is subjected to either compressive or tensile forces. If the dynamic response needs to be studied, it is subjected to either vibratory or impact loading. In holographic nondestructive testing (HNDT), it is essential to use the appropriate kind of loading. For example, debonds are easily seen when pressure or vacuum stressing is applied. Table 6.4 gives some areas of application of the five types of loading.

6.13 MEASUREMENT OF VERY SMALL VIBRATION AMPLITUDES

It has been shown that the first minimum of intensity in time-average HI occurs when $A(x, y) = (\lambda/4\pi)2.4048$. This is valid only when the directions of illumination and observation are along the surface normal. For illumination from a He–Ne laser ($\lambda = 632.8\,\text{nm}$), the amplitude of vibration $A(x, y)$ corresponds to $0.12\,\mu\text{m}$. Even smaller amplitudes of vibration can be monitored by observing the variation of intensity between the maximum and first zero of the Bessel function.

6.14 MEASUREMENT OF LARGE VIBRATION AMPLITUDES

6.14.1 Frequency Modulation of Reference Wave

The frequency of the reference wave is shifted by $n\omega$, where n is an integer and ω is the frequency of vibration, and a time-average hologram is recorded in the normal way. The recorded intensity distribution is given by

$$I(x,y) = \frac{1}{T}\int_0^T [O_0^2 + R_0^2 + OR^* + O^*R]\,dt \tag{6.27}$$

where the object wave and the reference wave are written explicitly as

$$\mathbf{O}(x, y; t) = O_0\,e^{i[\varpi t+\phi_o(x,y)+\delta(x,y;t)]} \tag{6.28a}$$

$$\mathbf{R}(x, y; t) = R_0\,e^{i[(\varpi-n\omega)t+\phi_R]} \tag{6.28b}$$

Further, assuming that the object is vibrating sinusoidally with an amplitude $A(x, y)$, we can write for the phase difference $\delta(x, y; t)$

$$\begin{aligned}\delta(x, y; t) &= \frac{2\pi}{\lambda}(\cos\theta_1 + \cos\theta_2)A(x, y)\sin\omega t\\ &= \frac{2\pi}{\lambda}2A(x, y)\sin\omega t\end{aligned}$$

when the directions of illumination and observation are along the surface normal.

TABLE 6.4
Some Applications of Holographic Testing

Loading/Problem Detected	Items Tested
Mechanical Loading	
Cracks	Bolt holes in steel channels, concrete, rocket nozzle liners, welded metal plates, turbine blades, glass
Debonds	Honeycomb panels, propellant liners
Fatigue damage	Composite materials
Deformation	Radial deformation in carbon cylinders and tubes
Pneumatic Loading	
Cracks	Welded joints, aluminum shells, bonded cylinders, ceramic heat exchanger tubes
Debonds/delaminations	Composite panels and tubes, honeycomb structures, bonded cylinders, tires, rocket launch tubes, rocket nozzle liners
Weld defects	Welded plastic pipes, honeycomb panels
Weakness (thinness)	Aluminum cylinders, pressure vessels, tubes, composite tubes
Structural flaws	Composite domes
Soldered joints	Medical implant devices
Bad brazes	Silicon pressure sensors
Thermal Loading	
Cracks	Glass/ceramic tubes, train wheels, turbofan blades
Debonds	Aircraft wing assemblies, honeycomb structures, laminate structures, rocket motor casings, rocket nozzle liners, rocket propellant liners, tires, bonded structures
Delaminations	Antique paintings
Various defects	Circuit boards and electronic modules
Vibrational Loading	
Cracks	Train wheels, ceramic bars
Debonds	Turbine blades, rocket propellant liners, sheet metal sandwich structures, laminate structures, honeycomb structures, fiberglass-reinforced plastics, helicopter rotors, ceramics bonded to composite plates, brake disks, adhesive joints, metallic interfaces, elastomer
Strength	CRTs
Impulse Loading	
Cracks	Steel plates, wing plank splices, turbine blades
Debonds	Foam insulation
Voids and thin areas	Aluminum plates

Since the various terms in the expression for the recorded intensity are separable, we pick up the term of interest as

$$\frac{1}{T}\int_0^T \mathbf{OR}^*\,dt = O_0 e^{i\phi_o} R_0 e^{-i\phi_R}\frac{1}{T}\int_0^T e^{(4\pi i/\lambda)A\sin\omega t}e^{-in\omega t}\,dt$$

$$= O_0 R_0 e^{i(\phi_o-\phi_R)}\frac{1}{T}\sum_{n'=-\infty}^{n'=\infty} J_{n'}\left(\frac{4\pi}{\lambda}A\right)\int_0^T e^{in'\omega t}e^{-in\omega t}\,dt \qquad (6.29)$$

The integral will vanish for all values of n' except $n' = n$. Thus, the amplitude distribution, on reconstruction, becomes proportional to $J_n(4\pi A/\lambda)$. The intensity distribution is then given by

$$I(x,y) = I_0 J_n^2 \left(\frac{4\pi}{\lambda} A\right) \tag{6.30}$$

Unlike the case of time-average HI, fringes of nearly equal intensity are formed, and hence large vibration amplitudes can be measured.

6.14.2 Phase Modulation of Reference Beam

The mirror in the reference arm is mounted on PZT, which is excited by a voltage signal at the frequency of vibration. The amplitude of the reference wave at the recording plane is

$$\mathbf{R}(x,y) = R_0\, e^{i\phi_R}\, e^{i\delta_R} \tag{6.31}$$

where

$$\delta_R = \frac{2\pi}{\lambda} 2A_R \sin(\omega t - \Delta)$$

with A_R and Δ representing the amplitude and phase of the reference mirror vibration. The object wave is expressed as

$$\mathbf{O}(x,y) = O_0\, e^{i\phi_o(x,y)} e^{i\delta} \tag{6.32}$$

where

$$\delta = \frac{2\pi}{\lambda} 2A(x,y) \sin \omega t$$

is the phase difference introduced by object vibration. It should be noted that the modulation in the reference wave is phase-shifted by Δ, which may be varied experimentally. A time-average hologram is now recorded. Again, we consider only the term of interest in the expression for the amplitude transmittance, and express the amplitude of the reconstructed wave as proportional to

$$\begin{aligned} \frac{1}{T}\int_0^T \mathbf{OR}^*\, dt &= O_0 R_0\, e^{i(\phi_o - \phi_R)} \frac{1}{T}\int_0^T e^{(4\pi i/\lambda)[A \sin \omega t - A_R \sin(\omega t - \Delta)]}\, dt \\ &= O_0 R_0\, e^{i(\phi_o - \phi_R)} J_0 \left(\frac{4\pi}{\lambda}(A^2 + A_R^2 - 2AA_R \cos \Delta)^{1/2}\right) \end{aligned} \tag{6.33}$$

The intensity distribution in the reconstructed image is given by

$$I(x,y) = I_0 \left[J_0 \left(\frac{4\pi}{\lambda}(A^2 + A_R^2 - 2AA_R \cos \Delta)^{1/2}\right)\right]^2 \tag{6.34}$$

As a special case, when the phase of modulation Δ of the reference wave is zero, the intensity distribution is given by

$$I(x,y) = I_0\left[J_0\left(\frac{4\pi}{\lambda}(A - A_R)\right)\right]^2 \tag{6.35}$$

Thus, control of the amplitude of the reference mirror vibration can be used to measure large vibration amplitudes. The zero-order fringe (the brightest maximum) will be formed at those regions where the vibration amplitude of the reference mirror matches that of the object. In this way, it is possible to extend the measurable range considerably, in practice up to about 10 μm. By varying the phase of the reference mirror, it is also possible to trace out areas of the object vibrating in the same phase as that of the mirror, thereby mapping the phase distribution of the object vibration.

6.15 STROBOSCOPIC ILLUMINATION/STROBOSCOPIC HI

The object is illuminated by laser pulses whose duration is much shorter than the period of vibration. The recorded hologram is thus like a double-exposure hologram. Let the first record be made when the object is in a state of vibration 1 and the second when the object is in state 2, as shown in Figure 6.9. The double-exposure hologram will display fringes corresponding to the displacement A between these two vibration states. By varying the pulse separation, holograms for different displacements can be recorded.

Alternatively, pulse separation may be adjusted to correspond to one-quarter of the time period, and double-exposure holograms may be recorded at various phases of the vibration cycle. The advantage of stroboscopic HI is that fringes of unit contrast, similar to those in double-exposure HI, are formed. In addition, the use of pulse illumination does not require vibration isolation of the holographic set-up.

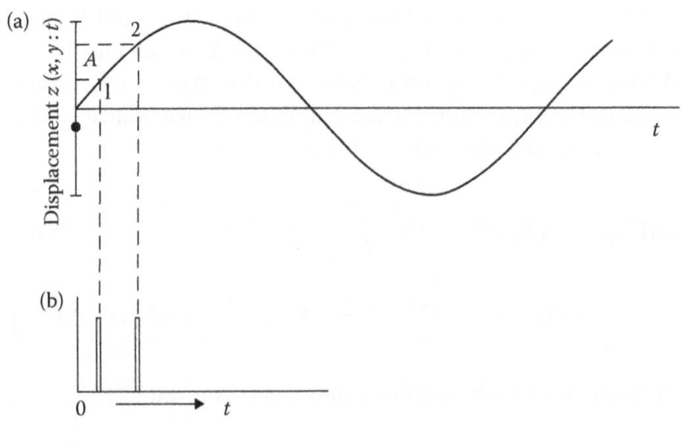

FIGURE 6.9 Stroboscopic HI of a vibrating object: (a) displacement of sinusoidally vibrating object; (b) pulse illumination.

6.16 SPECIAL TECHNIQUES IN HOLOGRAPHIC INTERFEROMETRY

6.16.1 Two-Reference-Beam HI

In the discussion of the phase-shift method of evaluation of the phase in an interferogram, it was mentioned that an additional phase is introduced that is independent of the phase to be measured. In real-time HI, this is provided by shifting the phase of the reference wave during reconstruction. However, it is not possible to achieve this in double-exposure HI, since the reconstruction of both the recorded waves is done with a single reference beam.

In order to have independent access to the two reconstructed waves, and introduce the desired phase difference between them, a holographic set-up with two reference waves is required (Figure 6.10a). The first exposure of a double-exposure hologram is performed with the object in its initial state and with the reference wave R_1 (the reference wave R_2 is blocked). The object is loaded, and the second exposure is made with the reference wave R_2 (the reference wave R_1 is blocked). The reconstruction of

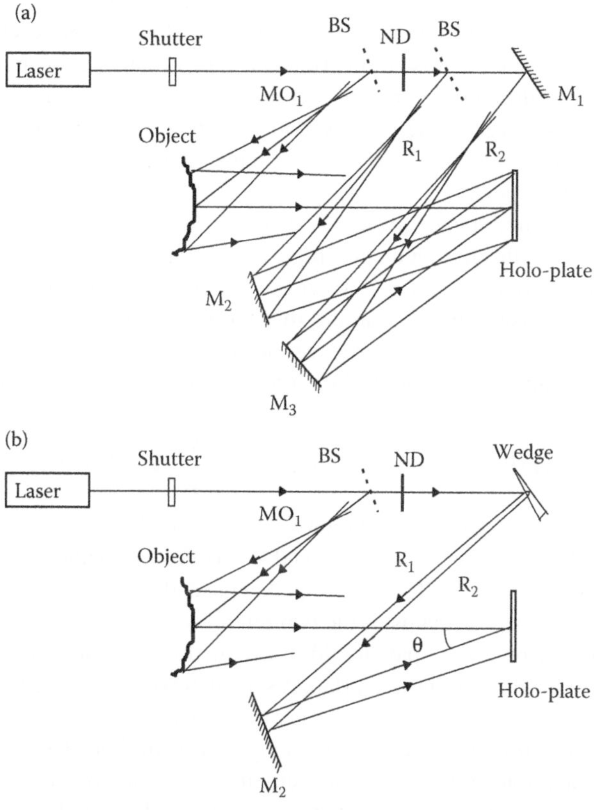

FIGURE 6.10 (a) Schematic of two-reference-wave HI. (b) Two-reference-wave HI with close waves.

the hologram is accomplished with both reference waves R_1 and R_2 simultaneously to generate the interference pattern between the initial and final states of the object: the interference pattern may be varied by changing one of the reference waves while leaving the other unaffected. Mathematically, the procedure is described as follows.

In the first exposure, an intensity distribution $I_1(x, y)$ is recorded, where

$$I_1(x, y) = O_0^2 + R_{10}^2 + \mathbf{O}\mathbf{R}_1^* + \mathbf{O}^*\mathbf{R}_1 \tag{6.36}$$

The intensity distribution recorded in the second exposure is given by

$$I_2(x, y) = O_0^2 + R_{20}^2 + \mathbf{O}'\mathbf{R}_2^* + \mathbf{O}'^*\mathbf{R}_2 \tag{6.37}$$

where $\mathbf{O}'(x, y) = \mathbf{O}(x, y)e^{i\delta(x,y)}$ is the object wave from the deformed state of the object. The total exposure is $T(I_1 + I_2)$, and the amplitude transmittance $t(x, y)$ of the doubly-exposed hologram is

$$t(x, y) = t_0 - \beta T(I_1 + I_2) \tag{6.38}$$

This hologram is illuminated by both reference waves. Therefore, the amplitude of the waves just behind the hologram is $(\mathbf{R}_1 + \mathbf{R}_2)t(x, y)$. This hologram generates a multiplicity of images, which may or may not overlap. However, $\mathbf{R}_1\mathbf{R}_1^*\mathbf{O}$ and $\mathbf{R}_2\mathbf{R}_2^*\mathbf{O}'$ overlap completely and give rise to interference fringes. Other undesired images can be separated from this interference pattern if the angle between the two reference waves is made very large. Generally, the two reference waves are taken on the same side of the object normal and have mutual angular separation much larger than the angular size of the object.

Although the desired images are separated by the large angular separation between the reference waves, the arrangement is highly sensitive to misalignment of the hologram during repositioning and also to any wavelength change between hologram recording and reconstruction. Assuming that the reference waves are plane waves propagating in directions given by propagation vectors $\mathbf{k}$ and $\mathbf{k}'$, the additional phase difference $\Delta\phi$ at any point $\mathbf{r}_H$ on the hologram due to misalignment and wavelength change is

$$\Delta\phi(\mathbf{r}_H) = [(\mathbf{k} - \mathbf{k}') \times \boldsymbol{\omega}] \cdot \mathbf{r}_H + \frac{\Delta\lambda}{\lambda}(\mathbf{k} - \mathbf{k}') \cdot \mathbf{r}_H \tag{6.39}$$

where $\boldsymbol{\omega} = (\Delta\xi, \Delta\eta, \Delta\chi)$ is the rotation vector for small rotations $\Delta\xi, \Delta\eta, \Delta\chi$ of the hologram about the x, y, and z axes, and $\Delta\lambda$ is the wavelength change between recording and reconstruction. It can be seen that the additional phase change $\Delta\phi$ will be small if $\mathbf{k} - \mathbf{k}'$ is small, that is, if the two reference waves propagate in nearly the same direction. The price to be paid for this is a reduction in fringe contrast, since all of the reconstructed waves will now overlap. A schematic of an arrangement to produce two close reference waves is shown in Figure 6.10b. This set-up reduces the influence of misalignment errors and wavelength-change error. In fact, if the recording and reconstruction are done at the same wavelength, there is no error due to wavelength change. However, if the recording is done with a pulsed laser and the reconstruction with a continuous-wave laser, the error due to wavelength change cannot be removed, but can be reduced using the above experimental arrangement.

6.16.2 Sandwich HI

HI records phase difference no matter what the source of this difference. However, when using HI for studying strains in an object, the main interest is in measuring local deformations on the object surface. Depending on the loading conditions, the object very often undergoes rigid-body movements (translations and tilts), and these are commonly much greater in magnitude than the local deformations. These rigid-body deformations may be too large to be measured by HI, or they may completely mask the local deformations of interest. Therefore, there is considerable interest in exploring techniques that can compensate for the influence of rigid-body movements. Fringe-control techniques have been developed, but they are applicable only with real-time HI. Sandwich HI offers an attractive alternative for compensating for rigid-body movements.

As the name suggests, the recording is done on two holographic plates H_1 and H_2, which form a sandwich as shown in Figure 6.11a. We assume a plane reference wave. The initial state of the object is recorded on H_1 (H_2 is replaced by a dummy) and the final state on H_2. A ray from an object point P hits plate H_1 at B and plate H_2 at A. Because of this geometry, an additional path difference AB − BD is recorded. From Figure 6.11a, $AB = s/\cos\alpha$ and $BD = AB\cos(\alpha+\beta)$, where s is the separation between the plates of the sandwich and α and β are the angles shown in Figure 6.11a. Therefore, the path difference AB − BD is equal to

$$\frac{s}{\cos\alpha}\,[1-\cos(\alpha+\beta)]$$

In other words, if only the initial state of the object is recorded on both the plates, the reconstruction will give a fringe-free image of the object. However, the path difference AB − BD can be varied by tilting the sandwich, resulting in the appearance of a fringe pattern.

We will now see how much change is introduced in the path difference by the tilt of the sandwich. Let us assume that the sandwich is tilted by a small angle ϕ so that the reference beam now makes an angle of β' with the normal to the sandwich as shown in Figure 6.11b. The corresponding points on the tilted sandwich are denoted by A_1 and B_1. The reconstructed rays from A_1 and B_1 are shown by the dotted lines. The path difference between these rays is $r\cos\phi - r\cos(\alpha+\beta')$, with $\beta' = \beta - \phi$ and $r = s/\cos\alpha$. Therefore, the net path difference is

$$\begin{aligned} & r[\cos\phi - \cos(\alpha+\beta') - 1 + \cos(\alpha+\beta)] \\ & \approx r\phi\sin(\alpha+\beta) = s\phi\tan(\alpha+\beta) \end{aligned} \tag{6.40}$$

It can be seen that the path difference change due to the tilt of the sandwich is proportional to ϕ.

Let us now assume that the object is tilted by a small angle $\Delta\xi$ as shown in Figure 11c as a result of loading. This introduces an additional path equal to $2x\Delta\xi$. For this tilt to be compensated, we must have

$$2x\Delta\xi = s\phi\tan(\alpha+\beta) \tag{6.41}$$

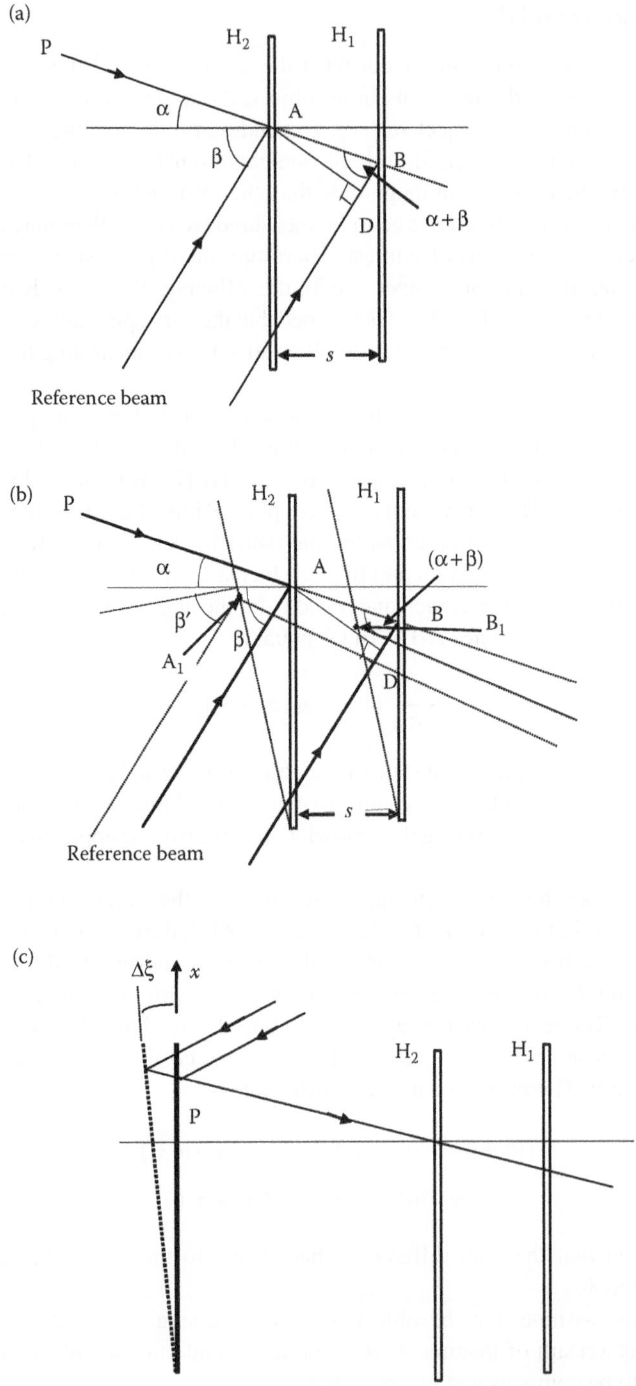

FIGURE 6.11 Sandwich HI. (a) Recording of a sandwich hologram. (b) Influence of the tilt of the sandwich. (c) Sandwich HI with an object tilted between exposures.

When the angles α and β are small and are also taken equal to each other (i.e., $\alpha = \beta$), Equation 6.41 becomes

$$2x\Delta\xi = 2\alpha s\phi \tag{6.42}$$

But the angle α can be expressed as $\alpha = x/z$, where z is the distance between the sandwich hologram and the object plane. Therefore,

$$\phi = \frac{z}{s}\Delta\xi \tag{6.43}$$

A tilt of the sandwich hologram of magnitude ϕ will fully compensate for a tilt of the object of magnitude $\Delta\xi$.

In order to fully utilize the strength of sandwich HI, a reference surface is placed by the side of the actual object—the reference surface should be free from fringes on reconstruction. However, when the fringe pattern on the object is compensated, a linear fringe pattern appears on the reference surface, which is used to determine the tilt given to the sandwich hologram. Alternatively, the reference surface is so placed that it is unaffected by loading but is influenced by bodily tilts and translations. Then, on reconstruction, the reference surface will also carry a fringe pattern, which should be fully compensated in order to obtain the correct deformation map of the object.

6.16.3 Reflection HI

Another method to eliminate the contributions of rigid-body movements and tilts to the measured phase difference is to clamp the holographic plate to the object under study as shown in Figure 6.12a. The photographic plate, with its emulsion side facing the

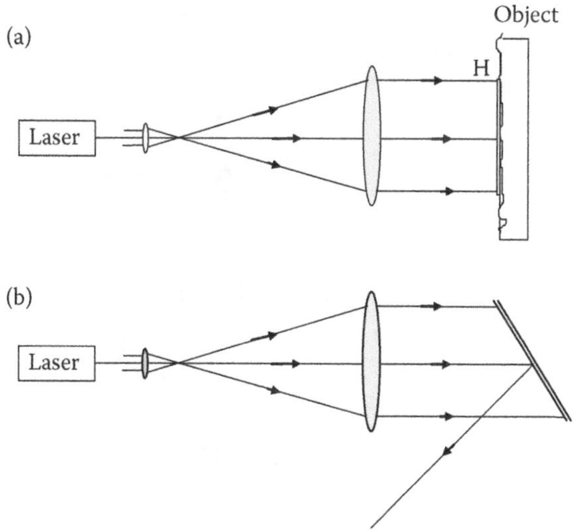

FIGURE 6.12 (a) Recording of a hologram in reflection HI. (b) Its reconstruction.

object, is illuminated by a plane wave at normal incidence. In the first pass through the plate, this wave acts as a reference wave, and the light scattered from the object forms the object wave. Therefore, a reflection hologram is recorded within the emulsion as interference planes between the incident wave and the one scattered by the object. The recording is made on a thick emulsion, say of about 10 μm thickness: Agfa-Gevaert 8E75 plates are suitable for recording with He–Ne lasers.

For interferometric comparison, two exposures with the object loaded in-between are made. The hologram is reconstructed with the laser light as shown in Figure 6.12b. As a result of emulsion shrinkage, reconstruction is possible only at a suitable shorter wavelength. Otherwise, the emulsion is swelled to its original thickness by soaking it in an aqueous solution of triethanolamine, $(CH_2OHCH_2)_3N$, followed by slow and careful drying in air to observe the fringe pattern at the recording wavelength.

The resulting hologram is insensitive to rigid-body translations parallel and perpendicular to the direction of illumination if plane-wave illumination is used during recording. Also, in-plane rotation has no influence in this case. This is because either no path difference or a constant path difference is introduced. However, additional fringes may be introduced if the rigid-body rotation is around an axis perpendicular to the direction of illumination and if the object and the holographic plate are separated as shown in Figure 6.13. We take a point P on the object, and consider recording, say, at point H of the hologram. Owing to the rotation, the hologram–object combination takes a new position, that is, P moves to P′ as the object rotates, and H moves to H′ as the plate rotates along with the object. In the first exposure, the optical path recorded

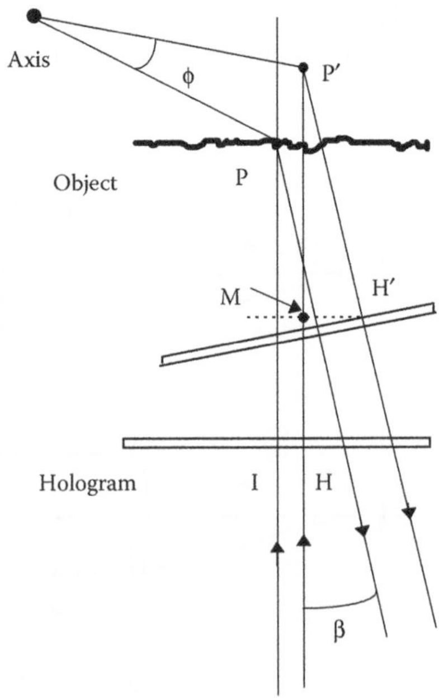

FIGURE 6.13 Calculation of path change when the hologram–object combination is rotated.

is IP + PH, and in the second recording, the path is MP′ + P′H′. The optical path difference Δ is IP + PH − MP′ − P′H′. But PH = P′H′. Therefore, the path difference Δ is IP − MP′ = PH $\cos\beta$ − P′H′ $\cos(\beta+\phi)$, where β is the direction of observation and ϕ is the angle of rotation. This may be rewritten as

$$\begin{aligned}\Delta &= \mathrm{PH}[\cos\beta - \cos(\beta+\phi)] = \mathrm{PH}\cos\beta\left[1 - \frac{\cos(\beta+\phi)}{\cos\beta}\right] \\ &= z_\mathrm{p}\left[1 - \frac{\cos\beta - \phi\sin\beta}{\cos\beta}\right], \quad \text{since } \phi \ll 1 \\ &= z_\mathrm{p}\phi\tan\beta \end{aligned} \tag{6.44}$$

where z_p is the distance between the object and the holographic plate. It can be seen that this path difference yields additional fringes. However, no additional fringes are introduced (i.e., $\Delta = 0$ for $\phi \neq 0$) when (i) $z_\mathrm{p} = 0$ and/or (ii) $\beta = 0$. Thus, placing the holographic plate in contact with the object eliminates the influence of rotation about an axis perpendicular to the illumination beam.

6.17 EXTENDING THE SENSITIVITY OF HI

6.17.1 HETERODYNE HI

It has been shown that the fringe separation usually corresponds to $\lambda/2$ in path difference. With phase-shifting techniques, deformations as small as $\lambda/30$ can be measured. If even smaller deformations are to be monitored, other means must be explored. Heterodyne HI is one such technique, being capable of giving a resolution of $\lambda/1000$. This is, however, achieved at the expense of a more complicated experimental set-up and slower data acquisition.

Figure 6.14 shows an experimental arrangement: it is a two-reference-wave geometry. One of the reference waves passes through two acousto-optical (AO) modulators. The first exposure is made with the reference beam R_1 (R_2 being blocked), and the second exposure, after the object has been loaded, is made with the reference wave R_2 (R_1 being blocked). Both exposures are made at the same wavelength, with the AO modulators turned off. The reconstruction is done with both reference waves; the AO modulators are turned on so that the wavelengths of the two reference waves are different. As an example, one AO modulator modulates the beam at 40 MHz; the frequency of the light wave diffracted in the first order is thus shifted by 40 MHz. This then passes through the second AO modulator, which modulates it at 40.08 MHz. We now consider diffraction in −1 diffraction order, to cancel the deviation of the beam and obtain a light wave that is frequency-shifted by 80 kHz (= 40.08 − 40.0 MHz). Thus, the two reference waves are now frequency-shifted by 80 kHz.

As has been pointed out earlier, a multiplicity of images are produced in two-reference-wave HI. However, we consider only terms of interest in the amplitude transmittance of the hologram, that is, $\mathbf{OR}_1^* + \mathbf{O}'\mathbf{R}_2^*$, where reference waves $\mathbf{R}_1$ and

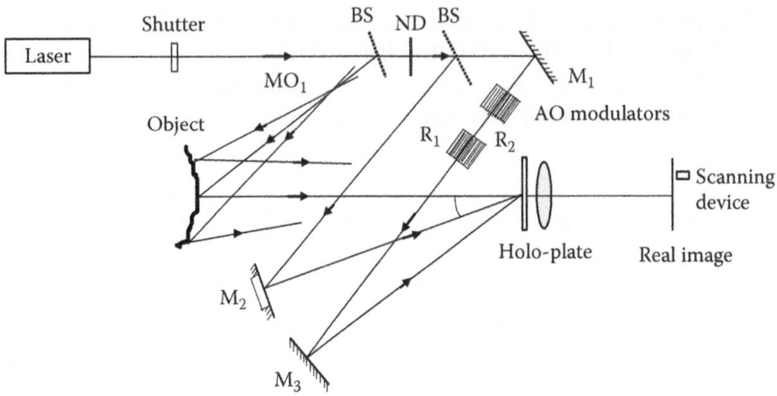

FIGURE 6.14 Schematic of heterodyne HI.

$\mathbf{R}_2$ have the same frequency. We write the reference and object waves explicitly as

$$\mathbf{R}_1 = R_{01}\, e^{i(\omega_1 t + \phi_{R1})} \tag{6.45a}$$

$$\mathbf{R}_2 = \mathrm{R}_{02}\, e^{i(\omega_1 t + \phi_{R2})} \tag{6.45b}$$

$$\mathbf{O} = O_0(x, y)\, e^{i(\omega_1 t + \phi_o)} \tag{6.45c}$$

$$\mathbf{O}' = \mathbf{O}\, e^{i\delta(x,y,z)} \tag{6.45d}$$

During reconstruction, $\mathbf{R}_1 = R_{01} e^{i(\omega_1 t + \phi_{R1})}$ and $\mathbf{R}_2 = \mathbf{R}_2'$, where $\mathbf{R}_2' = \mathrm{R}_{02} e^{i(\omega_2 t + \phi_{R2})}$, with $\omega_2 = \omega_1 \pm \Delta\omega$, $\Delta\omega$ being the frequency shift introduced by the AO modulators. Again considering just the terms of interest, the amplitude in the reconstructed image is proportional to

$$(\mathbf{R}_1 + \mathbf{R}_2')(\mathbf{O}\mathbf{R}_1^* + \mathbf{O}'\mathbf{R}_2^*)$$

and we obtain the intensity distribution in the image as

$$\left|\mathbf{O}\mathbf{R}_1\mathbf{R}_1^* + \mathbf{O}'\mathbf{R}_2'\mathbf{R}_2^*\right|^2 = a(x, y) + b(x, y)\cos[\delta(x, y) + \Delta\omega t] \tag{6.46}$$

where the various appropriate terms have been absorbed into $a(x, y)$ and $b(x, y)$. If the intensity at a point (x, y) is observed, then it is found to vary sinusoidally with time. The interference phase that is to be measured introduces a phase shift in this signal. Unfortunately, the phase $\delta(x, y) + \Delta\omega t$ carries no information. However, from the phase difference of the oscillating intensities at two different points, one obtains

$$\Delta\delta(x_1, y_1 : x_2, y_2) = [\delta(x_1, y_1) + \Delta\omega t] - [\delta(x_2, y_2) + \Delta\omega t] = \delta(x_1, y_1) - \delta(x_2, y_2)$$

The phase difference $\Delta\delta$, which is the difference between the interference phases at two points, can be measured with a very high degree of accuracy using an electronic phasemeter.

There are two ways to perform temporal heterodyne evaluation. In one approach, the phase difference is measured by keeping one detector fixed to a reference point and scanning the image with another detector. In this way, the phase difference modulo 2π with respect to the phase at the reference point is measured. The interference phase difference can be summed from point to point to yield the phase distribution along a line or a plane. In the second method, the real image is scanned using a pair, triplet, quadruplet, or quintuplet of photodetectors with known separation, and the phase differences are then measured. In practice, the detector consists of a photodiode to which a fiber is pigtailed. The other end of the fiber scans the real image.

Heterodyne HI cannot be implemented as real-time HI, because of the extremely high stability requirement of the experimental set-up and also of the surrounding environment.

6.18 HOLOGRAPHIC CONTOURING/SHAPE MEASUREMENT

For complete stress analysis, it is necessary to know the shape of the object under study. Three techniques that can be used to obtain the shape of an object are as follows:

- Change of wavelength between exposures—the dual-wavelength method
- Change of refractive index of the medium surrounding the object between exposures—the dual-refractive-index method
- Change of direction of the illumination beam between exposures—the dual-illumination method

6.18.1 Dual-Wavelength Method

We will describe the dual-wavelength method in detail. The technique requires that a holographic recording of the object be made on the same plate with two slightly different wavelengths λ_1 and λ_2. The processed plate, the hologram, is then illuminated with the reference wave at wavelength λ_1, so that the original object wave is reconstructed at λ_1 together with the distorted reconstruction of the object recorded at wavelength λ_2.

The waves from the original reconstruction and the distorted reconstruction interfere to yield a fringe pattern that is related to the shape of the object. In order to understand how this method works, we again go through the mathematics of recording and reconstruction. In the first exposure, we record an intensity distribution proportional to

$$O_0^2 + R_{01}^2 + \mathbf{O}_1\mathbf{R}_1^* + \mathbf{O}_1^*\mathbf{R}_1$$

The intensity recorded in the second exposure is proportional to

$$O_0^2 + R_{02}^2 + \mathbf{O}_2\mathbf{R}_2^* + \mathbf{O}_2^*\mathbf{R}_2$$

where the object and reference waves are defined as

$$\mathbf{O}_1 = O_0\, e^{i\mathbf{k}_1\cdot r} = O_0 \exp\left(\frac{2\pi i}{\lambda_1} z_o\right) \exp\left[\frac{2\pi i}{2\lambda_1 z_o}(x - x_o)^2\right] \tag{6.47a}$$

$$\mathbf{O}_2 = O_0\, e^{i\mathbf{k}_2\cdot r} = O_0 \exp(ik_2 z_o) \exp\left[\frac{k_2 i}{2z_o}(x - x_o)^2\right] \tag{6.47b}$$

$$\mathbf{R}_1 = R_0\, e^{ik_1 x \sin\theta} \tag{6.47c}$$

$$\mathbf{R}_2 = R_0\, e^{ik_2 x \sin\theta} \tag{6.47d}$$

For reconstruction, the hologram is illuminated with a reference wave at wavelength λ_1. Therefore, considering the terms of interest, the amplitude distribution in the virtual image is proportional to

$$\mathbf{R}_1\left(\mathbf{O}_1\mathbf{R}_1^* + \mathbf{O}_2\mathbf{R}_2^*\right) = \mathbf{O}_1 R_0^2 + \mathbf{O}_2 R_0^2 e^{ix\sin\theta(k_1 - k_2)} \tag{6.48}$$

The term $e^{ix\sin\theta(k_1-k_2)}$ represents a different direction of propagation of the object wave $\mathbf{O}_2$. This phase term will be zero provided that the angle of the reference beam R_2, before the second exposure, is adjusted such that $k_1 \sin\theta = k_2 \sin\phi$, where ϕ is the angle that the reference beam R_2 makes with the optical axis. Then, the amplitude distribution can be rewritten proportional to

$$R_0^2 O_o \left\{1 + e^{i(k_2-k_1)z_o} \exp\left[\frac{i(k_2 - k_1)}{2z_o}(x - x_o)^2\right]\right\}$$

The phase term

$$\frac{k_2 - k_1}{2z_o}(x - x_o)^2$$

is very small, since $x \ll z_o$ and $x_o \ll z_o$, and hence the amplitude distribution in the reconstructed virtual image is

$$R_0^2 O_0 (1 + e^{i(k_2-k_1)z_o})$$

The intensity distribution in the image can therefore be written as

$$I = I_o\{1 + \cos[(k_2 - k_1)z_o]\} = I_o\left\{\left[1 + \cos\frac{2\pi|\lambda_1 - \lambda_2|}{\lambda_1\lambda_2} z_o\right]\right\} \tag{6.49}$$

Bright fringes are formed wherever

$$\frac{2\pi\,|\lambda_1 - \lambda_2|}{\lambda_1\lambda_2} z_o(x, y) = 2m\pi$$

or

$$z_o(x, y) = \frac{m\lambda_1\lambda_2}{|\lambda_1 - \lambda_2|} \tag{6.50}$$

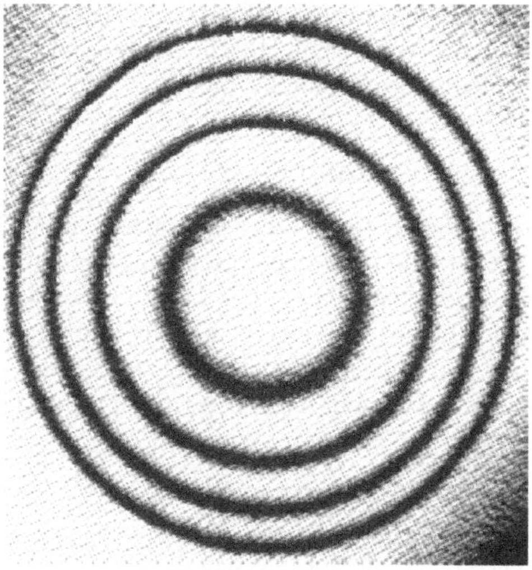

FIGURE 6.15 Fringes over a spherical surface obtained with dual-wavelength HI.

The contour interval is $\Delta z_o = \lambda_1\lambda_2/|\lambda_1 - \lambda_2|$. Essentially, the method produces interference planes perpendicular to the z axis and with a separation of $\lambda_1\lambda_2/|\lambda_1 - \lambda_2|$. The sensitivity of the contour fringes can be varied from 1 μm to several millimeters using two suitable wavelengths. Figure 6.15 is an interferogram of an object taken with two wavelengths. The object is a spherical surface, and hence the intersection by parallel planes yields circular contour lines.

6.18.2 Dual-Refractive-Index Method

Similar results are obtained when double-exposure HI is performed with a change in the refractive index of the medium surrounding the object. The object is placed in an enclosure with a transparent window, and the first exposure is made. The refractive index of the medium surrounding the object is changed, and then the second exposure is made on the same plate. The refractive index can be easily changed when, say, the water medium around the object is replaced by a mixture of water and alcohol. The hologram, when viewed, displays the object covered with interference planes.

If the vacuum wavelength of the recording light is λ_0, then the wavelength in a medium of refractive index n_1 is $\lambda_1 = \lambda_0/n_1$, and in the other medium of refractive index n_2 it is $\lambda_2 = \lambda_0/n_2$. Essentially, this is two-wavelength HI, and the contour interval is given by

$$\Delta z_o(x,y) = \frac{\lambda_1\lambda_2}{|\lambda_1 - \lambda_2|} = \frac{\lambda_0^2/n_1 n_2}{\left|\dfrac{\lambda_0}{n_1} - \dfrac{\lambda_0}{n_2}\right|} = \frac{\lambda_0}{|n_2 - n_1|} \tag{6.51}$$

There is no need to correct for the angle of the reference wave prior to the second exposure as was done in two-wavelength HI. The contour interval can be varied by a suitable choice of the media surrounding the object.

6.18.3 Dual-Illumination Method

The dual-illumination method is frequently used for contouring, because of its simplicity. A collimated beam illuminates the object at an angle θ with the optical axis, and a real-time hologram is recorded in the usual way. On reconstruction, a dark field is observed. Now, if the illumination beam is tilted by a small angle $\Delta\theta$ a system of equispaced interference planes intersects the object. These planes run parallel to the bisector of the angle $\Delta\theta$. The contour interval $\Delta z_0(x, y)$ is given by

$$\Delta z_0(x, y) = \frac{\lambda}{\sin\theta \sin\Delta\theta} \approx \frac{\lambda}{\sin\theta}\frac{1}{\Delta\theta} \tag{6.52}$$

Usually, the interference planes do not intersect the object perpendicular to the line of sight. Large objects are illuminated by a spherical wave from a point source that is translated laterally to produce the interference surfaces. Double-exposure HI can also be performed for contouring by shifting the point source between exposures.

6.19 HOLOGRAPHIC PHOTOELASTICITY

This section deals with the application of holographic interferometry to the study of photo-elastic models. The technique is described in detail in Chapter 8.

6.20 DIGITAL HOLOGRAPHY

6.20.1 Recording

Instead of using photographic emulsions or other recording media, a CCD is used as recording medium and the information is stored electronically. The reconstruction is performed numerically on the electronically stored data. The advantages of recording on CCD are that the hologram is recorded at video frequency and no chemical or physical process for development is necessary. CCD cameras, however, have resolutions of order 100 lines/mm, which is at least one order of magnitude smaller than that of the photographic emulsions commonly used in holography. For this reason, the maximum admissible angle between the object and the reference waves is about 1°.

Let us assume that the CCD contains $N \times N$ pixels of sizes Δx and Δy along the x and y directions. Assuming that the adjacent pixels have no space between them and no overlap, Δx and Δy are also pixel center distances. In digital holography, a Fresnel hologram is generated on the CCD target by the superposition of an object wave and a reference wave. The hologram is digitized, quantized, and stored in the memory of the image-processing system. Figure 6.16 shows schematics of the arrangements for recording a digital hologram. A plane reference wave is assumed for simplicity. The reconstruction is performed numerically. However, according to the sampling theorem, only spatial frequencies less than μ_{max} along the x direction and ν_{max} along

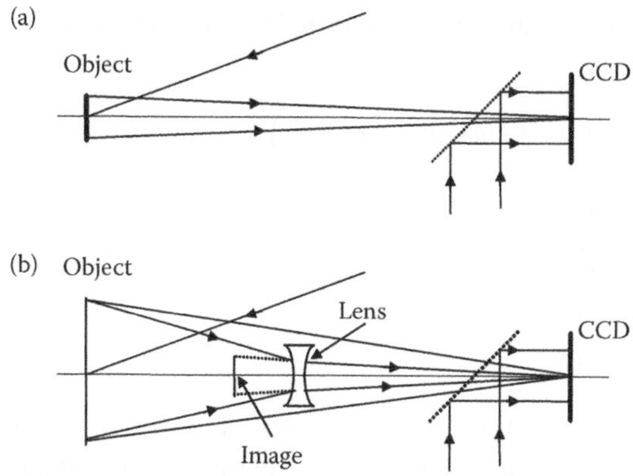

FIGURE 6.16 Schematic of arrangements for recording a digital hologram of (a) a small object and (b) a large object.

the y direction can be reliably reconstructed, where $\mu_{max} = (2\Delta x)^{-1}$ and $\nu_{max} = (2\Delta y)^{-1}$. This sets the maximum allowable angle between the object wave and the reference wave. The angle θ in the (x, z) plane can be written as $\theta = \sin^{-1}(\lambda\mu_{max})$. As a numerical example, consider a CCD of 1024×1024 pixels, each $6.8\,\mu m \times 6.8\,\mu m$ in size, and illumination of the object by a He–Ne laser beam at $0.633\,\mu m$. The maximum allowable angle is $2.67°$. Therefore, an object placed at a distance of 1 m from the CCD must be smaller than 4.3 cm along the x direction. Larger objects either have to be placed further away from the CCD plane or have their apparent size reduced by a lens.

6.20.2 Reconstruction

The reconstruction of the hologram is done by illuminating it with a plane reference wave. The amplitude in the real image can be expressed as

$$A(\xi, \eta) = \frac{iR}{\lambda z} \exp\left[-\frac{i\pi}{\lambda z}(\xi^2 + \eta^2)\right] \times \int_{-\infty}^{\infty}\int_{-\infty}^{\infty} t(x, y) \exp\left[-\frac{i\pi}{\lambda z}(x^2 + y^2)\right] \exp\left[\frac{2\pi i}{\lambda z}(x\xi + y\eta)\right] dx\, dy \tag{6.53}$$

where R is the amplitude of the reference wave. This equation is valid under the Fresnel approximation, namely,

$$z^3 \gg \frac{\pi}{4\lambda}\left[(x - \xi)^2 + (y - \eta)^2\right]^2$$

where z is the distance between the object and the hologram, and hence between the hologram and the real image.

Since the data is available in digital form, Equation 6.53 can be written as

$$A(m,n) = \frac{iR}{\lambda z} \exp\left[-\frac{i\pi}{\lambda z}\left(m^2\Delta\xi^2 + n^2\Delta\eta^2\right)\right]$$
$$\times \sum_{k=0}^{N-1}\sum_{l=0}^{N-1} t(k,l) \exp\left[-\frac{i\pi}{\lambda z}(k^2\Delta x^2 + l^2\Delta y^2)\right] \exp\left[2\pi i\left(\frac{km}{N} + \frac{ln}{N}\right)\right] \quad (6.54)$$

with $m = 0, 1, 2, \ldots, N-1; n = 0, 1, 2, \ldots, N-1$. Here $t(k, l)$ is an matrix of $N \times N$ (pixels in CCD) data that describes the digitally sampled amplitude transmittance of the hologram; Δx, Δy, and $\Delta\xi$, $\Delta\eta$ are the pixel sizes in the hologram plane and the plane of the real image, respectively. The amplitude distribution in the real image is obtained as the inverse Fourier transform of the product consisting of the hologram transmittance $t(k, l)$ multiplied by an exponential factor that contains a quadratic phase factor. The Fourier transform is performed by a fast Fourier transform (FFT) algorithm. The amplitude $A(\xi, \eta)$ in the reconstructed image is a complex function, and hence the intensity and the phase are computed at each pixel. The intensity $I(m, n)$

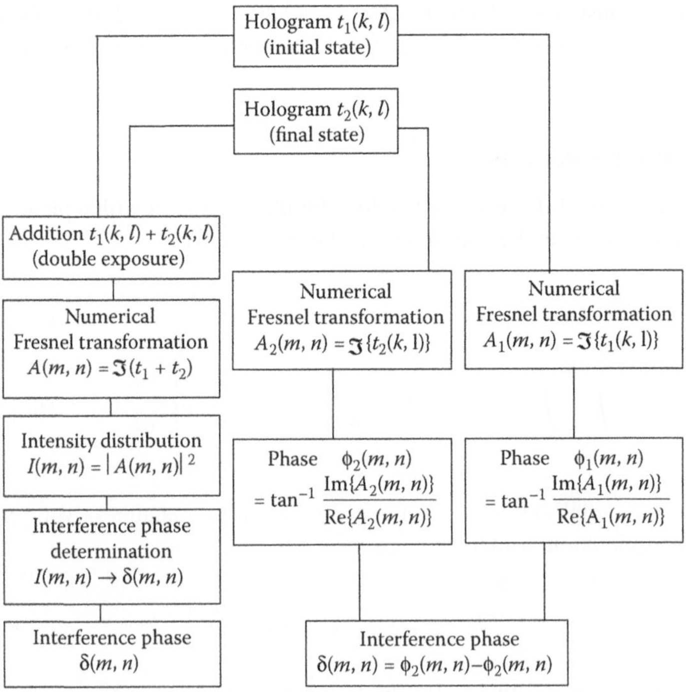

FIGURE 6.17 Procedures for obtaining digital holographic interferograms.

and the phase $\phi(m,n)$ are determined as follows:

$$I(m,n) = |A(m,n)|^2 = \mathrm{Re}\{A(m,n)\}^2 + \mathrm{Im}\{A(m,n)\}^2 \tag{6.55a}$$

$$\phi(m,n) = \tan^{-1}\frac{\mathrm{Im}\{A(m,n)\}}{\mathrm{Re}\{A(m,n)\}}, \quad \text{with value } -\pi \text{ to } \pi \tag{6.55b}$$

As a consequence of surface roughness, the phase $\phi(m,n)$ varies randomly. In digital holography, only the intensity variation on the object is of interest, and hence several calculations, such as that of $iR/\lambda z$, the phase $\phi(m,\ n)$, and the exponential term $e^{-(i\pi/\lambda z)(m^2\Delta\xi^2+n^2\Delta\eta^2)}$ are not carried out.

6.21 DIGITAL HOLOGRAPHIC INTERFEROMETRY

Comparison of states of the object due to loading can be performed by digital holographic interferometry in a manner similar to that of double-exposure HI. Two holograms with amplitude transmittances $t_1(x,y)$ and $t_2(x,y)$ belonging to the two states of the object are digitally recorded and stored. During reconstruction, there are two procedures that can be used to obtain the interferogram representing the change in the state of the object. In the first procedure, the transmittances of both holograms are added pixel by pixel and the Fresnel transform is taken. This reconstructs the

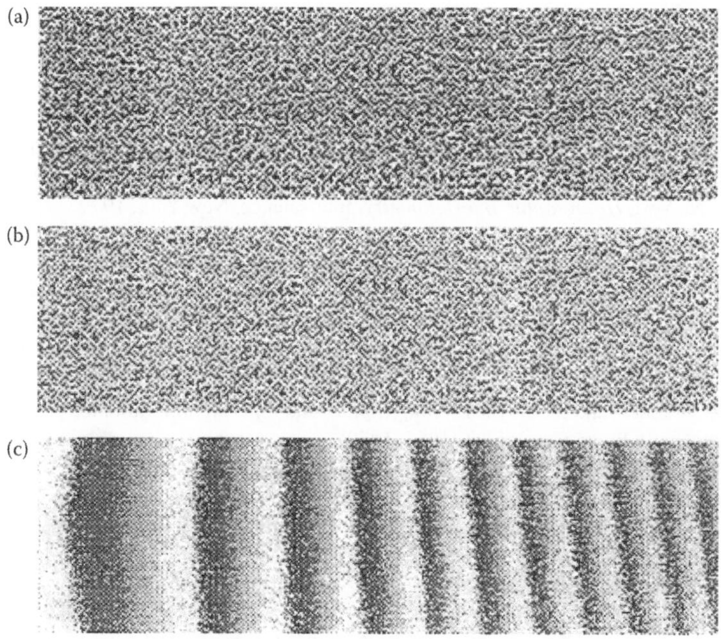

FIGURE 6.18 Numerically reconstructed phase of (a) an undeformed object and (b) a deformed object, and (c) phase difference between (a) and (b) producing an interferogram. (From U. Schnars, T. M. Kreis, and W. P. O. Jüptner, *Opt. Eng.*, 35, 977–982, 1996. With permission.)

sum of the two amplitudes. When the intensity distribution is calculated, it exhibits a cosine variation characteristic of the interference between the two waves. This interference pattern can be evaluated by any of the well-known procedures. In the second method, the holograms are reconstructed individually, and their phases $\phi_1(m,n)$ and $\phi_2(m,n)$ are calculated pixel-wise. Although $\phi_1(m,n)$ and $\phi_2(m,n)$ are random functions, their difference $\delta(m,n) = \phi_2(m,n) - \phi_1(m,n)]$ is deterministic, and gives the phase change due to loading. The phase change $\delta(m,n)$ is calculated as

$$\delta(m,n) = \begin{cases} \phi_2(m,n) - \phi_1(m,n) & \text{for } \phi_2(m,n) \geq \phi_1(m,n) \\ \phi_2(m,n) - \phi_1(m,n) + 2\pi & \text{for } \phi_2(m,n) < \phi_1(m,n) \end{cases} \tag{6.56}$$

These two approaches are summarized in Figure 6.17. Figures 6.18a and 6.18b show the numerically reconstructed phase of an undeformed and a deformed object, respectively, and Figure 6.18c shows the difference between these two as an interferogram.

Since digital HI can handle only small objects, it has potential applications in the testing and characterization of microsystems. Comparison of remote objects and also their responses to external agencies can be done conveniently with digital holography.

BIBLIOGRAPHY

1. R. K. Erf (Ed.), *Holographic Non-Destructive Testing*, Academic Press, New York, 1974.
2. Ju. I. Ostrovskii, M. M. Butusov, and G. V. Ostrovskaja, *GolografiCeskaja interferometrija* [Holographic Interferometry], Nauka, Moskva, 1977.
3. Yu. I. Ostrovsky, M. M. Butusov, and G. V. Ostrovskaya, *Interferometry by Holography*, Springer Series in Optical Sciences, Vol. 20, Springer-Verlag, Berlin/Heidelberg, 1980.
4. C. M. Vest, *Holographic Interferometry*, Interscience, New York, 1979.
5. W. Schumann and M. Dubas, *Holographic Interferometry*, Springer Series in Optical Sciences, Vol. 16, Springer-Verlag, Berlin/Heidelberg, 1979.
6. N. Abramson, *The Making and Evaluation of Holograms,* Academic Press, London, 1981.
7. G. Wernicke and W. Osten, *Holografische Interferometrie*, Physik-Verlag, Weinheim, 1982.
8. R. J. Jones and C. Wykes, *Holographic and Speckle Interferometry*, Cambridge University Press, Cambridge (1983, 1989).
9. W. Schumann, J. P. Zürcher and D. Cuche, *Holography and Deformation Analysis*, Springer Series in Optical Sciences, Vol. 46, Springer-Verlag, Berlin/Heidelberg, 1985.
10. Yu. I. Ostrovsky, V. P. Shchepinov, and V. V. Yakovlev, *Holographic Interferometry in Experimental Mechanics*, Springer Series in Optical Sciences, Vol. 60, Springer Verlag, Berlin/Heidelberg, 1991.
11. P. K. Rastogi (Ed.), *Holographic Interferometry, Principles and Methods*, Springer Series in Optical Sciences, Vol. 68, Springer-Verlag, Berlin/Heidelberg, 1994.
12. T. Kreis, *Holographic Interferometry, Principles and Methods*, Akademie Verlag Series in Optical Metrology, Vol. 1, Akademie Verlag, Berlin, 1996.
13. V. P. Shchepinov (Ed.), *Strain and Stress Analysis by Holographic and Speckle Interferometry*, Wiley, Chichester, 1996.

14. P. K. Rastogi (Ed.), *Optical Measurement Techniques and Applications*, Artech House, Boston, 1997.
15. R. S. Sirohi and K. D. Hinsch, *Selected Papers on Holographic Interferometry*, SPIE Milestone Series MS144, SPIE, Bellingham, WA (1998).
16. Thomas Kreis, *Handbook of Holographic Interferometry: Optical and Digital Methods*, Wiley-VCH, Weinheim, 2005.
17. W. Osten (Ed.), *Optical Inspection of Microsystems*, Taylor & Francis, New York, 2006.

ADDITIONAL READING

1. R. L. Powell and K. A. Stetson, Interferometric vibration analysis by wavefront reconstruction, *J. Opt. Soc. Am.*, 55, 1593–1598, 1965.
2. K. A. Stetson and R. L. Powell, Interferometric hologram evaluation and real-time vibration analysis of diffuse objects, *J. Opt. Soc. Am.*, 55, 1694–1695, 1965.
3. R. J. Collier, E. T. Doherty, and K. S. Pennington, Application of moiré techniques to holography, *Appl. Phys. Lett.*, 7, 223–225, 1965.
4. R. E. Brooks, L. O. Heflinger, and R. F. Wuerker, Interferometry with a holographically reconstructed comparison beam, *Appl. Phys. Lett.*, 7, 248–249, 1965.
5. M. H. Horman, An application of wavefront reconstruction to interferometry, *Appl. Opt.*, 4, 333–336, 1965.
6. L. H. Tanner, Some applications of holography in fluid mechanics, *J. Sci. Instrum.*, 43, 81–83, 1966.
7. B. P. Hildebrand and K. A. Haines, Interferometric measurements using wavefront reconstruction technique, *Appl. Opt.*, 5, 172–173, 1996.
8. K. A. Haines and B. P. Hildebrand, Surface-deformation measurement using the wavefront reconstruction technique, *Appl. Opt.*, 5, 595–602, 1966.
9. L. O. Heflinger, R. F. Wuerker, and R. E. Brooks, Holographic Interferometry, *J. Appl. Phys.*, 37, 642–649, 1966.
10. J. M. Burch, A. E. Ennos, and R. J. Wilson, Dual- and multiple-beam interferometry by wave front reconstruction, *Nature*, 209, 1015–1016, 1966.
11. B. P. Hildebrand and K. A. Haines, Multiple-wavelength and multiple-source holography applied for contour generation, *J. Opt. Soc. Am.*, 57, 155–162, 1967.
12. E. Archbold, J. M. Burch, and A. E. Ennos, The application of holography to the comparison of cylinder bores, *J. Sci. Instrum.*, 44, 489–494, 1967.
13. L. H. Tanner, The scope and limitations of three-dimensional holography of phase objects, *J. Sci. Instrum.*, 44, 1011–1014, 1967.
14. M. De and L. Sevigny, Three beam holographic interferometry, *Appl. Opt.*, 6, 1665–1671, 1967.
15. E. B. Aleksandrov and A. M. Bonch-Bruevich, Investigation of surface strains by the hologram technique, *Sov. Phys. Tech. Phys.*, 12, 258–265, 1967.
16. N. Shiotake, T. Tsuruta, Y. Itoh, J. Tsujiuchi, N. Takeya, and K. Matsuda, Holographic generation of contour map of diffusely reflecting surface by using immersion method, *Jpn J. Appl. Phys.*, 7, 661–662, 1968.
17. T. Tsuruta, N. Shiotake, and Y. Itoh, Hologram interferometry using two reference beams, *Jpn J. Appl. Phys.*, 7, 1092–1100, 1968.
18. J. S. Zelenka and J. R. Varner, A new method for generating depth contours holographically, *Appl. Opt.*, 7, 2107–2110, 1968.
19. E. Archbold and A. E. Ennos, Observation of surface vibration modes by stroboscopic hologram interferometry, *Nature*, 217, 942–943, 1968.

20. P. Shajenko and C. D. Johnson, Stroboscopic holographic interferometry, *Appl. Phys.*, 13, 44–46, 1968.
21. B. M. Watrasiewicz and P. Spicer, Vibration analysis by stroboscopic holography, *Nature*, 217, 1142–1143, 1968.
22. J. W. C. Gates, Holographic phase recording by interference between reconstructed wavefronts from separate holograms, *Nature*, 220, 473–474, 1968.
23. G. S. Ballard, Double exposure holographic interferometry with separate reference beams, *J. Appl. Phys.*, 39, 4846–4848, 1968.
24. W. G. Gottenberg, Some applications of holographic interferometry, *Exp. Mech.*, 8, 405–410, 1968.
25. M. A. Monahan and K. Bromley, Vibration analysis by holographic interferometry, *J. Acouts. Soc. Am.*, 44, 1225–1231, 1968.
26. A. E. Ennos, Measurement of in-plane surface strain by hologram interferometry, *J. Sci. Instrum.: Phys. E*, 1, 731–734, 1968.
27. O. Bryngdahl, Shearing interferometry by wavefront reconstruction, *J. Opt. Soc. Am.*, 58, 865–871, 1968.
28. C. C. Aleksoff, Time average holography extended, *Appl. Phys. Letts.*, 14, 23–24, 1969.
29. F. M. Mottier, Time average holography with triangular phase modulation of the reference wave, *Appl. Phys. Lett.*, 15, 285–287, 1969.
30. R. M. Grant and G. M. Brown, Holographic non-destructive testing (HNDT), *Mater. Eval.*, 27, 79–84, 1969.
31. G. M. Brown, R. M. Grant and G. W. Stroke, Theory of holographic interferometry, *J. Acoust. Soc. Am.*, 45, 1166–1179, 1969.
32. K. Matsumoto, Holographic multiple beam interferometry, *J. Opt. Soc. Am.*, 59, 777–778, 1969.
33. N.-E. Molin and K. A. Stetson, Measuring combination mode vibration patterns by hologram interferometry, *J. Sci. Instrum.: Phys. E*, 2, 609–612, 1969.
34. I. Yamaguchi and H. Saito, Application of holographic interferometry to the measurement of Poisson's ratio, *Jpn J. Appl. Phys.*, 8, 768–771, 1969.
35. W. Van Deelan and P. Nisenson, Mirror blank testing by real-time holographic interferometry, *Appl. Opt.*, 8, 951–955, 1969.
36. N. Abramson, The holodiagram: A practical device for making and evaluating holograms, *Appl. Opt.*, 8, 1235–1240, 1969.
37. A. A. Friesem and C. M. Vest, Detection of micro fractures by holographic interferometry, *Appl. Opt.*, 8, 1253–1254, 1969.
38. J. S. Zelenka and J. R. Varner, Multiple-index holographic contouring, *Appl. Opt.*, 8, 1431–1434, 1969.
39. J. E. Sollid, Holographic interferometry applied to measurements of small static displacements of diffusely reflecting surfaces, *Appl. Opt.*, 8, 1587–1595, 1969.
40. W. T. Welford, Fringe visibility and localization in hologram interferometry, *Opt. Commun.*, 1, 123–125, 1969.
41. T. Tsuruta, N. Shiotake and Y. Itoh, Formation and localization of holographically produced interference fringes, *Opt. Acta*, 16, 723–733, 1969.
42. K. A. Stetson, A rigorous treatment of the fringes of hologram interferometry, *Optik*, 29, 386–400, 1969.
43. S. Mallick and M. L. Roblin, Shearing interferometry by wavefront reconstruction using single exposure, *Appl. Phys. Lett.*, 14, 61–62, 1969.
44. S. Walles, On the concept of homologous rays in holographic interferometry of diffusely reflecting surfaces, *Opt. Acta*, 17, 899–913, 1970.

45. R. A. Jeffries, Two wavelength holographic interferometry of partially ionised plasmas, *Phys. Fluids*, 13, 210–212, 1970.
46. A. D. Wilson, Holographically observed torsion in a cylindrical shaft, *Appl. Opt.*, 9, 2093–2097, 1970.
47. R. C. Sampson, Holographic interferometry applications in experimental mechanics, *Exp. Mech.*, 10, 313–320, 1970.
48. A. D. Wilson, Characteristic functions for time-average holography, *J. Opt. Soc. Am.*, 60, 1068–1071, 1970.
49. K. A. Stetson, Effect of beam modulation on fringe loci and localization in time-average hologram interferometry, *J. Opt. Soc. Am.*, 60, 1378–1384, 1970.
50. D. B. Neumann, C. F. Jacobson, and G. M. Brown, Holographic technique for determining the phase of vibrating objects, *Appl. Opt.*, 9, 1357–1362, 1970.
51. R. Pryputniewicz and K. A. Stetson, Determination of sensitivity vectors in hologram interferometry from two known rotations of the object, *Appl. Opt.*, 19, 2201–2205, 1970.
52. K. S. Mustafin, V. A. Seleznev, and E. I. Shtyrkov, Use of the non-linear properties of a photoemulsion for enhancing the sensitivity of holographic interferometry, *Opt. Spectrosc.*, 28, 638–640, 1970.
53. K. Matsumoto and M. Takashima, Phase difference amplification by non-linear holography, *J. Opt. Soc. Am.*, 60, 30–33, 1970.
54. C. H. F. Velzel, Small phase differences in holographic interferometry, *Opt. Commun.*, 2, 289–291, 1970.
55. A. D. Wilson, Computed time-average holographic interferometric fringes of a circular plate vibrating simultaneously in two rationally or irrationally related modes, *J. Opt. Soc. Am.*, 61, 924–929, 1971.
56. F. Weigl, A generalized technique of two wavelength non diffuse holographic interferometry, *Appl. Opt.*, 10, 187–192, 1971.
57. G. V. Ostrovskaya and Yu. I. Ostrovsky, Two-wavelength hologram method for studying the dispersion properties of phase objects, *Sov. Phys. Tech. Phys.*, 15, 1890–1892, 1971.
58. R. D. Matulka and D. J. Collins, Determination of three-dimensional density fields from holographic interferograms, *J. Appl. Phys.*, 42, 1109–1119, 1971.
59. L. A. Kersch, Advanced concepts of holographic nondestructive testing, *Mater. Eval.*, 29, 125–129, 1971.
60. R. A. Ashton, D. Slovin, and H. I. Gerritsen, Interferometric holography applied to elastic stress and surface corrosion, *Appl. Opt.*, 10, 440-441, 1971.
61. A. D. Wilson, In-plane displacement of a stressed membrane with a hole measured by holographic interferometry, *Appl. Opt.*, 10, 908–912, 1971.
62. C. C. Aleksoff, Temporally modulated holography, *Appl. Opt.*, 10, 1329–1341, 1971.
63. J. P. Waters, Holographic inspection of solid propellant to liner bonds, *Appl. Opt.*, 10, 2364–2365, 1971.
64. T. D. Dudderar and R. O'Regan, Measurement of the strain field near a crack tip in polymethylmethacrylate by holographic interferometry, *Exp. Mech.*, 11, 49–56, 1971.
65. A. D. Wilson, C. H. Lee, H. R. Lominac, and D. H. Strope, Holographic and analytic study of semi-clamped rectangular plate supported by struts, *Exp. Mech.*, 11, 229–234, 1971.
66. W. Aung and R. O'Regan, Precise measurement of heat transfer using holographic interferometry, *Rev. Sci. Instrum.*, 42, 1755–1759, 1971.
67. P. M. Boone, Determination of three orthogonal displacement components from one double-exposure hologram, *Opt. Laser Technol.*, 4, 162–167, 1972.

68. C. H. Agren and K. A. Stetson, Measuring the resonances of treble-viol plates by hologram interferometry, *J. Acoust. Soc. Am.*, 51, 1971–1983, 1972.
69. P. Larinov, A. V. Lukin, and K. S. Mustafin, Holographic inspection of shapes of unpolished surfaces, *Sov. J. Opt. Technol.*, 39, 154–155, 1972.
70. B. U. Achia and D. W. Thompson, Real-time hologram-moiré interferometry for liquid flow visualization, *Appl. Opt.*, 11, 953–954, 1972.
71. W. P. Chu, D. M. Robinson and J. H. Goad, Holographic non-destructive testing with impact excitation, *Appl. Opt.*, 11, 1644–1645, 1972.
72. R. E. Rowlands and I. M. Daniel, Application of holography to anisotropic composite plates, *Exp. Mech.*, 12, 75–82, 1972.
73. S. K. Dhir and J. P. Sikora, An improved method for obtaining the general-displacement field from a holographic interferogram, *Exp. Mech.*, 12, 323–327, 1972.
74. T. R. Hsu and R. G. Moyer, Application of holography in high-temperature displacement measurements, *Exp. Mech.*, 12, 431–432, 1972.
75. K. S. Mustafin and V. A. Seleznev, Holographic interferometry with variable sensitivity, *Opt. Spectrosc.*, 32, 532–535, 1972.
76. T. Matsumoto, K. Iwata and R. Nagata, Measuring accuracy of three dimensional displacements in holographic interferometry, *Appl. Opt.*, 12, 961–967, 1973.
77. P. Hariharan and Z. S. Hegedus, Simple multiplexing technique for double-exposure hologram interferometry, *Opt. Commun.*, 9, 152–155, 1973.
78. C. A. Sciammarella and J. A. Gilbert, Strain analysis of a disk subjected to diametral compression by means of holographic interferometry, *Appl. Opt.*, 12, 1951–1956, 1973.
79. D. W. Sweeny and C. M. Vest, Reconstruction of three-dimensional refractive index field from multidirectional interferometric data, *Appl. Opt.*, 12, 2649–2664 (1973).
80. N. Abramson and H. Bjelkhagen, Industrial holographic measurements, *Appl. Opt.*, 12, 2792–2796, 1973.
81. K. Grunwald, D. Kaletsch, V. Lehmann, and H. Wachuta, Holographische Interferometrie und deren quantitative Auswertung, demonstriet am Beispiel zylindrischer GfK-Rohre, *Optik*, 37, 102–109, 1973.
82. A. D. Luxmoore, Holographic detection of cracks in concrete, *Nondestr. Test.*, 6, 258–263, 1973.
83. R. Dandliker, B. Ineichen, and F. M. Mottier, High resolution hologram interferometry by electronic phase measurement, *Opt. Commun.*, 9, 412–416, 1973.
84. Y. Doi, T. Komatsu and T. Fujimoto, Shearing interferometry by holography, *Jpn J. Appl. Phys.*, 12, 1036–1042, 1973.
85. N. Abramson, Sandwich hologram interferometry: A new dimension in holographic comparison, *Appl. Opt.*, 13, 2019–2025, 1974.
86. M. Dubas and W. Schumann, Sur la détermination holographique de l'état de déformation á la surface d'un corps non-transparent, *Opt. Acta*, 21, 547–562, 1974.
87. I. Prikryl, Localization of interference fringes in holographic interferometry, *Opt. Acta*, 21, 675–681, 1974.
88. K. A. Stetson, Fringe interpretation for hologram interferometry of rigid-body motions and homogeneous deformations, *J. Opt. Soc. Am.*, 64, 1–10, 1974.
89. R. Jones and D. A. Bijl, Holographic interferometric study of the end effects associated with the four point bending technique for measuring Poisson's ratio, *J. Phys. E: J. Sci. Instrum.*, 7, 357–358, 1974.
90. J. P. Sikora and F. T. Mendenhall, Holographic vibration study of a rotating propeller blade, *Exp. Mech.*, 14, 230–232, 1974.

91. T. R. Hsu, Large-deformation measurements by real-time holographic interferometry, *Exp. Mech.*, 14, 408–411, 1974.
92. E. R. Robertson and W. King, The technique of holographic interferometry applied to the study of transient stresses, *J. Strain Anal.*, 9, 44–49, 1974.
93. A. R. Luxmoore, Holographic detection of cracks, *J. Strain Anal.*, 9, 50–51, 1974.
94. T. Matsumoto, K. Iwata, and R. Nagata, Measurement of deformation in a cylindrical shell by holographic interferometry, *Appl. Opt.*, 13, 1080–1084, 1974.
95. V. Fossati Bellani and A. Sona, Measurement of three-dimensional displacements by scanning a double-exposure hologram, *Appl. Opt.*, 13, 1337–1341, 1974.
96. S. Amadesi, F. Gori, R. Grella, and G. Guattari, Holographic methods for painting diagnostics, *Appl. Opt.*, 13, 2009–2013, 1974.
97. P. R. Wendendal and H. I. Bjelkhagen, Dynamics of human teeth in function by means of double-pulsed holography: An experimental investigation, *Appl. Opt.*, 13, 2481–2485, 1974.
98. M. Dubas and W. Schumann, The determination of strain at a nontransparent body by holographic interferometry, *Opt. Acta*, 21, 547–562, 1974.
99. H. J. Raterink and R. L. van Renesse, Investigation of holographic interferometry, applied for the detection of cracks in large metal objects, *Optik*, 40, 193–200, 1974.
100. H. Spetzler, C. H. Scholz, and C. P. J. Lu, Strain and creep measurements on rocks by holographic interferometry, *Pure Appl. Geophys.*, 112/113, 571–582, 1974.
101. C. A. Sciammarella and T. Y. Chang, Holographic interferometry applied to the solution of shell problem, *Exp. Mech.*, 14, 217–224, 1974.
102. E. Roth, Holographic interferometric investigations of plastic parts, In *Laser 75, Proceedings of International Conference on Optoelectronics*, 168–174, 1975.
103. M. C. Collins and C. E. Watterson, Surface-strain measurements on a hemispherical shell using holographic interferometry, *Exp. Mech.*, 15, 128–132, 1975.
104. K. Grunwald, W. Fritzsch, and H. Wachuta, Quantitative holographische Verformungsmessungen an Kunststoff- und GFK Bauteilen, *Materialprufung*, 17, 69–72, 1975.
105. Y. Chen, Holographic interferometry applied to rotating disk, *Trans. ASME*, 42, 499–512, 1975.
106. F. Able, A. L. Dancer, H. Fagot, R. B. Franke, and P. Smigielski, Holographic interferometry applied to investigations of tympanic-membrane displacements in guinea-pig ears subjected to acoustical pulses, *J. Acoust. Soc. Am.*, 58, 223–228, 1975.
107. J. D. Trolinger, Flow visualization holography, *Opt. Eng.*, 14, 470–481, 1975.
108. R. J. Radley Jr, Two-wavelength holography for measuring plasma electron density, *Phys. Fluids*, 18, 175–179, 1975.
109. C. A. Sciammarella and J. A. Gilbert, Holographic interferometry applied to the measurement of displacements of interior points of transparent bodies, *Appl. Opt.*, 15, 2176–2182, 1976.
110. C. G. H. Foster, Accurate measurement of Poisson's ratio in small samples, *Exp. Mech.*, 16, 311–315, 1976.
111. H. D. Meyer and H. A. Spetzler, Material properties using holographic interferometry, *Exp. Mech.*, 16, 434–438, 1976.
112. D. Dameron and C. M. Vest, Fringe sharpening and diffraction in nonlinear two-exposure holographic interferometry, *J. Opt. Soc. Am.*, 66, 1418–1421, 1976.
113. N. Takai, M. Yamada and T. Idogawa, Holographic interferometry using a reference wave with a sinusoidally modulated amplitude, *Opt. Laser Technol.*, 8, 21–23, 1976.
114. J. R. Crawford and R. Benson, Holographic interferometry of circuit board components, *Appl. Opt.*, 15, 24–25, 1976.

115. P. Greguss, Holographic interferometry in biomedical sciences, *Opt. Laser Technol.*, 8, 153–159, 1976.
116. N. Abramson, Holographic contouring by translation, *Appl. Opt.*, 15, 1018–1022, 1976.
117. P. M. de Larminant and R. P. Wie, A fringe compensation technique for stress analysis by reflection holographic interferometry, *Exp. Mech.*, 16, 241–248, 1976.
118. R. Dandliker, E. Marom and F. M. Mottier, Two-reference beam holographic interferometry, *J. Opt. Soc. Am.*, 66, 23–30, 1976.
119. K. A. Stetson, Use of fringe vectors in hologram interferometry to determine fringe localization, *J. Opt. Soc. Am.*, 66, 627–628, 1976.
120. P. Hariharan and Z. S. Hegedus, Two-hologram interferometry: A simplified sandwich technique, *Appl. Opt.*, 15, 848–849, 1976.
121. G. E. Sommargren, Double-exposure holographic interferometry using common-path reference waves, *Appl. Opt.*, 16, 1736–1741, 1977.
122. N. Abramson, Sandwich hologram interferometry: Holographic studies of two milling machines, *Appl. Opt.*, 16, 2521–2531, 1977.
123. D. C. Holloway, A. M. Patacca and W. L. Fourney, Application of holographic interferometry to a study of wave propagation in rock, *Exp. Mech.*, 17, 281–289, 1977.
124. G. Wernicke and G. Frankowski, Untersuchung des Fliessbeginns einsatzgeharter und gaskarbonitrierter Stahl mit Hilfe der holographischen Interferometrie, *Die Technik*, 32, 393–396, 1977.
125. C. M. Vest and D. W. Sweeney, Applications of holographic interferometry to non-destructive testing, *Int. Adv. Nondestr. Test.*, 5, 17–21, 1977.
126. M. D. Mayer and T. E. Katyanagi, Holographic examination of a composite pressure vessel, *J. Test. Eval.*, 5, 47–52, 1977.
127. S. Toyooka, Holographic interferometry with increased sensibility for diffusely reflecting objects, *Appl. Opt.*, 16, 1054–1057, 1977.
128. H. Kohler, General formulation of the holographic interferometric evaluation methods, *Optik*, 47, 469–475, 1977.
129. N. Abramson and H. Bjelkhagan, Pulsed sandwich holography, 2. Practical application, *Appl. Opt.*, 17, 187–191, 1978.
130. K. A. Stetson and I. R. Harrison, Computer-aided holographic vibration analysis for vectorial displacements of bladed disks, *Appl. Opt.*, 17, 1733–1738, 1978.
131. D. Nobis and C. M. Vest, Statistical analysis of errors in holographic interferometry, *Appl. Opt.*, 17, 2198–2204, 1978.
132. P. M. De Laminat and R. P. Wei, Normal surface displacement around a circular hole by reflection holographic interferometry, *Exp. Mech.*, 18, 74–80, 1978.
133. K. N. Petrov and Yu P. Presnyakov, Holographic interferometry of the corrosion process, *Opt. Spectrosc.*, 44, 179–181, 1978.
134. J. M. Caussignac, Application of holographic interferometry to the study of structural deformations in civil engineering, *Proc. SPIE*, 136, 136–142, 1978.
135. A. Felske, Holographic analysis of oscillations in squealing disk brakes, *Proc. SPIE*, 136, 148–155, 1978.
136. J. D. Dubourg, Application of holographic interferometry to testing of spun structures, *Proc. SPIE*, 136, 186–191, 1978.
137. P. Meyrueis, M. Pharok, and J. Fontaine, Holographic interferometry in osteosynthesis, *Proc. SPIE*, 136, 202–205, 1978.
138. S. Toyooka, Computer mapping of the first and the second derivatives of plate deflection by using modulated diffraction gratings made by double-exposure holography, *Opt. Acta*, 25, 991–1000, 1978.

139. Hans Rottenkolber and Hans Steinbichler, Method for the optical determination and comparison of shapes and positions of objects, and arrangement for practicing said method, US Patent, 4,111,557, 1978.
140. K. A. Stetson, The use of an image-derotator in hologram interferometry and speckle photography of rotating objects, *Exp. Mech.*, 18, 67–73, 1978.
141. K. D. Hinsch, Holographic interferometry of surface deformations of transparent fluids, *Appl. Opt.*, 17, 3101–3108, 1978.
142. F. T. S. Yu, A. Tai, and H. Chen, Multi-wavelength rainbow holographic interferometry, *Appl. Opt.*, 18, 212–218, 1979.
143. C. Shakher and R. S. Sirohi, Fringe control techniques applied to holographic non-destructive testing, *Can. J. Phys.*, 57, 2155–2160, 1979.
144. J. C. McBain, I. E. Horner, and W. A. Stange, Vibration analysis of a spinning disk using image-derotated holographic interferometry, *Exp. Mech.*, 19, 17–22, 1979.
145. G. von Bally, Otological investigations in living man using holographic interferometry, In *Holography in Medicine and Biology* (ed. G. von Bally), Vol. 18, 198–205, Springer-Verlag, Berlin/Heidelberg, 1979.
146. S. Vikram and K. Vedam, Holographic interferometry of corrosion, *Optik*, 55, 407, 1980.
147. S. Vikram, K. Vedam, and E. G. Buss, Non-destructive evaluation of the strength of eggs by holography, *Poultry Sci.*, 59, 2342, 1980.
148. J. C. McBain, W. A. Stange, and K. G. Harding, Real-time response of a rotating disk using image-derotated holographic interferometry, *Exp. Mech.*, 21, 34–40, 1981.
149. W. F. Fagan, M.-A. Beeck, and H. Kreitlow, The holographic vibration analysis of rotating objects using a reflective image derotator, *Opt. Lasers Eng.*, 2, 21–33, 1981.
150. C. A. Sciammarella, Holographic moiré, an optical tool for the determination of displacements, strains, contours, and slopes of surfaces, *Opt. Eng.*, 21, 447–457, 1982.
151. M. Yonemura, Holographic contour generation by spatial frequency modulation, *Appl. Opt.*, 21, 3652–3658, 1982.
152. H. J. Tiziani, Real-time metrology with BSO crystals, *Opt. Acta*, 29, 463–470, 1982.
153. M. J. Marchant and M. B. Snell, Determination of flexural stiffness of thin plates from small deflection measurements using optical holography, *J. Strain Anal.*, 17, 53–61, 1982.
154. A. E. Ennos and M. S. Virdee, Application of reflection holography to deformation measurement problems, *Exp. Mech.*, 22, 202–209, 1982.
155. P. Hariharan, B. F. Oreb and N. Brown, A digital phase-measurement system for real-time holographic interferometry, *Opt. Commun.*, 41, 393–396, 1982.
156. T. Yatagai, S. Nakadate, M. Idesawa and H. Saito, Automatic fringe analysis using digital image processing techniques, *Opt. Eng.*, 21, 432–435, 1982.
157. K. A. Stetson, Method of vibration measurements in heterodyne interferometry, *Opt. Lett.*, 7, 233–234, 1982.
158. H. G. Leis, Vibration analysis of an 8-cylinder V-engine by time-average holographic Interferometry, *Proc. SPIE*, 398, 90–94, 1983.
159. R. Dandliker and R. Thalmann, Deformation of 3D displacement and strain by holographic interferometry for non-plane objects, *Proc. SPIE*, 398, 11–16, 1983.
160. J. Geldmacher, H. Kreitlow, P. Steinlein, and G. Sepold, Comparison of vibration mode measurements on rotating objects by different holographic methods, *Proc. SPIE*, 398, 101–110, 1983.
161. G. Schonebeck, New holographic means to exactly determine coefficients of elasticity, *Proc. SPIE*, 398, 130–136, 1983.

162. S. Amadesi, A. D'Altorio, and D. Paoletti, Single-two hologram interferometry: A combined method for dynamic tests on painted wooden statues, *J. Optics*, 14, 243–246, 1983.
163. A. A. Antonov, Inspecting the level of residual stresses in welded joints by laser interferometry, *Weld Prod.*, 30, 29–31, 1983.
164. J. L. Goldberg, A method of three-dimensional strain measurement on non-ideal objects using holographic interferometry, *Exp. Mech.*, 23, 59–73, 1983.
165. H. P. Rossmanith, Determination of dynamic stress-intensity factors by holographic interferometry, *Opt. Lasers Eng.*, 4, 129–143, 1983.
166. P. Hariharan, B. F. Oreb, and N. Brown, Real-time holographic interferometry: A microcomputer system for the measurement of vector displacement, *Appl. Opt.*, 22, 876–880, 1983.
167. Z. Füzessy and F. Gyimesi, Difference holographic interferometry: Displacement measurement, *Opt. Eng.*, 23, 780–783, 1984.
168. P. Lam, J. D. Gaskill, and J. C. Wyant, Two-wavelength holographic interferometer, *Appl. Opt.*, 23, 3079–3081, 1984.
169. L. Wang, J. D. Hovanesian, and Y. Y. Hung, A new fringe carrier method for the determination of displacement derivatives in hologram interferometry, *Opt. Lasers Eng.*, 5, 109–120, 1984.
170. P. Hariharan, Quasi-heterodyne hologram interferometry, *Opt. Eng.*, 24, 632–638, 1985.
171. R. Thalmann and R. Dandliker, Heterodyne and quasi-heterodyne holographic interferometry, *Opt. Eng.*, 24, 824–831, 1985.
172. R. J. Pryputniewicz, Heterodyne holography, applications in studies of small components, *Opt. Eng.*, 24, 843–848, 849–854, 1985.
173. R. Thalmann and R. Dandliker, Holographic contouring using electronic phase detection, *Opt. Eng.*, 24, 930–935, 1985.
174. D. L. Mader, Holographic interferometry on pipes: Precision interpretation by least square fitting, *Appl. Opt.*, 24, 3784–3790, 1985.
175. D. B. Neumann, Comparative holography: A technique for eliminating background fringes in holographic interferometry, *Opt. Eng.*, 24, 625–627, 1985.
176. A. Stimpfling and P. Smigielski, New method for compensating and measuring any motion of three-dimensional objects in holographic interferometry, *Opt. Eng.*, 24, 821–823, 1985.
177. R. Pryputniewicz, Time average holography in vibration analysis, *Opt. Eng.*, 24, 843–848, 1985.
178. R. Pryputniewicz, Heterodyne holography, applications in study of small components, *Opt. Eng.*, 24, 849–854, 1985.
179. W. Osten, Some consideration on the statistical error analysis in holographic interferometry with application to an optimized interferometry, *Opt. Acta*, 32, 827–838, 1985.
180. Z. Fuzessy, F. Geimesi, and J. Kornis, Comparison of two filament lamps by difference hologram interferometry, *Opt. Laser Technol.*, 18, 318–320, 1986.
181. D. V. Nelson and J. T. McCrickerd, Residual-stress determination through combined use of holographic interferometry and blind-hole drilling, *Exp. Mech.*, 26, 371–378 (1986).
182. S. Takemoto, Application of laser holographic techniques to investigate crustal deformations, *Nature*, 322, 49–51, 1986.
183. T. Kreis, Digital holographic phase measurement using Fourier transform method, *J. Opt. Soc. Am.*, A3, 847–855, 1986.

184. D. Paoletti, G. Schirripa-Spagnolo, and A. D'Altorio, Sandwich hologram for displacement derivatives, *Opt. Commun.*, 56, 325–329, 1986.
185. D. W. Watt and C. M. Vest, Digital interferometry for flow visualization, *Exp. Fluids*, 5, 401–406, 1987.
186. P. Del Vo and M. L. Rizzi, Vibrational testing of an X-ray concentrator by holographic interferometry, *Proc. SPIE*, 814, 357–364, 1987.
187. H. Kreitlow, T. Kreis and W. Jüptner, Holographic interferometry with reference beams modulated by the object motion, *Appl. Opt.*, 26, 4256–4262, 1987.
188. D. E. Cuche, Determination of Poisson's ratio by the holographic moiré technique, *Proc. SPIE*, 1026, 165–170, 1988.
189. P. J. Bryanston-Cross and J. W. Gardner, Application of holographic interferometry to the vibrational analysis of the harpsichord, *Opt. Laser Technol.*, 20, 199–204, 1988.
190. L. Wang and J. Ke, The measurement of residual stresses by sandwich holographic interferometry, *Opt. Lasers in Eng.*, 9, 111–119, 1988.
191. H. Kasprzak, H. Podbielska, and G. von Bally, Human tibia rigidly examined in bending and torsion loading by using double-exposure holographic interferometry, *Proc. SPIE*, 1026, 196v201, 1988.
192. P. K. Rastogi, P. Jacquot, and L. Pflug, Holographic interferometry applied at subfreezing temperatures: Study of damage in concrete exposed to frost action, *Opt. Eng.*, 27, 172–178, 1988.
193. T. E. Carlsson, G. Bjarnholt, N. Abramson, and D. C. Holloway, Holographic interferometry applied to a model study of ground vibrations produced from blasting, *Opt. Eng.*, 27, 923–927, 1988.
194. R. Dandliker, R. Thalmann, and D. Prongue, Two-wavelength laser interferometry using super-heterodyne detection, *Opt. Lett.*, 13, 339–341, 1988.
195. G. Lai and T. Yatagai, Dual-reference holographic interferometry with a double pulsed laser, *Appl. Opt.*, 27, 3855–3858, 1988.
196. S. Fu and J. Chen, Phase difference amplification of double-exposure holograms, *Opt. Commun.*, 67, 417–420, 1988.
197. N. V. Kukhtarev and V. V. Muravev, Dynamic holographic interferometry in photorefractive crystals, *Opt. Spectros.*, 64, 656–659, 1988.
198. M. Kujawinska and D. W. Robinson, Multichannel phase-stepped holographic interferometry, *Appl. Opt.*, 27, 312–320, 1988.
199. I. N. Odintsev, V. P. Shchepinov and V. V. Yakovlev, Material elastic constants measurement by holographic compensation technique, *Zhur. Tech. Fiz.*, 58, 108–113, 1988.
200. J. Gryzagoridis, Holographic non-destructive testing of composites, *Opt. Laser Technol.*, 21, 113–116, 1989.
201. E. Simova and V. Sainov, Comparative holographic moiré interferometry for nondestructive testing: Comparison with conventional holographic interferometry, *Opt. Eng.*, 28, 261–266, 1989.
202. A. J. Decker, Holographic interferometry with an injection sealed Nd : Yag laser and two reference beams, *Appl. Opt.*, 29, 2697–2700, 1990.
203. M. Takeda, Spatial-carrier fringe-pattern analysis and its application to precision interferometry and profilometry: An overview, *Indust. Metrol.*, 1, 79–99, 1990.
204. K. A. Stetson, Use of sensitivity vector variations to determine absolute displacements in double-exposure hologram interferometry, *Appl. Opt.*, 29, 502–504, 1990.
205. V. V. Balalov, V. S. Pisarev, V. P. Shchepinov, and V. V. Yakovlev, Holographic interference measurements of 3D displacement fields and their use in stress determination, *Opt. Spectros.*, 68, 75–78, 1990.

206. D. W. Watt, T. S. Gross and S. D. Hening, Three illumination beam phase-shifted holographic interferometry study of thermally induced displacements on a printed wiring board, *Appl. Opt.*, 30, 1617–1623, 1991.
207. Z. Fuzessy, Application of double pulse holography for the investigations of machines and systems, *Physical Res.*, 15, 75–107, 1991.
208. M. R. Sajan, T. S. Radha, B. S. Ramprasad, and E. S. R. Gopal, Measurement of corrosion rate of aluminum in sodium hydroxide using holographic interferometry, *Opt. Lasers Eng.*, 15, 183–188, 1991.
209. P. K. Rastogi, M. Barillot, and G. H. Kaufmann, Comparative phase-shifting holographic interferometry, *Appl. Opt.*, 30, 722–728, 1991.
210. R. C. Troth and J. C. Dainty, Holographic interferometry using anisotropic self-diffraction in $Bi_{12}SiO_{20}$, *Opt. Lett.*, 16, 53–55, 1991.
211. S. L. Sochava, R. C. Troth, and S. I. Stepanov, Holographic interferometry using −1 order diffraction in photorefractive $Bi_{12}SiO_{20}$ and $Bi_{12}TiO_{20}$ crystals, *J. Opt. Soc. Am. B*, 9, 1521–1527, 1992.
212. A. M. Lyalikov and A. F. Tuev, Multiwavelength single-exposure holographic interferometry, *Opt. Spectros.*, 73, 227–229, 1992.
213. C. S. Vikram and W. K. Witherow, Critical needs of fringe order accuracies in two-color holographic interferometry, *Exp. Mech.*, 32, 74–77, 1992.
214. M.-A. Beeck, Pulsed holographic vibration analysis on high speed rotating objects: fringe formation, recording techniques and practical applications, *Opt. Eng.*, 31, 553–561, 1992.
215. W. Jüptner, J. Geldmacher, Th. Bischof, and Th. Kreis, Measurement of the deformation of a pressure vessel above a weld point, *Proc. SPIE*, 1756, 98–105, 1992.
216. G. Brown, J. W. Forbes, Mitchell M. Marchi, and R. R. Wales, Hologram interferometry in automotive component vibration testing, *Proc. SPIE*, 1756, 146–152, 1992.
217. S. W. Biederman and R. P. Pryputniewicz, Holographic study of vibrations of a wing section, *Proc. SPIE*, 1756, 153–163, 1992.
218. J. Woisetschlaeger, D. B. Sheffer, H. Mikati, K. Somasundaram, C. W. Loughry, S. K. Chawla, and P. J. Wesolowski, Breast cancer detection by holographic interferometry, *Proc. SPIE*, 1756, 176–183, 1992.
219. X. Wang, R. Magnusson, and A. Haji-Sheikh, Real-time interferometry with photorefractive reference holograms, *Appl. Opt.*, 32, 1983–1986, 1993.
220. P. Zanetta, G. P. Solomos, M. Zurn, and A.C. Lucia, Holographic detection of defects in composites, *Opt. Laser Technol.*, 25, 97–102, 1993.
221. M. M. Ratnam and W. T. Evans, Comparison of measurement of piston deformation using holographic interferometry and finite elements, *Exp. Mech.*, 12, 336–342, 1993.
222. C. M. E. Holden, S. C. J. Parker, and P. J. Bryanston-Cross, Quantitative three-dimensional holographic interferometry for flow field analysis, *Opt. Lasers Eng.*, 19, 285–298, 1993.
223. J. D. Trolinger and J. C. Hsu, Flowfield diagnostics by holographic interferometry and tomography, In *Proceedings of FRINGE'93*, 423–439, 1993.
224. A. Kreuttner, B. Lau, A. Mann, E. Mattes, R. Miller, M. Stuber, and K. Stocker, Holographic interferometry for display of shock wave induced deformations and vibrations—A contribution to laser lithotripsy, *Lasers Med. Sci.*, 8, 211–220, 1993.
225. Q. Huang, J. A. Gilbert, and H. J. Caulfield, Holographic interferometry using substrate guided waves, *Opt. Eng.*, 33, 1132–1137, 1994.
226. P. K. Rastogi, Holographic comparison between waves scattered from two physically distinct rough surfaces, *Opt. Eng.*, 33, 3484–3485, 1994.

227. U. Schnars and W. Jüptner, Digital recording and reconstruction of holograms in hologram interferometry and shearography, *Appl. Opt.*, 33, 4373–4377, 1994.
228. S. V. Miridonov, A. A. Kamshilin, and E. Barbosa, Recyclable holographic interferometer with a photorefractive crystal: Optical scheme optimization, *J. Opt. Soc. Am. A*, 11, 1780–1788, 1994.
229. D. Dirksen and G. von Bally, Holographic double-exposure interferometry in near real time with photorefractive crystals, *J. Opt. Soc. Am. A*, 11, 1858–1863, 1994.
230. U. Schnars, Direct phase determination in hologram interferometry with use of digitally recorded holograms, *J. Opt. Soc. Am. A*, 11, 2011–2015, 1994.
231. H. Kasprzak, W. N. Forester, and G. von Bally, Holographic measurement of changes of the central corneal curvature due to interocular pressure differences, *Opt. Eng.*, 33, 198–203, 1994.
232. N. Ninane and M. P. George, Holographic interferometry using two-wavelength holography: The measurement of large deformations, *Appl. Opt.*, 34, 1923–1928, 1995.
233. Th. Kreis, R. Biederman, and W. Jüptner, Evaluation of holographic interference patterns by artificial neural networks, *Proc. SPIE*, 2544, 11–24, 1995.
234. B. Pouet and S. Krishnaswamy, Dynamic holographic interferometry by photorefractive crystals for quantitative deformation measurements, *Appl. Opt.*, 35, 787–794, 1996.
235. V. I. Vlad, D. Malacara-Hernandez, and A. Petris, Real-time holographic interferometry using optical phase conjugation in photorefractive materials and direct spatial phase modulation, *Opt. Eng.*, 35, 1383–1387, 1996.
236. J. Zhu, Optical nondestructive examination of glass-fiber reinforced plastic honeycomb structures, *Opt. Lasers Eng.*, 25, 133–143, 1996.
237. M. F. Ralea and N. N. Rosu, Laser holographic interferometry for investigations of cylindrical transparent tubes, *Opt. Eng.*, 35, 1393–1395, 1996.
238. T. J. McIntyre, M. J. Wegener, A. I. Bishop, and H. Rubinsztein-Dunlop, Simultaneous two-wavelength holographic interferometry in a superorbital expansion tube facility, *Appl. Opt.*, 36, 8128–8134, 1997.
239. S. Seebacher, W. Osten, and W. Juptner, Measuring shape and deformation of small objects using digital holography, *Proc. SPIE*, 3479, 104–115, 1998.
240. G. Pedrini, P. Fröning, H. Fessler, and H. J. Tiziani, In-line digital holographic interferometry, *Appl. Opt.*, 37, 6262–6269, 1998.
241. C. Herman, E. Kang, and M. Wetzel, Expanding the applications of holographic interferometry to the quantitative visualization of oscillatory thermofluid processes using temperature as tracer, *Exp. Fluids*, 24, 431–446, 1998.
242. J. T. Sheridan and R. Patten, Holographic interferometry and the fractional Fourier transformation, *Opt. Lett.*, 25, 448–450, 2000.
243. C. Wagner, W. Osten, and S. Seebacher, Direct shape measurement by digital wavefront reconstruction and multiwavelength contouring, *Opt. Eng.*, 39, 79–85, 2000.
244. G. Cedilnik, M. Esselbach, A. Kiessling, and R. Kowarschik, Real-time holographic interferometry with double two-wave mixing in photorefractive crystals, *Appl. Opt.*, 39, 2091–2100, 2000.
245. V. Palero, N. Andrés, M. P. Arroyo, and M. Quintanilla, Holographic interferometry versus stereoscopic PIV for measuring out-of-plane velocity fields in confined flows, *Meas. Sci. Technol.*, 11, 655–666, 2000.
246. S. Herrmann, H. Hinrichs, K.D. Hinsch, and C. Surmann, Coherence concepts in holographic particle image velocimetry. *Exp. Fluids*, 29, S108–S116, 2000.

247. S. Seebacher, W. Osten, T. Baumbach, and W. Jüptner, The determination of material parameters of microcomponents using digital holography, *Opt. Lasers Eng.*, 36, 103–126, 2001.
248. V. S. Pisarev, V. V. Balalov, V. S. Aistov, M. M. Bondarenko, and M. G. Yustus, Reflection hologram interferometry combined with hole drilling technique as an effective tool for residual stresses fields investigation in thin-walled structures, *Opt. Lasers Eng.*, 36, 551–597, 2001.
249. W. Osten, T. Baumbach, and W. Jüptner, Comparative digital holography, *Opt. Lett.*, 27, 1764–1766, 2002.
250. S. R. Guntaka, V. Toal, and S. Martin, Holographically recorded photopolymer diffractive optical element for holographic and electronic speckle-pattern interferometry, *Appl. Opt.*, 41, 7475–7479, 2002.
251. K. D. Hinsch, Holographic particle image velocimetry. *Meas. Sci. Technol.*, 13, R61–R72, 2002.
252. W. Steinchen, G. Kupfer, and P. Mäckel, Full field tensile strain shearography of welded specimens, *Strain*, 38, 17–26, 2002.
253. N. Demoli, D. Vukicevic, and M. Torzynski, Dynamic digital holographic interferometry with three wavelengths, *Opt. Exp.*, 11, 767–774, 2003.
254. L. Zipser, H. Franke, E. Olsson, N.-E. Molin, and M. Sjödahl, Reconstruction of two-dimensional acoustic object fields by use of digital phase conjugation of scanning laser vibrometry recordings, *Appl. Opt.*, 42, 5831–5838, 2003.
255. F. Pinard, B. Laine, and H. Vach, Musical quality assessment of clarinet reeds using optical holography, *J. Acoust. Soc. Am.*, 113, 1736–1742, 2003.
256. N. Demoli and D. Vukicevic, Detection of hidden stationary deformations of vibrating surfaces by use of time-averaged digital holographic interferometry, *Opt. Lett.*, 29, 2423–2425, 2004.
257. V. S. Pisarev, V. S. Aistov, V. V. Balalov, M. M. Bondarenko, S. V. Chumak, V. D. Grigoriev, and M. G. Yustus, Metrological justification of reflection hologram interferometry with respect to residual stresses determination by means of blind hole drilling, *Opt. Eng.*, 41, 353–410, 2004.
258. S. R. Guntaka, V. Toal, and S. Martin, Holographic and electronic speckle pattern interferometry using a photopolymer recording material, *Strain*, 40, 79–81, 2004.
259. J. Fernández-Sempere, F. Ruiz-Beviá, and R. Salcedo-Díaz, Measurements by holographic interferometry of concentration profiles in dead-end ultrafiltration of polyethylene glycol solutions, *J. Membr. Sci.*, 229, 187–197, 2004.
260. Y. Morimoto, T. Nomura, M. Fujigaki, S. Yoneyama, and I. Takahashi, Deformation measurement by phase-shifting digital holography, *Exp. Mech.*, 45, 65–70, 2005.
261. R. Ambu, F. Aymerich, F. Ginesu, and P. Priolo, Assessment of NDT interferometric techniques for impact damage detection in composite laminates, *Comp. Sci. Technol.*, 66, 199–205, 2006.
262. T. Baubach, W. Osten, C. Von Kopylow, and W. Jüptner, Remote metrology by comparative digital holography, *Appl. Opt.*, 45, 925–934, 2006.
263. D. Pantelić, L. Blažić, S. Savić-Šević, and Bratimir Panić, Holographic detection of a tooth structure deformation after dental filling polymerization, *J. Biomed. Opt.*, 12, 024026, 2007.

7 Speckle Metrology

7.1 THE SPECKLE PHENOMENON

Illumination of a diffuse object by coherent light produces a grainy structure in space. This grainy light distribution is known as a speckle pattern. It arises as a result of self-interference of numerous waves from scattering centers on the surface of the diffuse object, as shown in Figure 7.1: the amplitudes and phases of these scattered waves are random variables. We assume that (i) the amplitude and phase of each scattered wave are statistically independent variables, and also independent of the amplitudes and the phases of all other waves, and (ii) the phases of these waves are uniformly distributed between $-\pi$ and π. Such a speckle pattern is fully developed. The resultant complex amplitude $\mathbf{u}(x, y) = \mathrm{u}(x, y)e^{i\phi}$ is given by

$$u(x,y)e^{i\phi} = \frac{1}{\sqrt{N}}\sum_{k=1}^{N} u_k = \frac{1}{\sqrt{N}}\sum_{k=1}^{N} a_k e^{i\phi_k} \tag{7.1}$$

where a_k and ϕ_k are the amplitude and phase of the wave from the kth scatterer. For such a speckle pattern, the complex amplitude of the resultant $\mathbf{u}(x, y)$ obeys Gaussian statistics.

The probability density function $p(I)$ of speckle intensity is given by

$$p(I) = \frac{e^{-I/\bar{I}}}{\bar{I}} \tag{7.2}$$

where $\bar{I}$ is the average intensity. This gives the probability that the intensity at a point in the speckle pattern will have the value I. The probability density function in the speckle pattern follows a negative exponential law. The most probable intensity value is zero. A measure of the contrast in the speckle pattern is the ratio $c = \sigma/\bar{I}$, where σ is the standard deviation of the speckle intensity. The contrast in the fully developed linearly polarized speckle pattern is unity.

7.2 AVERAGE SPECKLE SIZE

The grains or the speckles in the pattern are not well defined but have a structure. However, we can associate with the speckle an average size. We consider two cases.

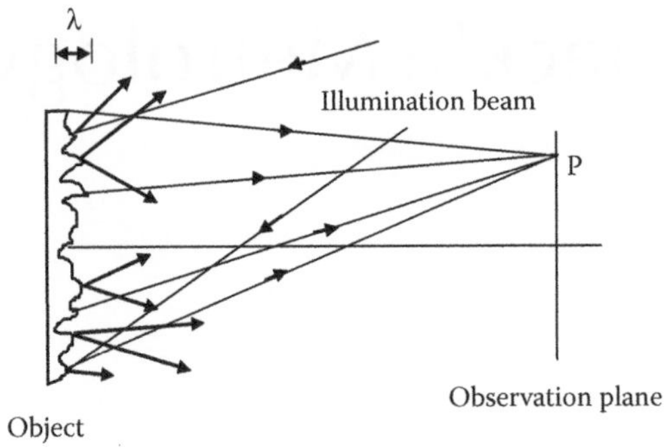

FIGURE 7.1 Formation of a speckle pattern.

7.2.1 Objective Speckle Pattern

It was pointed out earlier that a speckle pattern is formed in space when a diffuse object is illuminated by a coherent wave. This pattern, resulting from free-space propagation, is termed an objective speckle pattern. The speckle size in an objective speckle pattern is given by

$$\sigma_{\text{ob}} = \frac{\lambda z}{D} \tag{7.3}$$

where D is the size of the illuminated area of the object and z is the distance between the object and the observation plane (Figure 7.2a). The size is governed by the interference between the waves from the extreme scattering points on the object. The relationship is the same as that expected from Young's double-slit experiment, the slits being at the extreme positions in the illuminated area. The speckle size increases linearly with the separation between the object and the observation plane.

7.2.2 Subjective Speckle Pattern

A speckle pattern formed at the image plane of a lens is called a subjective speckle pattern. This arises owing to interference of waves from several scattering centers in the resolution element (area) of the lens: in the image of this resolution area, the randomly dephased impulse response functions are added, resulting in a speckle. Therefore, the speckle size is governed by the well-known Airy formula. The diameter of the Airy disk is given approximately by

$$\sigma_{\text{s}} \approx \frac{\lambda b}{D'} \tag{7.4}$$

where D' is the diameter of the lens and b is the image distance (Figure 7.2b). Here again, the speckle size is determined by the maximum aperture of the lens. The objective speckle pattern immediately in front of the lens is transmitted through it and appears on the other side of the lens. The speckles at the periphery of this pattern determine the speckle size at the image plane. By introducing the f-number of the

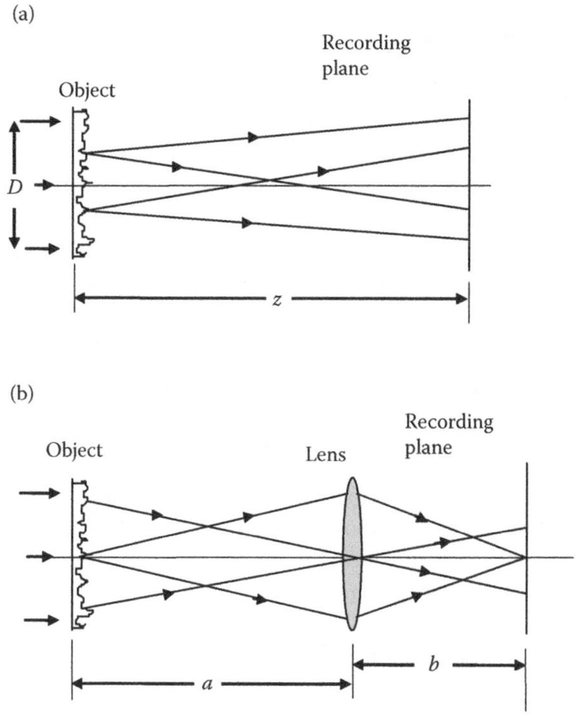

FIGURE 7.2 (a) Objective speckle pattern. (b) Subjective speckle pattern.

lens, $F\# = f/D'$, where f is the focal length of the lens, the average speckle size can be expressed as

$$\sigma_s = (1 + m)\lambda F\# \tag{7.5}$$

Here we have introduced the magnification of the lens, $m = b/a = (b - f)/f$. It can thus be seen that the speckle size can be controlled by (i) the magnification m, and (ii) the $F\#$ of the lens. Control of speckle size by $F\#$ is often used in speckle metrology to match the speckle size with the pixel size of charge-coupled device (CCD) array detectors.

7.3 SUPERPOSITION OF SPECKLE PATTERNS

Speckle patterns can be added on either an amplitude basis or an intensity basis. An example of the addition of speckle patterns on an amplitude basis arises in shear speckle interferometry, where the two speckle patterns are shifted and then superposed. In such a speckle pattern, the statistics of the resultant speckle pattern remains unchanged. However, when the speckle patterns are added on an intensity basis—for example, when two speckle records are made on the same plate—the speckle statistics is completely modified and is governed by the correlation coefficient.

7.4 SPECKLE PATTERN AND OBJECT SURFACE CHARACTERISTICS

In the explanation of a fully developed linearly polarized speckle pattern, it must be stressed that the phases are uniformly distributed between $-\pi$ and π and that there are large number of scatterers participating in speckle formation. This results in a unit-contrast speckle pattern. However, if the surface is smoother than is necessary to satisfy this condition, the speckle contrast decreases. The speckle contrast depends on surface roughness and coherence of light. In fact, speckle contrast has been employed to measure surface roughness over a large range using both monochromatic and polychromatic illumination.

7.5 SPECKLE PATTERN AND SURFACE MOTION

7.5.1 Linear Motion in the Plane of the Surface

Let us assume a translucent object that is translated by a distance d in its own plane: the objective speckle pattern also translates by the same magnitude in the same direction (Figure 7.3a). However, the structure of the speckle pattern begins to change (i.e., decorrelation sets in) when some of the scattering centers go out of the illumination beam. For the subjective speckle pattern, the speckle motion is in the direction opposite to that of the object and its magnitude is md, where m is the magnification (Figure 7.3b).

7.5.2 Out-of-Plane Displacement

Let us consider that a speckle is formed at a position P(r,0), as shown in Figure 7.4a. This speckle is due to the superposition of all the waves from the object. When the object translates axially by a small distance ε, all of these waves are phase-shifted by nearly the same amount. If this phase shift is a multiple of 2π, then a similar state of speckle will exist at the point $P'(r - \Delta r, 0)$; that is, the speckle will have shifted

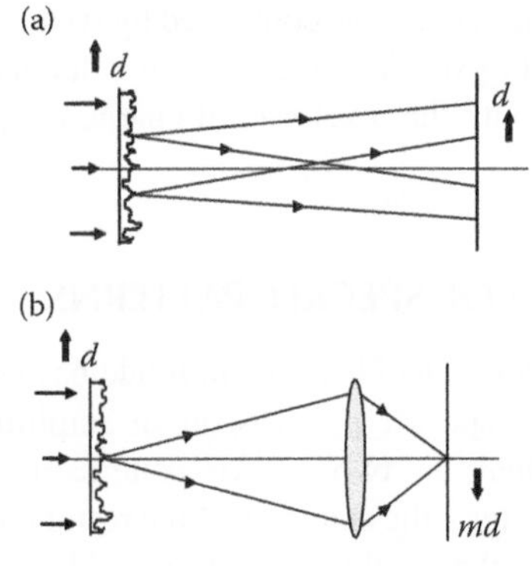

FIGURE 7.3 In-plane displacement: (a) objective speckle pattern; (b) subjective speckle pattern.

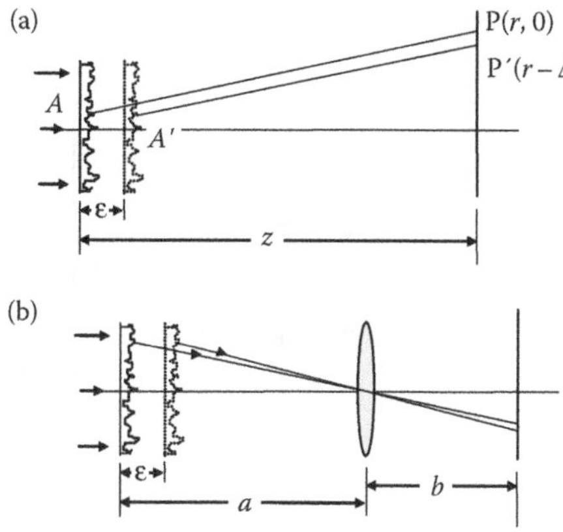

FIGURE 7.4 Out-of-plane displacement: (a) objective speckle pattern; (b) subjective speckle pattern.

radially by Δr. We therefore have

$$\frac{r}{z} = \frac{r - \Delta r}{z - \varepsilon}$$

or

$$|\Delta r| = \varepsilon \frac{r}{z} \tag{7.6}$$

A similar situation holds for a subjective speckle pattern, as shown in Figure 7.4b. We therefore obtain

$$\frac{a}{a - \varepsilon} = \frac{r}{r + \Delta r}$$

or

$$|\Delta r| = \varepsilon \frac{r}{a} \tag{7.7}$$

where a is the object distance. Thus, for axial translation of the object, the speckle pattern shifts radially—it either expands or contracts, depending on the direction of the shift of the object. It may be noticed that the radial movement of the speckle pattern requires a rather large out-of-plane displacement of the object owing to the presence of the r/z or r/a factor, which is very small.

7.5.3 Tilt of the Object

Let the object be illuminated in a direction that makes an angle α with the z axis (taken in the direction of the surface local normal) and let the speckle pattern be observed at a point P in the direction making an angle β, as shown in Figure 7.5a. When the object is tilted by a small angle $\Delta\phi$, the speckle pattern shifts to a new position, which is shifted angularly by $\Delta\psi$, where $\Delta\psi = (1 + \cos\alpha/\cos\beta)\Delta\phi$. Only for the

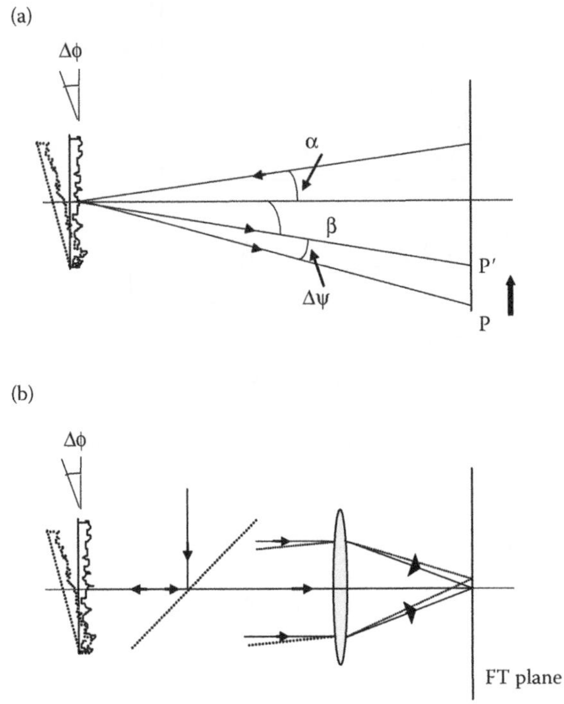

FIGURE 7.5 Motion of the speckle due to the tilt of the object: (a) at a far plane (objective speckle pattern); (b) at the FT plane.

case of normal illumination and observation directions (very small angles α and β) is the angular shift of the speckle pattern twice the tilt of the object. The situation is quite different when we consider the subjective speckle pattern. In fact, there is no shift at the image plane due to the tilt of the object. Between the image plane and the focal plane [or Fourier-transform (FT) plane], speckle shift is due to both in-plane and tilt contributions. At the FT plane, the speckle pattern does not shift owing to in-plane translation, but shifts by $\Delta x_f = f\Delta\phi$ when the object is tilted by $\Delta\phi$ (Figure 7.5b). When the object is illuminated by a divergent wave, one can find a plane that is sensitive only to the in-plane motion and another plane sensitive to tilt alone.

It can thus be seen that a speckle pattern undergoes changes when the object is either translated or tilted. However, for deformation measurement, we are interested in measuring changes/shifts at various points of an object, and hence we employ only the subjective speckle pattern. In this way, a correspondence between the object and the image points is maintained. In other words, the local changes in the object cause changes locally in the speckle pattern rather than over the whole speckle pattern plane. The changes in a speckle pattern due to object deformation are (i) positional changes accompanied by irradiance changes and decorrelation and (ii) phase changes, which are made visible by adding a specular reference wave or speckled reference wave at the image plane. Often, both changes occur simultaneously, but one may dominate over the other.

The speckle methods used for deformation measurement and vibration analysis can be placed in the following four categories:

- Speckle photography
- Speckle interferometry
- Speckle shear interferometry
- Electronic speckle pattern interferometry (ESPI) and shear ESPI.

The first three of these methods employ a photographic medium, photorefractive crystals, and so on for recording; the ESPI methods employ electronic detection. Correlation, as in speckle photography, can also be accomplished digitally.

7.6 SPECKLE PHOTOGRAPHY

The object may be illuminated obliquely or normally. We make an image of the object, as shown in Figure 7.6a, on a photographic plate capable of resolving the speckle pattern, and the first exposure is recorded. The object is then loaded, and another record of the displaced speckle pattern is made on the same plate. In this way, we have recorded two speckle patterns, one of them translated locally by **d**. We need to find out **d** at various locations on the plate, and then generate the deformation map. It was pointed out earlier that the speckle displacement has poor sensitivity for axial (out-of-plane) displacements. Hence, speckle photography is used mostly to measure in-plane displacements and in-plane vibration amplitudes.

Let us first examine the specklegram (negative film or plate) realized by making a single exposure. The intensity recorded is given by $\mathrm{I}(x, y) = |\mathbf{u}(x, y)|^2$. The amplitude transmittance of this negative (specklegram) is expressed as

$$t(x,\ y) = t_0 - \beta T I(x,\ y) \tag{7.8}$$

where t_0 is the bias transmittance, β is a constant, and T is the exposure time. As the speckle pattern consists of a grainy structure, each grain being identified by a δ-function, the intensity $I(x, y)$ can also be expressed as

$$I(x, y) = \iint I(x', y')\delta(x - x', y - y')\, dx'dy' \tag{7.9}$$

When this specklegram is placed in a set-up as shown in Figure 7.6b, and illuminated by a parallel beam of light, the amplitude transmitted is given by $\mathbf{u}_0(x, y)t(x, y)$, where $\mathbf{u}_0(x, y)$ is the amplitude of the illuminating plane wave. The specklegram will diffract the light over a reasonably large cone, depending on the speckle size.

We collect this diffracted light with a lens and make an observation at the back focal plane of the lens. Essentially, we are taking the Fourier transform of the amplitude transmittance $t(x, y)$ of the speckle record. The amplitude at the FT plane, assuming

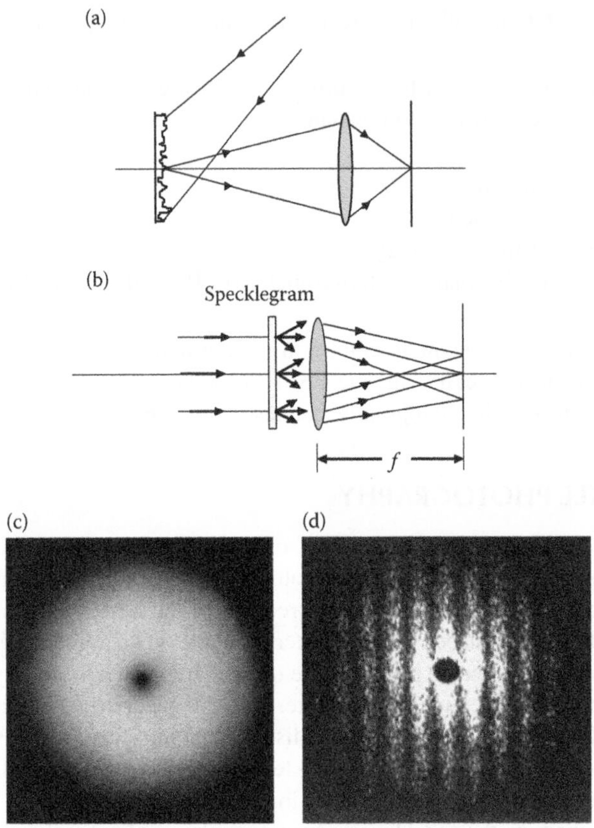

FIGURE 7.6 Speckle photography: (a) bodily in-plane displacement; (b) set-up to observe diffracted filed from the specklegram at the focal plane of a lens; (c) halo from a single-exposure specklegram; (d) fringes in the halo from a double-exposure specklegram.

a unit-amplitude illumination wave, is given by

$$\begin{aligned} U(\mu,\nu) &= \iint t(x,y)\, e^{-2\pi i(\mu x+\nu y)} dx\, dy \\ &= \iint t_0\, e^{-2\pi i(\mu x+\nu y)} dx\, dy - \beta T \iint I(x,y) e^{-2\pi i(\mu x+\nu y)} dx\, dy \\ &= t_0 \delta(\mu,\nu) - \beta T \Im[I(x,y)] \end{aligned} \quad (7.10)$$

where $\Im[\ ]$ signifies the Fourier transform. The intensity distribution at the FT plane, $|U(\mu,\nu)|^2$, consists of a strong central peak and a light distribution around called the halo. This is shown in Figure 7.6c; the central portion is blocked while recording the halo. The halo contains a range of spatial frequencies between 0 and $1/\sigma_s$, where σ_s is the average size in the subjective speckle pattern. The diameter of the halo is $2f\lambda/\sigma_s$. The halo distribution is given by the autocorrelation of the aperture function of the imaging lens. A physical insight into the formation of the halo can be obtained if we consider a specklegram as having a large number of sinusoidal gratings of continuously varying pitch and random orientations. When the specklegram is illuminated, these gratings diffract the beam in various directions, forming the halo at the back focal plane of the lens.

Let us now consider a double-exposure specklegram. In the first exposure, an intensity distribution $I_1(x, y)$ is recorded. The object is then deformed, and the second exposure $I_2(x, y)$ is recorded on the same plate. Owing to the deformation, the speckle pattern shifts locally. Therefore, the intensity distribution $I_2(x, y)$ can be expressed as

$$I_2(x,y) = \iint I(x', y')\delta(x + d_x - x', y + d_y - y')\, dx'dy' \tag{7.11}$$

where d_x and d_y are the components of $\mathbf{d}$ along the x and y directions respectively. The total intensity recorded is

$$\begin{aligned} I_t(x,y) &= I_1(x,y) + I_2(x,y) \\ &= \iint I(x',y')\left[\delta(x - x',\, y - y') + \delta(x + d_x - x',\, y + d_y - y')\right] dx'dy' \end{aligned} \tag{7.12}$$

Again, if this double-exposure specklegram is illuminated by a collimated beam, then one obtains, at the focal plane of the lens, a central order and the superposition of halos belonging to the initial and the final states of the object. The amplitude transmittance of the double-exposure specklegram is given by

$$t(x,y) = t_0 - \beta T[I_1(x,y) + I_2(x,y)] \tag{7.13}$$

The amplitude at the FT plane is given by

$$\Im[t(x,y) = t_0\delta(\mu,\nu) - \beta T\Im[I_1(x,y) + I_2(x,y)] \tag{7.14}$$

For simplicity, we now confine ourselves to one dimension, and hence write the total intensity as

$$I_t(x) = \iint I(x')\left[\delta(x - x') + \delta(x + d_x - x')\right] dx'dy' \tag{7.15}$$

which is the convolution of $I(x)$ with $\delta(x) + \delta(x + d_x)$. Therefore,

$$\Im[I_t(x)] = \Im[I_1(x,y) + I_2(x,y)] = \Im[I(x)]\Im[\delta(x) + \delta(x + d_x)] \tag{7.16}$$

The Fourier transform of $\delta(x)$ is a plane wave propagating on-axis, and that of $\delta(x + d_x)$ is also a plane wave propagating inclined with the axis; that is,

$$\begin{aligned} \Im[\delta(x) + \delta(x + d_x)] &= \int \delta(x)e^{-2\pi i\mu x}\, dx + \int \delta(x + d_x)e^{-2\pi i\mu x}\, dx \\ &= c + ce^{2\pi i\mu d_x} \end{aligned} \tag{7.17}$$

where c is a constant, being the Fourier transform of a δ-function. The spatial frequency μ is defined as $\mu = x_f/f\lambda$, with x_f being the x coordinate on the FT plane and f the focal length of the lens. Suppressing the central order, we can write the intensity distribution in the FT plane (in the halo) as

$$I(\mu) \propto 2\beta^2T^2c^2(1 + \cos 2\pi\mu d_x) = I_0(\mu)\cos^2 \pi\mu d_x \tag{7.18}$$

It is therefore seen that the halo, given by $I_0(\mu)$, is modulated by the $\cos^2(\pi\mu d_x)$ term. In other words, a fringe pattern appears in the halo. Figure 7.6d shows the fringe pattern in the halo. This fringe pattern is similar to the Young's double-aperture fringe pattern. Therefore the fringes are termed Young's fringes. In the $\cos^2(\pi\mu d_x)$ term, there are two variables, namely μ, (the coordinate along the x axis) and d_x, which is a function of local coordinates if the deformation is not constant over the whole specklegram. For constant d_x, the interpretation is simple and straightforward: the halo distribution is modulated by the cosinusoidal fringes, whose spacing is $\Delta x_f = f\lambda/d_x$. Therefore, the magnitude of d_x may be obtained from the fringe width measurement. The direction of the displacement d_x is always along the normal to the fringes. However, if d_x is not constant, then each value of d_x will produce its own fringe pattern, which on superposition may completely wash out this intensity variation. In such a situation, the specklegram is interrogated with a narrow or unexpanded laser beam to extract displacement information from each region of interrogation. The displacement in the region of interrogation should be constant.

7.7 METHODS OF EVALUATION

A double-exposure specklegram contains two shifted speckle patterns, and this shift is to be determined at a number of locations on the specklegram in order to generate the deformation map. It was remarked earlier that Young's-type fringes are formed when the specklegram is illuminated by a narrow beam. These fringes give the direction and magnitude of the displacement at a point. The process of extracting the information from a specklegram is called filtering. The filtering is done both at the plane of the specklegram and at its FT plane. These methods are called pointwise filtering and wholefield filtering.

Another method of filtering, usually applicable to out-of-plane displacement measurement, is known as Fourier filtering. It can also be used in other cases; the fringes are generally localized on the specklegram.

7.7.1 POINTWISE FILTERING

It was mentioned earlier that if the displacement of the speckles is nonuniform, no fringe pattern may be formed at the FT plane when the whole specklegram is illuminated. However, it is safe to assume that the speckle movement over a very small region of the image (specklegram) is uniform. Therefore, if the double-exposure specklegram is illuminated by a narrow (unexpanded) beam and the observation is made at a plane sufficiently far away (far field), as shown in Figure 7.7a, then a system of Young's fringes will be formed. The fringes always run perpendicular to the direction

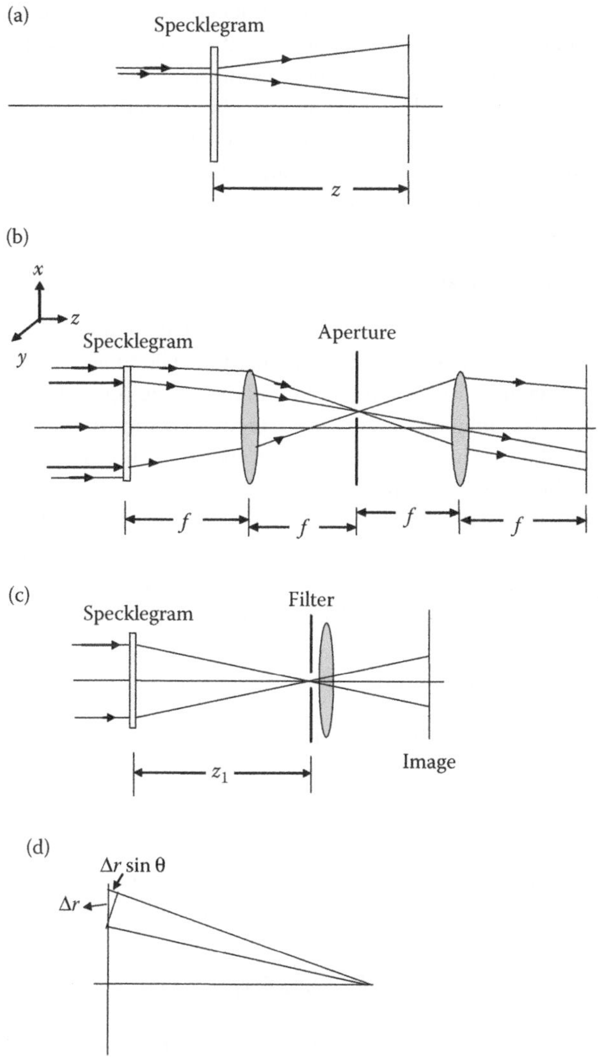

FIGURE 7.7 (a) Pointwise filtering arrangement. (b) Wholefield filtering arrangement. (c) Fourier filtering arrangement. (d) Path difference from a speckle pair.

of displacement, and the fringe width $\bar{p}$ is inversely proportional to the displacement; that is, $\bar{p} = \lambda z/|\mathbf{d}|$. Thus, both the direction and magnitude of the displacement at each interrogation region on the specklegram are obtained. The sign ambiguity is still not resolved. It can be resolved by giving a linear known displacement to the photographic plate before the second exposure. By obtaining the magnitude and direction of the displacement at a large number of points on the specklegram, a displacement map on the specklegram is generated, which is then translated to the object surface through the magnification of the imaging system.

The contrast of the Young's fringes is influenced by several factors, the most important of which are nonuniform displacement and missing speckle pairs in the region of illumination. Further, it must be kept in mind that the halo distribution is nonuniform and the background intensity may not be constant. Therefore, the Young's fringes do not have unit contrast and the positions of their maxima and minima are shifted. This shift introduces errors in the calculation of the displacement. Methods have been developed to correct for this.

7.7.2 Wholefield Filtering

We consider an arrangement shown in Figure 7.7b. This is a $4f$ configuration. The filtering is carried out at the FT plane by an aperture. The image of the specklegram is formed at unit magnification on the output plane by the light allowed through the filtering aperture. The image contains the fringes that represent constant in-plane displacements in the direction of the filtering aperture. In order to understand the working of the wholefield filtering technique, let us place an aperture of appropriate size at a position x_f along the x direction. All of the identical pair scatterers (speckles) on the specklegram will diffract light in phase in the direction of the filtering aperture if their separation is such that the waves diffracted by these point scatterers have a path difference of integral multiples of λ; that is,

$$d_x \sin\theta = m\lambda, \quad m = 0, \pm 1, \pm 2, \pm 3, \ldots \tag{7.19}$$

where d_x is the x component of the displacement vector and $\sin\theta = x_\mathrm{f}/f$. Thus, we obtain

$$d_x = m\lambda f/x_\mathrm{f} \tag{7.20a}$$

These areas therefore appear bright in the image, forming bright fringes that are loci of constant d_x. Similarly, when the filtering aperture is placed at y_f along the y axis, we obtain d_y:

$$d_y = m'\lambda f/y_\mathrm{f} \tag{7.20b}$$

These fringes are loci of constant d_y. In other words, the fringes represent the contours of constant in-plane displacement components, which are separated by incremental displacements Δd_x and Δd_y, where

$$\Delta d_x = \lambda f/x_\mathrm{f} \tag{7.21a}$$

$$\Delta d_y = \lambda f/y_\mathrm{f} \tag{7.21b}$$

This is a wholefield method: the fringes are formed over the whole object surface. In order to obtain d_x and d_y, we need to know m and m', and hence regions on the object where no displacement has taken place. It can also be seen that the sensitivity of the method is variable, being maximum when the filtering aperture is placed at the periphery of the diffraction halo; the increase in sensitivity follows the decrease in the available light for image formation.

7.7.3 Fourier Filtering: Measurement of Out-of-Plane Displacement

It has been shown that a longitudinal or axial displacement ε of the object results in a radial displacement of the speckle pattern, $\Delta r = \varepsilon r/z$. The displacement magnitude ε is obtained by Fourier filtering. The specklegram is illuminated by a parallel beam, as shown in Figure 7.7c, and an aperture is placed on the optical axis for filtering. A lens placed just behind the aperture images the specklegram at the observation plane. All of these point pairs (identical scatterers) will diffract the incident light in phase at the filtering plane if $\Delta r \sin\theta = m\lambda$, where $\sin\theta = r/z_1$ and λ is the wavelength of light used for filtering (Figure 7.7c). Thus,

$$\Delta r \sin\theta = (\varepsilon r/z)(r/z_1) = m\lambda$$

or

$$\varepsilon r^2/zz_1 = m\lambda \tag{7.22}$$

This indicates that a circular fringe pattern is observed at the observation plane. The displacement ε is obtained by measuring the radii of fringes of different orders as follows:

$$\varepsilon = \frac{n\lambda zz_1}{r_{m+n}^2 - r_m^2} \tag{7.23}$$

where r_{m+n} and r_m are the radii of the $(m+n)$th- and mth-order circular fringes, respectively.

7.8 Speckle Photography with Vibrating Objects: In-Plane Vibration

Let us consider an object that is executing an in-plane vibration with amplitude $A(x, y)$. The object is imaged on a photographic plate. Owing to the in-plane vibration, a speckle stretches to a line of length $2A(x, y)m$, where m is the magnification. As mentioned earlier, a sinusoidally vibrating object spends more time at the positions of maximum displacements. Therefore, the speckle line has a nonuniform brightness distribution. However, if the recording is made on a high-contrast photographic plate, the speckle line will have a uniform density after development. We therefore assume that the speckles are uniformly elongated to $2A(x, y)m$. When such a specklegram (time-average recording) is pointwise-filtered, the halo distribution is modulated by $\text{sinc}^2(2\mu mA)$. The zeros of this function occur where

$$2\mu mA = n \quad \text{for } n = 0, \pm1, \pm2, \pm3, \ldots$$

or

$$A(x, y) = n\lambda z/2mx_{fn} \tag{7.24}$$

where x_{fn} is the position of the nth fringe. From this expression, the amplitude of vibration at any location on the specklegram can be found.

In fact, when an object vibrates, the speckles on the regions that are moving smear out and the contrast of the speckles is reduced. The nodal regions where there is no movement have unit contrast. Therefore, one observes a low-contrast vibration mode pattern, with the nodal lines appearing dark. A visual speckle interferometer was proposed by Burch and later modified by Stetson. This interferometer uses a reference beam to visualize out-of-plane vibration amplitudes.

7.9 SENSITIVITY OF SPECKLE PHOTOGRAPHY

The halo size is governed by the speckle size, which, in turn, is governed by the *F#* of the imaging lens. For the speckle displacement to be measurable, at least one fringe should be formed within a halo. In other words, the fringe width must be equal to or less than the halo diameter. Indeed, one fringe will be formed within the halo when the speckle shifts by an amount equal to its average size. This probably is the lower limit of the displacement that speckle photography can sense. The upper limit is also set by the speckle size. When the fringe width becomes equal to the speckle size, the fringes will not be discernible. In fact, one must choose a fringe width of approximately 10 times the speckle size for it to be discernible. This therefore sets an upper limit for speckle displacement measurement. For wholefield filtering, similar limits apply. The quality of fringes in wholefield filtering is considerably poorer than in pointwise filtering.

One of the serious drawbacks of speckle photography is that the positions of the maxima of the fringes in the fringe pattern are shifted owing to nonuniform halo distribution. The same is true for the minima positions when some background is present. This effect introduces errors in the measured values of the displacements. Methods are available to correct for these errors.

7.10 PARTICLE IMAGE VELOCIMETRY

Speckle photography of seeded flows carried out with pulsed lasers or scanned laser beams provides information about the flow (spatial variation of the velocity of flow—in fact the velocity of seeds). The technique is known as particle image velocimetry (PIV). First, a record is made with a short-duration laser pulse so that the motion is frozen, and then the second exposure is made a short time later. The seeds (particles) will have moved during this time to new locations depending on their velocities. The double-exposure record is pointwise-filtered to generate the velocity map.

7.11 WHITE-LIGHT SPECKLE PHOTOGRAPHY

Most objects are good candidates for white-light speckle photography, since the surface structure is quite adequate for this. However, this property is enhanced by coating the surface with a retro-reflective paint. The images of the embedded glass balls in retro-reflective paint act as speckles. Deformation studies can therefore be carried out using white-light speckles based on principles similar to those applied to laser speckles.

7.12 SHEAR SPECKLE PHOTOGRAPHY

It is obvious that speckle photography is essentially a tool to measure in-plane displacement or the in-plane component of deformation, since it has very low sensitivity to the out-of-plane component. It is also used to measure bodily tilts. A stress analyst is often interested in strain rather than displacement. Strain is obtained from displacement data by numerical differentiation, which is error-prone. However, strain can be obtained by optical differentiation, that is, by obtaining displacement values at two closely separated points. With a double-exposure specklegram, this can be achieved by interrogating it with two parallel-displaced narrow beams. Each beam produces Young's-type fringes in the diffraction halo, and the superposition of the two patterns, from each illumination beam, will generate a moiré pattern, from which strain can be calculated (Chapter 9). Alternatively, two states of the object can be recorded on two separate films/plates. During filtering, one plate is displaced with respect to the other, to introduce shear. Again, a moiré pattern is formed, from which strain is calculated. Both of these techniques provide a variable shear by varying the beam separation or specklegram displacement.

When two narrow beams are used for illumination, the diffraction halos are spatially displaced. For low-angle diffraction, the overlap region may be very small, to provide a meaningful moiré pattern. On the other hand, when two double-exposure films/plates are used, the underlying assumption that the object has been identically loaded for each double-exposure record may be difficult to realize. It is therefore desirable to record a single-shear double-exposure specklegram. For this purpose, a pair of wedge plates is used in front of the imaging lens: each wedge plate carries a sheet polarizer. The transmission axes of the polarizers are crossed. The object is illuminated either with circularly polarized light or with linearly polarized light with its azimuth at 45° to the polarizer axis. A double-exposure record, with object loaded in between, can be recorded. Interrogation of such a record with a narrow beam will generate a moiré pattern. The drawback of this method is that it has a fixed shear determined by the wedge angles. In addition, the Young's fringes have a nonuniform intensity, are noisy,

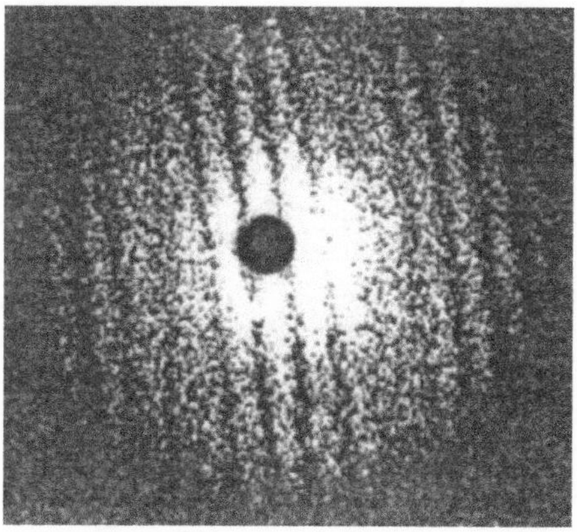

FIGURE 7.8 Moiré pattern due to superposition of two Young's fringe patterns.

and are few in number, and hence the moiré fringes are of poor contrast. Figure 7.8 shows a moiré pattern obtained using such an arrangement.

7.13 SPECKLE INTERFEROMETRY

The speckle phenomenon itself is essentially an interference phenomenon. However, when a reference beam is added to the speckle pattern to code its phase, the technique is then termed speckle interferometry. Speckle interferometry was first applied to measure in-plane displacement by Leendertz. The sensitivity of the measurement could be increased over that of speckle photography, which was limited by the speckle size. The basic theory was borrowed from holographic interferometry, since the phase difference introduced by deformation is governed by the same equation, namely, $\delta = (\mathbf{k}_2 - \mathbf{k}_1) \cdot \mathbf{d}$. When the object is illuminated with two beams with directions symmetric with respect to the object normal (also the optical axis) and observation is made along the optical axis, the arrangement generates fringes that are contours of constant in-plane displacement. The fringes are called correlation fringes—the reason for this name will become obvious in later sections.

The schematic of the set-up to measure in-plane displacement is shown in Figure 7.9. The object is illuminated by two plane waves incident symmetrically at angles θ and $-\theta$ with respect to the optical axis. An image of the object is made on the recording material, say a photographic plate. The object is deformed, and a second exposure is made on the same plate. This double-exposure record, on filtering, generates fringes representing the in-plane component along the x direction. The whole process can be explained mathematically as follows.

Let us consider that the object is illuminated by unit-amplitude plane waves, which are represented by $e^{ikx\sin\theta}$ and $e^{-ikx\sin\theta}$ at the plane of the object. The net amplitude of the waves at the plane of the object is $e^{ikx\sin\theta} + e^{-ikx\sin\theta}$. Let the scattering process from the object be represented by $\Re(x, y)$, which is a complex function and includes the surface characteristics. Thus, the field on the object surface just after scattering is

$$\Re(x, y)(e^{ikx\sin\theta} + e^{-ikx\sin\theta})$$

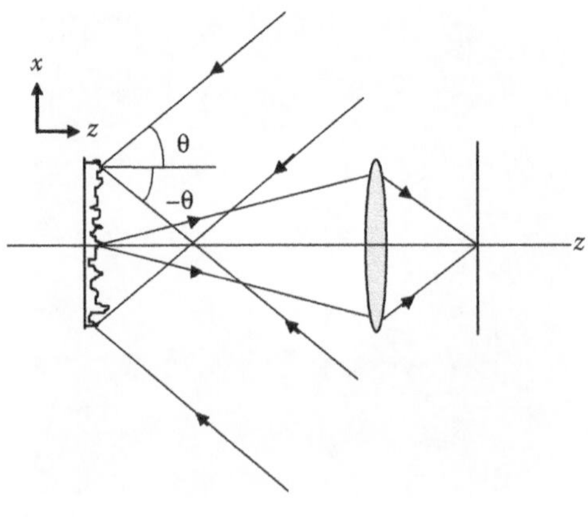

FIGURE 7.9 Configuration for in-plane displacement measurement.

The amplitude at the image plane is obtained through the superposition integral; that is,

$$u(x_i, y_i) = \iint \Re(x, y)(e^{ikx\sin\theta} + e^{-ikx\sin\theta})h(x_i - mx, y_i - my)\,dx\,dy$$
$$= a_1(x_i, y_i)e^{i\phi_1(x_i, y_i)} + a_2(x_i, y_i)e^{i\phi_2(x_i, y_i)} \quad (7.25)$$

where $h(x, y)$ is the impulse function and m is the magnification of the imaging lens. $a_1(x_i, y_i)$, $\phi_1(x_i, y_i)$, $a_2(x_i, y_i)$, $\phi_2(x_i, y_i)$ are the amplitudes and phases of the speckled waves at the image plane due to each of the illuminating waves. When the object is displaced by d_x in the x direction, the amplitude in the image may be written as

$$u'(x_i, y_i) = \iint \Re(x + d_x, y)(e^{ik(x+d_x)\sin\theta} + e^{-ik(x+d_x)\sin\theta})h(x_i - mx, y_i - my)\,dx\,dy$$
$$= a_1(x_i, y_i)e^{i\phi_1(x_i, y_i)}\,e^{ikd_x\sin\theta} + a_2(x_i, y_i)e^{i\phi_2(x_i, y_i)}e^{-ikd_x\sin\theta} \quad (7.26)$$

This equation has been derived under the tacit assumption that $\Re(x + d_x, y) = \Re(x, y)$. This assumption implies that there is no speckle displacement at the object surface due to the motion and that the surface characteristics do not change over a distance d_x. In fact, there is a speckle displacement, but we assume that it is much smaller than the speckle size.

Now, we can write the intensities recorded in the two exposures as

$$I_1(x_i, y_i) = a_1^2 + a_2^2 + 2a_1a_2\cos(\phi_1 - \phi_2) = a_1^2 + a_2^2 + 2a_1a_2\cos\phi \quad (7.27a)$$

$$I_2(x_i, y_i) = a_1^2 + a_2^2 + 2a_1a_2\cos(\phi + 2kd_x\sin\theta) \quad (7.27b)$$

For subsequent analyses of the various techniques, we will be writing the intensity distribution in the second exposure as

$$I_2(x_i, y_i) = a_1^2 + a_2^2 + 2a_1a_2\cos(\phi + \delta) \quad (7.28)$$

where δ is the phase introduced by the deformation. In this case,

$$\delta = 2kd_x\sin\theta = \frac{4\pi}{\lambda}d_x\sin\theta$$

Later, we will use different arguments to prove that δ is indeed equal to $2kd_x\sin\theta$ for the experimental configuration shown in Figure 7.9. For the present, we mention that ϕ_1, ϕ_2, a_1, a_2 are random variables, since $\Re(x, y)$ is a random variable. Therefore, $\phi = \phi_1 - \phi_2$ is also a random variable.

The total intensity recorded will be

$$I_1 + I_2 = 2a_1^2 + 2a_2^2 + 4a_1a_2\cos(\phi + \delta/2)\cos(\delta/2) \quad (7.29)$$

In fact, examination of Equations 7.27a and 7.28 for I_1 and I_2 reveals that when $\delta = 2m\pi$, the equations are identical—the speckles in the two exposures are fully correlated. When $\delta = (2m + 1)\pi$, the speckles are uncorrelated. This correlation provides the basis for fringe formation.

It should be noted that the contrast of the fringes is very poor in this arrangement, mainly because of the very strong granular bias term $2(a_1^2 + a_2^2)$. We will discuss other arrangements in which this bias term has been isolated, thereby improving the contrast of the fringes. First, let us prove that the phase change δ due to an in-plane displacement d_x is indeed given by $2kd_x \sin\theta$. The phase change δ due to a displacement **d** can be expressed as

$$\delta = \delta_2 - \delta_1 = (\mathbf{k}_2 - \mathbf{k}_1)\cdot\mathbf{d} - (\mathbf{k}_2 - \mathbf{k}_1')\cdot\mathbf{d} = (\mathbf{k}_1' - \mathbf{k}_1)\cdot\mathbf{d} \tag{7.30}$$

where $\mathbf{d} = d_x\mathbf{i} + d_y\mathbf{j} + d_z\mathbf{k}$, and $\mathbf{k}_1'$ and $\mathbf{k}_1$ are the propagation vectors of the illumination beams. The equation for the phase difference $\delta_2 = (\mathbf{k}_2 - \mathbf{k}_1)\cdot\mathbf{d}$ was derived in Chapter 5, and is also applicable to speckle interferometry. From the geometry of the experimental configuration of Figure 7.9, $\mathbf{k}_1'$ and $\mathbf{k}_1$ are expressed as

$$\mathbf{k}_1' = \frac{2\pi}{\lambda}(\sin\theta\mathbf{i} - \cos\theta\mathbf{k}) \tag{7.31a}$$

$$\mathbf{k}_1 = \frac{2\pi}{\lambda}(-\sin\theta\mathbf{i} - \cos\theta\mathbf{k}) \tag{7.31b}$$

Thus, the vector $\mathbf{k}_1' - \mathbf{k}_1 = (2\pi/\lambda)2\sin\theta\mathbf{i}$ lies in the plane of the object. Hence,

$$\delta = (\mathbf{k}_1' - \mathbf{k}_1)\cdot\mathbf{d} = \left(\frac{2\pi}{\lambda}2\sin\theta\mathbf{i}\right)\cdot(d_x\mathbf{i} + d_y\mathbf{j} + d_z\mathbf{k}) = \frac{2\pi}{\lambda}2d_x\sin\theta$$

The phase δ is thus governed by the in-plane component d_x of the displacement **d**. Bright fringes are formed when

$$\frac{2\pi}{\lambda}2d_x\sin\theta = 2m\pi$$

Therefore,

$$d_x = \frac{m\lambda}{2\sin\theta} \tag{7.32}$$

The arrangement is sensitive only to the in-plane component of displacement lying in the plane of the illumination beams. Consecutive in-plane fringes differ by $\lambda/(2\sin\theta)$. Obviously, the sensitivity can be varied over a very wide range from 0 to $\lambda/2$ per fringe by varying the interbeam angle. Furthermore, this particular arrangement has the following features:

- As can be seen from the theoretical analysis, the contribution of the out-of-plane displacement component d_z has been fully compensated for collimated illumination. This is due to the fact that both waves suffer equal phase changes arising from out-of-plane displacement, and hence the net phase change is zero.

- The arrangement is not sensitive to the y component d_y. However, d_y can be measured by rotating the experimental arrangement through 90°.
- It offers variable sensitivity.

The disadvantage is that the fringes obtained are of low contrast.

7.14 CORRELATION COEFFICIENT IN SPECKLE INTERFEROMETRY

The correlation coefficient of two random variables X and Y is defined by

$$\rho_{XY} = \frac{\langle XY\rangle - \langle X\rangle\langle Y\rangle}{\sigma_X \sigma_Y} \tag{7.33}$$

where $\sigma_X^2 = \langle X^2\rangle - \langle X\rangle^2$, $\sigma_Y^2 = \langle Y^2\rangle - \langle Y\rangle^2$, and the angular brackets indicate the ensemble average. If X and Y are uncorrelated, then $\langle XY\rangle = \langle X\rangle\langle Y\rangle$ and the correlation coefficient is zero, as expected. The correlation coefficient of the random variables $I_1(x_i, y_i)$ and $I_2(x_i, y_i)$, the intensities recorded in the first and second exposures, is given by

$$\rho(\delta) = \frac{\langle I_1 I_2\rangle - \langle I_1\rangle\langle I_2\rangle}{[\langle I_1^2\rangle - \langle I_1\rangle^2]^{1/2}[\langle I_2^2\rangle - \langle I_2\rangle^2]^{1/2}} \tag{7.34}$$

where the intensities I_1 and I_2 are expressed as follows:

$$I_1(x, y) = a_1^2 + a_2^2 + 2a_1 a_2 \cos\phi = i_1 + i_2 + 2\sqrt{i_1 i_2}\cos\phi \tag{7.35a}$$

$$I_2(x, y) = i_1 + i_2 + 2\sqrt{i_1 i_2}\cos(\phi + \delta) \tag{7.35b}$$

In these expressions, the intensities i_1 and i_2 refer to the intensities at the image plane due to individual illuminating beams. Equation 7.34 for the correlation coefficient is evaluated by noting the following:

- The intensities i_1, i_2, and the phase ϕ are independent random variables, and hence can be averaged separately.
- $\langle\cos\phi\rangle = \langle\cos(\phi + \delta)\rangle = 0$
- $\langle i_1^2\rangle = 2\langle i_1\rangle^2$, and.

When the intensity values I_1 and I_2 are substituted into Equation 7.34 for the correlation coefficient and the averages are taken, we obtain

$$\rho(\delta) = \frac{\langle i_1^2\rangle + \langle i_2^2\rangle + 2\langle i_1\rangle\langle i_2\rangle\cos\delta}{(\langle i_1\rangle + \langle i_2\rangle)^2} \tag{7.36}$$

The correlation coefficient depends on the intensities of the beams and the phase introduced by deformation. If we assume that $\langle i_1\rangle = r\langle i_2\rangle$, that is, one beam is r

times stronger than the other, the correlation coefficient takes the simpler form

$$\rho(\delta) = \frac{1 + r^2 + 2r\cos\delta}{(1+r)^2} \tag{7.37}$$

This has a maximum value of unity when $\delta = 2m\pi$ and a minimum value of $(1-r)^2/(1+r)^2$, when $\delta = (2m+1)\pi$. The minimum value will be zero when $r = 1$, that is, when the average intensities of the beams are equal. The correlation coefficient then varies between 0 and 1 as the value of δ varies over the record. This situation is completely at variance with holographic interferometry, where the reference beam is taken as being stronger than the object beam.

7.15 OUT-OF-PLANE SPECKLE INTERFEROMETER

An interferometer for measuring out-of-plane displacement is shown in Figure 7.10. This is a Michelson interferometer in which one of the mirrors has been replaced by the object under study. It therefore has a reference wave that is smooth or specular. The lens L_2 makes an image of the object at the recording plane. The record consists of an interference pattern between the smooth reference wave and the speckle field in the image of the object. The second exposure is recorded on the same plate after the object has been loaded. This double-exposure specklegram (interferogram), when filtered, yields fringes that are contours of constant out-of-plane displacement. As before, the intensity distribution in the first exposure is

$$I_1(x_i, y_i) = a_1^2 + r_0^2 + 2a_1 r_0 \cos(\phi_1 - \phi_r) = a_1^2 + r_0^2 + 2a_1 r_0 \cos\phi \tag{7.38}$$

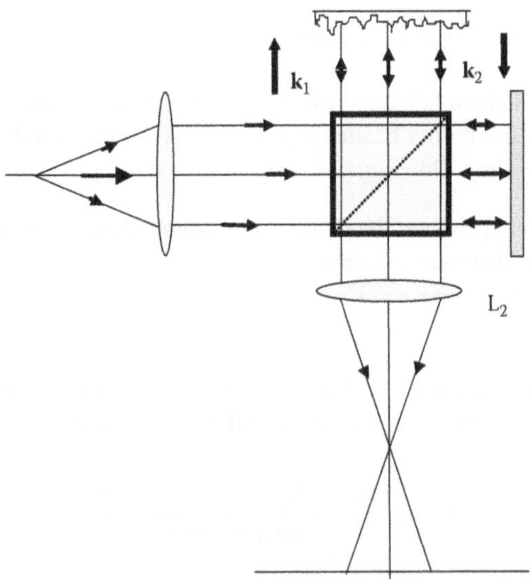

FIGURE 7.10 Configuration for out-of-plane displacement measurement.

where a_1, ϕ_1, ϕ are random variables. The second exposure records the intensity distribution given by

$$I_2(x_i, y_i) = a_1^2 + r_0^2 + 2a_1 r_0 \cos(\phi + \delta) \tag{7.39}$$

where the phase difference δ, introduced by the deformation, is given by

$$\delta = (\mathbf{k}_2 - \mathbf{k}_1) \cdot \mathbf{d} = \frac{2\pi}{\lambda} 2d_z$$

The phase difference depends only on the out-of-plane displacement component d_z. Bright fringes are formed wherever

$$\frac{2\pi}{\lambda} 2d_z = 2m\pi$$

or

$$d_z = m\lambda/2 \tag{7.40}$$

Thus, consecutive fringes are separated by out-of-plane displacements of $\lambda/2$. Again, the fringes are of low contrast. However, it may be noted that, as a consequence of the imaging geometry and customized configurations, only one of the components of deformation is sensed, unlike in holographic interferometry, where the phase difference δ introduced by deformation is invariably dependent on all three components. Further, the correlation fringes are localized at the plane of the specklegram, unlike in holographic interferometry, where the fringe pattern is localized in space in general.

7.16 IN-PLANE MEASUREMENT: DUFFY'S METHOD

In the method due to Leendertz described earlier, the object is illuminated symmetrically with respect to the surface normal, and observation is made along the bisector of the angle enclosed by the illuminating beams. This method, which is also called the dual-illumination, single-observation-direction method, has very high sensitivity but yields poor-contrast fringes.

It is also possible to illuminate the object along one direction and make an observation along two different directions symmetric with respect to the optical axis, which is also along the local normal to the object. This method is due to Duffy, and was developed in the context of moiré gauging. It is also known as the single-illumination, dual-observation-direction method. Figure 7.11a shows a schematic of the experimental arrangement. The object is illuminated at an angle θ, and a two-aperture mask is placed in front of the lens. The apertures enclose an angle of 2α at the object distance. The lens makes an image of the object via each aperture—these images are perfectly superposed.

Each wave passing through the aperture generates a speckled image with a speckle size $\lambda b/D$, where D is the aperture size. These waves are superposed obliquely, and hence each speckle is modulated by a fringe pattern when recorded. The fringe spacing

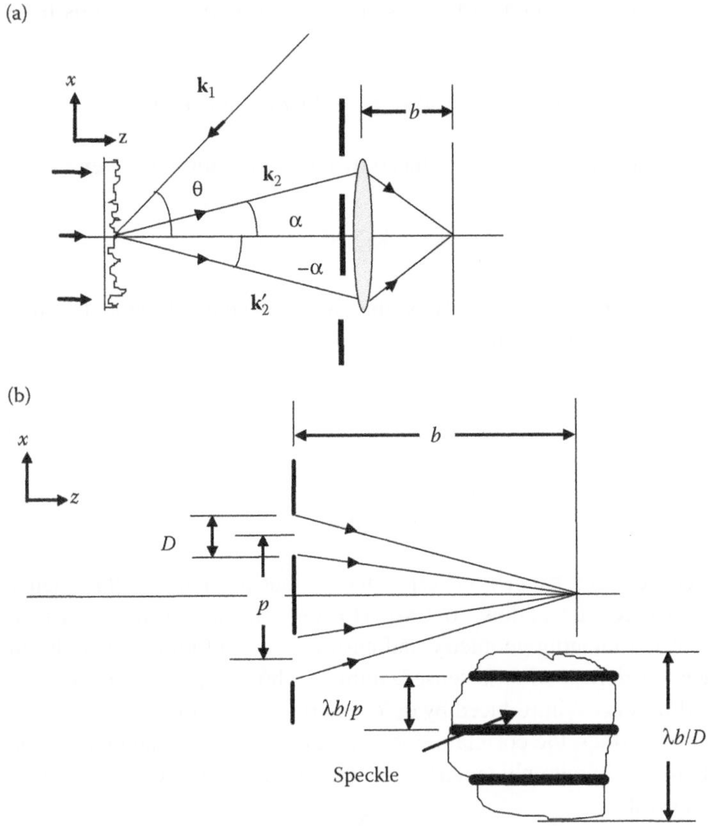

FIGURE 7.11 (a) Duffy's two-aperture arrangement. (b) Modulation of a speckle.

in the speckle is $\lambda b/p$, where p is the separation between the two apertures. This is shown in Figure 7.11b. When the object is deformed, these fringes shift in the speckle. When the deformation is such that a fringe moves by one period or a multiple thereof, it is then exactly superposed on the earlier recorded position: the fringe contrast is then high. The region will diffract strongly on filtering, and hence these areas will appear bright. On the other hand, if the deformation is such that the fringe pattern moves by half a fringe width or an odd multiple of a half-period, the new pattern will fall midway in the earlier recorded pattern, resulting in almost complete washout of the fringe pattern. These regions will not diffract, or will diffract poorly, and hence will appear dark on filtering. Therefore, bright and dark fringes correspond to regions where the displacement is an integral multiple of $\lambda b/p = \lambda/(2 \sin \alpha)$ and an odd integral multiple of $\lambda b/2p$, respectively.

We write the amplitudes of the waves via each aperture as

$$a_{11} = a_0\, e^{i\phi_{11}} \tag{7.41a}$$

$$a_{12} = a_0\, e^{i(\phi_{12}+2\pi\beta x)} \tag{7.41b}$$

where $\beta = p/\lambda b$ is the spatial frequency of the fringe pattern in the speckle pattern. The intensity recorded in the first exposure is

$$I_1(x,y) = |a_{11} + a_{12}|^2 = a_0^2 + a_0^2 + 2a_0^2 \cos(\phi_1 + 2\pi\beta x), \phi_1 = \phi_{12} - \phi_{11} \quad (7.42)$$

Similarly, when the object is loaded, the waves, acquire additional phase changes δ_1 and δ_2, respectively; that is, they can be expressed as

$$a_{21} = a_0 \, e^{i(\phi_{11}+\delta_1)} \quad (7.43a)$$

$$a_{22} = a_0 \, e^{i(\phi_{12}+2\pi\beta x+\delta_2)} \quad (7.43b)$$

The intensity distribution recorded in the second exposure is then

$$I_2(x,y) = |a_{21} + a_{22}|^2 = a_0^2 + a_0^2 + 2a_0^2 \cos(\phi_1 + 2\pi\beta x + \delta) \quad (7.44)$$

where $\delta = \delta_2 - \delta_1$ is the phase difference introduced by the deformation. The specklegram is now ascribed a transmittance $t(x,y)$, given by

$$t(x,y) = t_0 - \beta T(I_1 + I_2) \quad (7.45)$$

where β is a constant and T the exposure time. Information about the deformation is extracted by filtering.

7.17 FILTERING

The specklegram is placed in a set-up as shown in Figure 7.12 and is illuminated by a collimated beam, say, a unit-amplitude wave. The field at the FT plane is $\Im[t(x,y)]$. This consists of a halo distribution with a very strong central peak, as shown in the figure. Owing to the grating-like structure in each speckle, the halo distribution is modified: it has a central halo (zero order) and ± 1-order halos. The zero order arises as a consequence of the terms $a_0^2 + a_0^2$, which do not carry any information. Since the speckle size is now larger, owing to the smaller apertures used for imaging, the halo size (zero-order halo) shrinks. The filtering is done by choosing any one of the first-order halos. A fringe pattern of almost unit contrast is obtained, since the halo (zero order), which carries no information, has been isolated.

7.17.1 Fringe Formation

The phase differences δ_2 and δ_1, due to the deformation, experienced by the waves passing through the apertures can be expressed as

$$\delta_2 = (\mathbf{k}_2' - \mathbf{k}_1) \cdot \mathbf{d}$$

$$\delta_1 = (\mathbf{k}_2 - \mathbf{k}_1) \cdot \mathbf{d}$$

Hence, the phase difference $\delta = \delta_2 - \delta_1$ is given by

$$\delta = (\mathbf{k}_2' - \mathbf{k}_2) \cdot \mathbf{d} \quad (7.46)$$

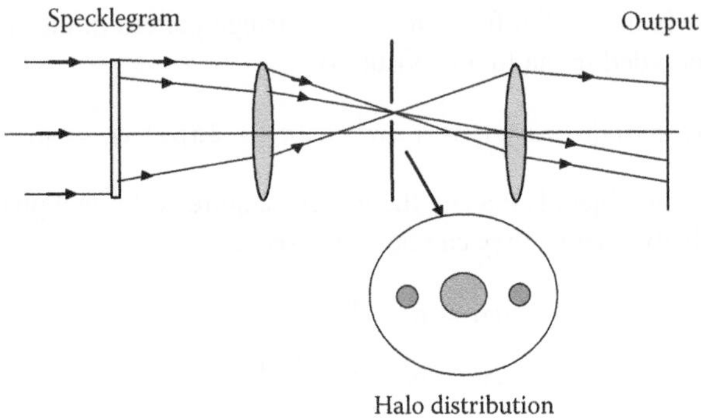

FIGURE 7.12 Wholefield filtering of a specklegram recorded with Duffy's arrangement.

This phase difference generates an interference pattern. Since the illumination and observation beams lie in the (x, z) plane, the wave vectors $\mathbf{k}_2'$ and $\mathbf{k}_2$ can be expressed as

$$\mathbf{k}_2' = \frac{2\pi}{\lambda}(\sin\alpha\mathbf{i} + \cos\alpha\mathbf{k})$$

$$\mathbf{k}_2 = \frac{2\pi}{\lambda}(-\sin\alpha\mathbf{i} + \cos\alpha\mathbf{k})$$

Thus,

$$\delta = (\mathbf{k}_2' - \mathbf{k}_2) \cdot \mathbf{d} = \frac{2\pi}{\lambda}2d_x \sin\alpha$$

Bright fringes are formed wherever

$$\frac{2\pi}{\lambda}2d_x \sin\alpha = 2m\pi$$

or

$$d_x = \frac{m\lambda}{2\sin\alpha} \tag{7.47}$$

This result is similar to that obtained earlier for the Leendertz method, except that the angle θ is replaced by the angle α. Obviously, α cannot take very large values—the magnitude of α is determined by the lens aperture or by $F\#$. Therefore, the method has intrinsically poor sensitivity. At the same time, the speckle size is very large compared with that in the Leendertz method, and hence the range of in-plane displacement measurement is large. The method yields high-contrast fringes owing to the removal of the unwanted speckled field by the grating-like structure formed during recording.

It is indeed very simple and easy to extend Duffy's method to measure both components of the in-plane displacement simultaneously. We describe here two configurations. In one, an aperture configuration as shown in Figure 7.13a is used. The apertures are located at the four corners of a square. In another configuration,

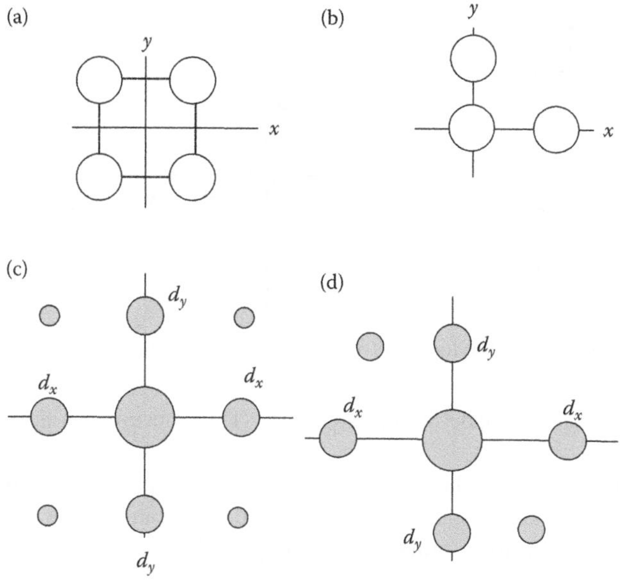

FIGURE 7.13 (a, b) Aperture configurations. (c, d) Corresponding halo distributions.

shown in Figure 7.13b, only three apertures are used: these apertures lie on the vertices of a right-angled triangle. The corresponding halo distributions at the FT plane are shown in Figure 7.13c and 7.13d, respectively.

Filtering through the halos indicated as d_x or d_y yields the component d_x or d_y. It is thus seen that both components, either with the same sensitivity (when the apertures are placed at equal distances along the x and y directions) or with different sensitivities (when the aperture separations are different) can be obtained from a single double-exposure specklegram.

Leendertz's and Duffy's methods can be combined: the object is illuminated by symmetric collimated beams and the specklegram is recorded using an apertured lens. This combination extends the range and sensitivity of in-plane displacement measurement from a small value dictated by Leendertz's method to a large value governed by Duffy's method. On filtering the double-exposure specklegram, both systems of fringes are observed simultaneously.

7.18 OUT-OF-PLANE DISPLACEMENT MEASUREMENT

A schematic of a configuration to measure the out-of-plane component of deformation is shown in Figure 7.14. The object is illuminated by collimated beam at an angle θ. A two-aperture mask is placed in front of the imaging lens. One of the apertures carries a ground-glass plate, which is illuminated by the unexpanded laser beam to generate a diffuse reference wave. The object is imaged via the other aperture. At the image plane, an interference pattern between the diffuse reference wave and the speckled object image is recorded. This constitutes the first exposure. The object is loaded, and

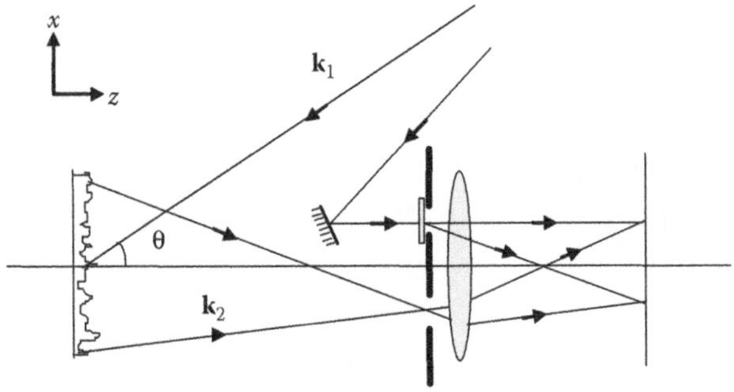

FIGURE 7.14 Configuration for out-of-plane displacement measurement when $\theta = 0$.

the second exposure is made on the same plate. This constitutes a double-exposure specklegram.

On filtering via one of the first-order halos, a fringe pattern depicting contours of constant out-of-plane displacement is obtained. This can be seen as follows. The phase difference δ due to deformation is again given by

$$\delta = (\mathbf{k}_2 - \mathbf{k}_1) \cdot \mathbf{d}$$

where

$$\mathbf{k}_2 = \frac{2\pi}{\lambda}\mathbf{k}, \qquad \mathbf{k}_1 = \frac{2\pi}{\lambda}(-\sin\theta\mathbf{i} + \cos\theta\mathbf{k})$$

Therefore,

$$\delta = \frac{2\pi}{\lambda}[\sin\theta\, d_x + (1 + \cos\theta)d_z] \tag{7.48}$$

The arrangement senses both components, d_x and d_z. However, when the object is illuminated normally (i.e., $\theta = 0$), the phase difference $\delta = (2\pi/\lambda)2d_z$. The phase difference depends only on the out-of-plane component. Bright fringes are formed wherever

$$\frac{2\pi}{\lambda}2d_z = 2m\pi$$

or

$$d_z = m\lambda/2 \tag{7.49}$$

Consecutive bright fringes are separated by $\lambda/2$ in out-of-plane displacement changes.

7.19 SIMULTANEOUS MEASUREMENT OF OUT-OF-PLANE AND IN-PLANE DISPLACEMENT COMPONENTS

By a judicious choice of aperture configuration, it is possible to obtain in-plane and out-of-plane displacement components from a single double-exposure specklegram. One such aperture configuration, along with the halo distribution, is shown

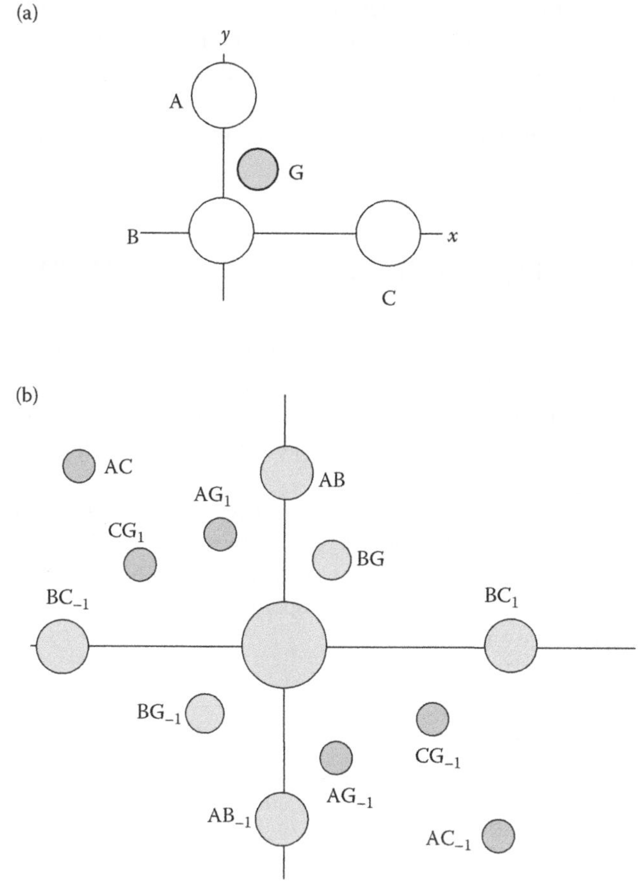

FIGURE 7.15 Simultaneous measurement of in-plane and out-of-plane displacement components: (a) aperture configuration; (b) halo distribution.

in Figure 7.15. It can be seen that the halo AB or AB_{-1} yields the y component of the in-plane displacement, and the halo BC_1 or BC_{-1} yields the x component. The halos BG or BG_{-1}, or AG_1 or AG_{-1}, or CG_1 or CG_{-1} yield the z component. Obviously, all three components of the displacement vector can be retrieved from a single double-exposure specklegram. In fact, redundancy is built into the arrangement.

7.20 OTHER POSSIBILITIES FOR APERTURING THE LENS

In addition to obtaining all components of the deformation vector from a single double-exposure specklegram, aperturing can be used to multiplex the information record; that is, several states of the object can be stored and the information retrieved from the specklegram. The multiplexing can be done in two ways: (i) frequency modulation, where the apertures are shifted laterally on the lens after each exposure;

(ii) θ-multiplexing (modulation), where the apertures are shifted angularly. These methods can also be combined. The amount of information that can be recorded and retrieved depends on lens size, aperture size, the spatial frequency content of the object, and the dynamic range of the recording material. In some experiments, in addition to measurement of displacement components, slope information is also recorded and retrieved later.

7.21 DUFFY'S ARRANGEMENT: ENHANCED SENSITIVITY

The sensitivity of Duffy's arrangement is limited by lens aperture. This limitation, however, can be overcome by modification of the recording set-up, as shown in Figure 7.16. The object is illuminated normally, and is observed along two symmetric directions: the beams are folded and directed to a pair of mirrors and then onto a two-aperture mask in front of the lens. The image of the object is thus made via these folded paths.

In this arrangement, the speckle size is governed by the aperture diameter (as in the case with the other methods); the fringe frequency is determined by the aperture separation; and the sensitivity is governed by the angle θ, which can be varied over a large range, and is not restricted by the lens aperture.

The phase difference δ is given by

$$\begin{aligned}\delta = \delta_2 - \delta_1 = \delta &= (\mathbf{k}_2 - \mathbf{k}_1) \cdot \mathbf{d} - (\mathbf{k}_2' - \mathbf{k}_1) \cdot \mathbf{d} \\ &= (\mathbf{k}_2 - \mathbf{k}_2') \cdot \mathbf{d} = \frac{2\pi}{\lambda} 2d_x \sin\theta \end{aligned} \tag{7.50}$$

The in-plane fringes are extracted by filtering using one of the first-order halos. The technique introduces perspective errors, and also shear when large objects are studied at larger angles. The perspective error, however, can be reduced using a pair of prisms to decrease the convergence.

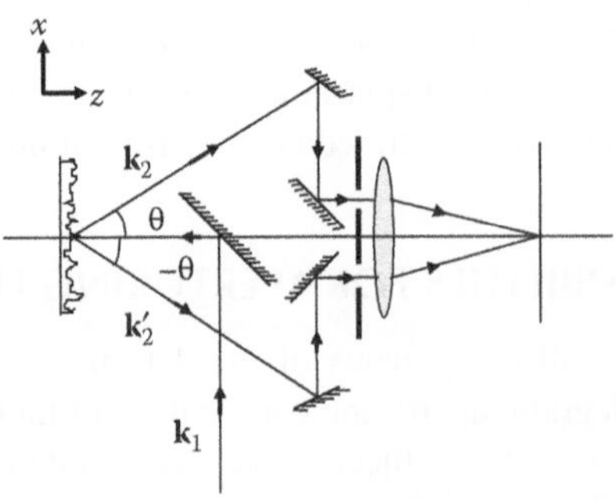

FIGURE 7.16 Configuration for in-plane measurement with enhanced sensitivity.

7.22 SPECKLE INTERFEROMETRY—SHAPE MEASUREMENT/CONTOURING

The following methods are available for contouring using speckle interferometry:

- Change of direction of the illumination beam between exposures
- Change of wavelength between exposures: dual-wavelength technique
- Change of refractive index of the surrounding medium between exposures: dual-refractive-index technique
- Rotation of the object between exposures in an in-plane sensitive configuration

These methods will be discussed in Section 7.28, where we present a technique that allows the contour fringes to be obtained in real time.

7.23 SPECKLE SHEAR INTERFEROMETRY

So far, we have discussed techniques that measure the displacement components only. As mentioned earlier, a stress analyst is usually interested in strains rather than displacements. The strain is obtained by fitting the displacement data numerically and then differentiating it. This procedure could lead to large errors. Therefore, methods have been investigated that can yield fringe patterns representing the derivatives of the displacement. This is achieved with speckle shear interferometry. Since all speckle techniques for displacement measurement use subjective speckles (i.e., image plane recordings), we restrict ourselves to shear at the image plane. The shear methods are grouped under the five categories listed in Table 7.1.

7.23.1 The Meaning Of Shear

Shear essentially means shift. When an object is imaged via two identical paths, as in a two-aperture arrangement, the images are perfectly superposed; there is no shear, even though there are two images. Since the imaging is via two independent paths, the two images can be manipulated independently. In linear shear, one image is

TABLE 7.1
Shear Methods Used in Speckle Interferometry

Shear Types	Phase Difference Leading to Fringe Formation
Lateral shear or linear shear	$\delta(x + \Delta x, y + \Delta y) - \delta(x, y)$
Rotational shear	$\delta(r, \theta + \Delta\theta) - \delta(r, \theta)$
Radial shear	$\delta(r \pm \Delta r, \theta) - \delta(r, \theta)$
Inversion shear	$\delta(x, y) - \delta(-x, -y)$
Folding shear	$\delta(x, y) - \delta(-x, y)$: folding about y axis
	$\delta(x, y) - \delta(x, -y)$: folding about x axis

laterally shifted in any desired direction by an appropriate amount. In interferometry, one image acts as a reference for the other, and hence there is no need to supply an additional reference wave. With shear, we compare the response of an object to the external agencies at any point with that of a shifted point.

In rotational shear, one image is rotated, usually about the optical axis, by a small angle with respect to the other image. In radial shear, one image is either radially contracted or radially expanded with respect to the other image. The inversion shear allows a point at (x, y) to be compared with another point at $(-x, -y)$. This is equivalent to a rotational shear of π. In folding shear, a point is compared with its mirror image: the image may be taken about the y axis or the x axis.

7.24 METHODS OF SHEARING

One of the most commonly used methods of shearing employs the Michelson interferometer, where the object is seen via two independent paths, OABAD and OACAD, as shown in Figure 7.17a. When the mirrors M_1 and M_2 are normal to each other, and at equal distances from A, the two images are perfectly superposed. Tilting one of the mirrors, as shown, displaces one image in the direction of the arrow: the images

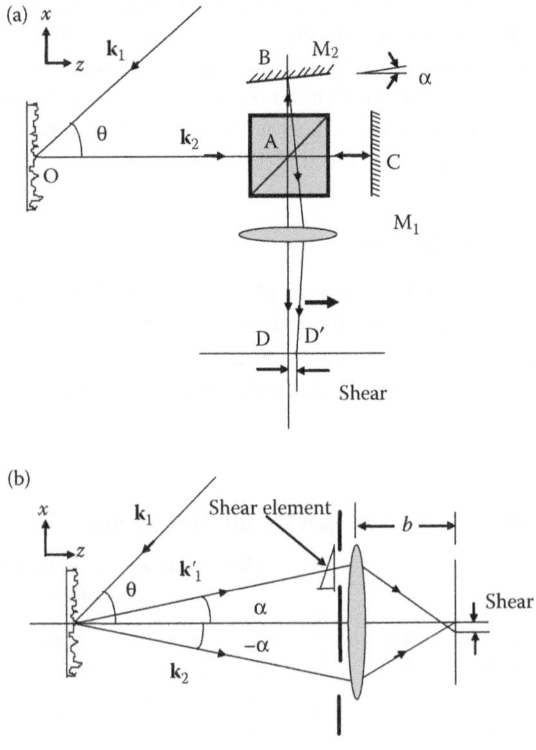

FIGURE 7.17 Shearing with (a) a Michelson interferometer and (b) Duffy's arrangement with a wedge.

are linearly separated. A detailed analysis of this arrangement reveals that a large spherical aberration is introduced owing to the use of a beam-splitting cube as the beams travel in glass a distance of three times the size of the cube.

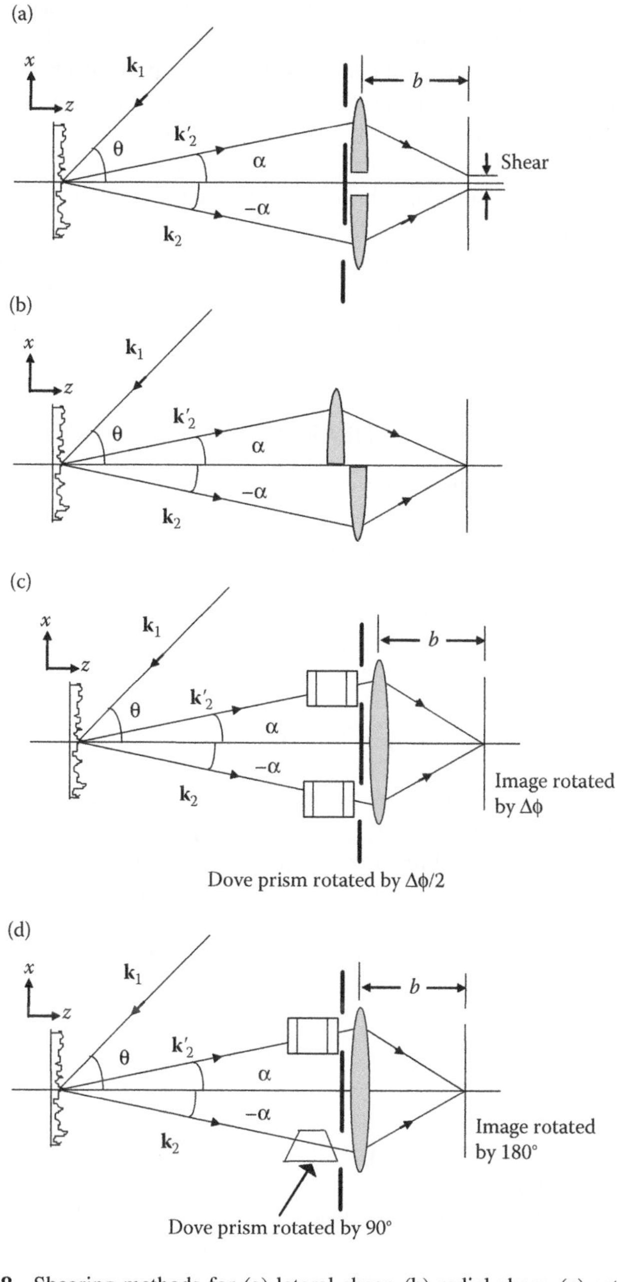

FIGURE 7.18 Shearing methods for (a) lateral shear, (b) radial shear, (c) rotational shear, and (d) inversion shear.

All of the shear methods described in Table 7.1 can be conveniently implemented when an aperture mask is placed in front of an imaging lens. For linear shear, a pair of plane parallel plates, wedges (Figure 7.17b), or a biprism have been used. Shearing has also been done with gratings. If the imaging lens is cut into two halves that can be translated in the lens own plane or along the axis, it makes an excellent shearing device: both functions of shearing and imaging are performed by the same device. In fact, a diffractive optical element can be designed that performs both functions of imaging and shearing. Figure 7.18 shows the schematics of introducing various shear types using an aperture mask and additional optical components.

We now present two configurations based on the Michelson interferometer that can be used to introduce both lateral shear and folding shear. Figure 7.19a shows a conventional arrangement except that the mirrors are replaced by right-angled prisms. Lateral shear is introduced by translation of one of the prisms. The configuration shown in Figure 7.19b is used for folding shear. It uses only one right-angled prism. Depending on the orientation of the prism, folding about the x or the y axis can be achieved.

7.25 THEORY OF SPECKLE SHEAR INTERFEROMETRY

As pointed out earlier, in shear interferometry, a point on the object is imaged as two points or two points on the object are imaged as a single point. One therefore obtains either object plane shear or image plane shear: these are related through the magnification of the imaging lens.

Let a_1 and a_2 be the amplitudes at any point on the image plane due to two points (x_o, y_o) and $(x_o + \Delta x_o, y_o + \Delta y_o)$ at the object plane. The intensity distribution recorded at the image plane is given by

$$I_1(x, y) = a_1^2 + a_2^2 + 2a_1a_2 \cos\phi, \quad \phi = \phi_2 - \phi_1 \tag{7.51a}$$

After the object has been loaded, the deformation vectors at the two points are represented by $\mathbf{d}(x_o, y_o)$ and $\mathbf{d}(x_o + \Delta x_o, y_o + \Delta y_o)$, respectively. Therefore, the waves from these two points arrive at the image point with a phase difference $\delta = \delta_2 - \delta_1$. The intensity distribution in the second exposure can therefore be expressed as

$$I_2(x, y) = a_1^2 + a_2^2 + 2a_1a_2 \cos(\phi + \delta), \quad \delta = \delta_2 - \delta_1 \tag{7.51b}$$

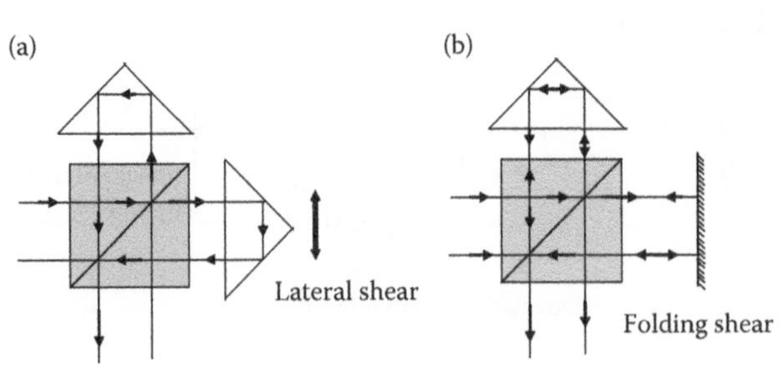

FIGURE 7.19 Michelson interferometer for (a) lateral shear and (b) folding shear.

The total intensity recorded is $I_1(x, y) + I_2(x, y)$, and hence the amplitude transmittance of the double-exposure specklegram is

$$t(x, y) = t_0 - \beta T[I_1(x, y) + I_2(x, y)] \tag{7.52}$$

On filtering, a fringe pattern is obtained representing the derivatives of the displacement components, as will be shown later.

7.26 FRINGE FORMATION

7.26.1 THE MICHELSON INTERFEROMETER

The phase difference δ can be expressed, assuming shear only along the x direction (Figure 7.17a), as

$$\begin{aligned} \delta &= (\mathbf{k}_2 - \mathbf{k}_1) \cdot \mathbf{d}(x_o + \Delta x_o, y_o)(\mathbf{k}_2 - \mathbf{k}_1) \cdot \mathbf{d}(x_o, y_o) \\ &\approx (\mathbf{k}_2 - \mathbf{k}_1) \cdot \frac{\partial \mathbf{d}}{\partial x} \Delta x_o \end{aligned} \tag{7.53}$$

Now, substituting

$$\mathbf{k}_2 = \frac{2\pi}{\lambda}\mathbf{k}, \quad \mathbf{k}_1 = \frac{2\pi}{\lambda}(-\sin\theta\mathbf{i} - \cos\theta\mathbf{k})$$

into Equation 7.53, we obtain

$$\delta \approx \frac{2\pi}{\lambda}\left[\sin\theta\frac{\partial d_x}{\partial x} + (1 + \cos\theta)\frac{\partial d_z}{\partial x}\right]\Delta x_o \tag{7.54}$$

Bright fringes are formed wherever

$$\sin\theta\frac{\partial d_x}{\partial x} + (1 + \cos\theta)\frac{\partial d_z}{\partial x} = \frac{m\lambda}{\Delta x_o}, \quad m = 0, \pm 1, \pm 2, \pm 3, \ldots \tag{7.55}$$

The fringe pattern has contributions from both the strain $\partial d_x/\partial x$ and the slope $\partial d_z/\partial x$. However, when the object is illuminated normally (i.e., $\theta = 0$), the fringe pattern represents a partial x-slope pattern only; that is,

$$\frac{\partial d_z}{\partial x} = \frac{m\lambda}{2\Delta x_o} \tag{7.56}$$

The fringe pattern corresponding to partial y-slope $\partial d_z/\partial y$ is obtained when a shear is applied along the y direction.

7.26.2 The Apertured Lens Arrangement

The phase difference δ can again be expressed, assuming shear only along the x direction (Figure 7.17b), as

$$\begin{aligned}\delta &= (\mathbf{k}_2 - \mathbf{k}_1)\cdot\mathbf{d}(x_o + \Delta x_o, y_o) - (\mathbf{k}_2' - \mathbf{k}_1)\cdot\mathbf{d}(x_o, y_o)\\ &\approx (\mathbf{k}_2 - \mathbf{k}_1)\cdot\mathbf{d}(x_o, y_o) + (\mathbf{k}_2 - \mathbf{k}_1)\cdot\frac{\partial\mathbf{d}}{\partial x}\Delta x_o - (\mathbf{k}_2' - \mathbf{k}_1)\cdot\mathbf{d}(x_o, y_o)\\ &\approx (\mathbf{k}_2 - \mathbf{k}_2')\cdot\mathbf{d}(x_o, y_o) + (\mathbf{k}_2 - \mathbf{k}_1)\cdot\frac{\partial\mathbf{d}}{\partial x}\Delta x_o\end{aligned}$$

$$\delta \approx \frac{2\pi}{\lambda}2d_x\sin\alpha + \frac{2\pi}{\lambda}\left[\sin\theta\frac{\partial d_x}{\partial x} + (1+\cos\theta)\frac{\partial d_z}{\partial x}\right]\Delta x_o \tag{7.57}$$

Comparison with the expression for the Michelson interferometer reveals that there is an in-plane component-dependent term in addition to the usual expression. This term arises owing to the two apertures separated by a distance—an arrangement that has been shown to be inherently sensitive to the in-plane component. An interesting aspect of aperturing of a lens, as has been pointed out earlier, is its ability to measure simultaneously in-plane and out-of-plane displacement components and their derivatives. Figure 7.20 shows photographs of out-of-plane displacement, partial

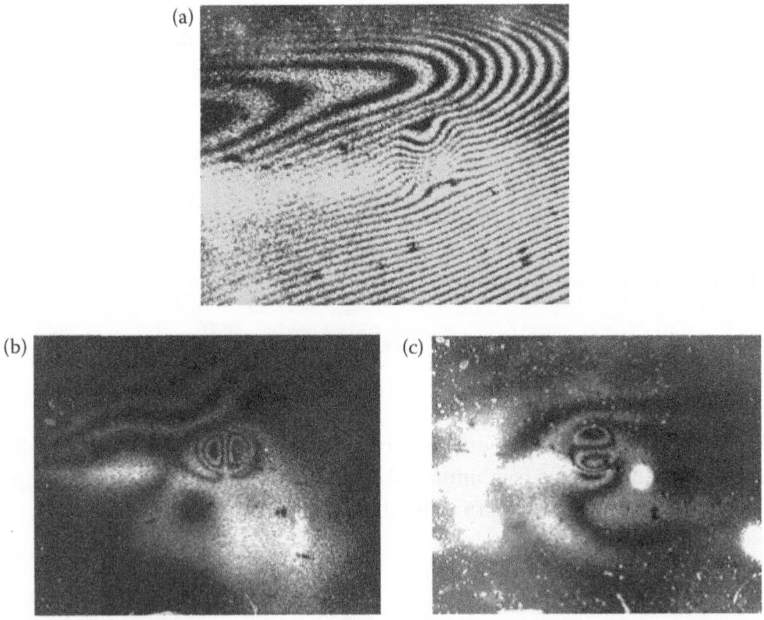

FIGURE 7.20 Interferograms from a double-exposure specklegram: (a) out-of-plane displacement fringe pattern; (b) partial x-slope fringe pattern; (c) partial y-slope fringe pattern.

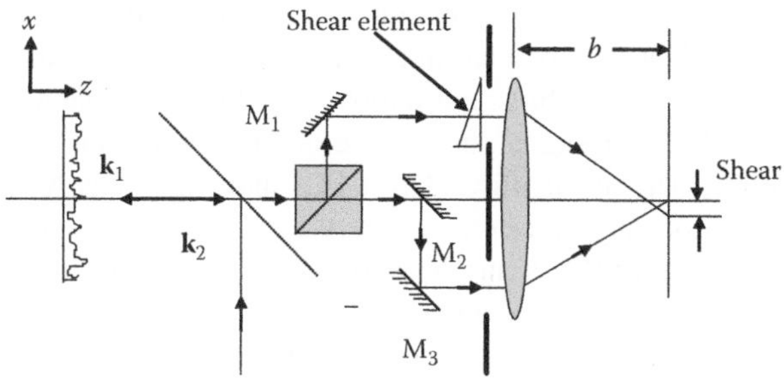

FIGURE 7.21 An experimental arrangement in shear interferometry that is insensitive to in-plane displacement.

x-slope, and partial y-slope fringe patterns of a defective pipe obtained from the same double-exposure specklegram.

7.27 SHEAR INTERFEROMETRY WITHOUT INFLUENCE OF THE IN-PLANE COMPONENT

It has been shown earlier that shear interferometry performed with an aperture mask in front of the imaging lens always yields a fringe pattern that is due to the combined effect of the in-plane displacement component and the derivatives of the displacement. At the same time, it is desirable to have an aperture mask for obtaining high-contrast fringe patterns. In order to retain this advantage and eliminate the in-plane displacement-component sensitivity, the configuration is modified as shown in Figure 7.21.

The object is illuminated normally and viewed axially. The object beam is divided and arranged to pass through the apertures to form the two images. Shear is introduced by placing a wedge plate in front of an aperture. It could just as well be introduced by tilting any one of the mirrors M_1, M_2, or M_3. The mirror combination M_2, M_3 compensates for any extra path difference. Fringe formation is now governed by

$$\frac{\partial d_z}{\partial x} = \frac{m\lambda}{2\Delta x_o} \tag{7.58}$$

This arrangement gives a pure partial x-slope fringe pattern.

7.28 ELECTRONIC SPECKLE PATTERN INTERFEROMETRY

The speckle size in speckle interferometry can be controlled by the *F#* of the imaging lens. Further, the size can be doubled by adding a reference beam axially. It is thus possible to use electronic detectors, which have limited resolution, for recording instead of photographic emulsions. The use of electronic detectors avoids the messy wet development process. Further, the processing is done at video rates, making the technique almost real-time. Photographic emulsions, as mentioned earlier, integrate the light intensity falling on them. In speckle techniques with photographic recording, the two exposures were added in succession, and then techniques were developed to remove the undesired DC component. In electronic detection, the two exposures are handled independently, and subtraction removes the DC component. Phase-shifting techniques are easily incorporated, and hence deformation maps can be presented almost in real-time. The availability of fast PCs and high-density CCD detectors makes the technique of electronic detection very attractive. In fact, electronic speckle pattern interferometry (ESPI) is an alternative to holographic interferometry, and perhaps will replace it in industrial environments.

It might be argued that all of the techniques discussed under speckle interferometry and speckle shear interferometry can be adopted simply by replacing the recording medium by an electronic detector. However, this is not the case, since the resolution of electronic detectors is limited to the range of 50–100 lines/mm, and hence to speckle sizes in the range of 10–20 μm are desired.

7.28.1 Out-of-Plane Displacement Measurement

Figure 7.22 shows one of several configurations used for measuring the out-of-plane component. The reference beam is added axially such that it appears to emerge from the center of the exit pupil of the imaging lens. The speckle size is matched with the pixel size by controlling the lens aperture. The intensities of both the object and the reference beams at the CCD plane are adjusted to be equal. The first frame is stored in the frame-grabber, and the second frame, captured after loading of the object, is subtracted pixel by pixel. The difference signal is rectified and then sent to the monitor to display the fringe pattern.

This procedure can be described mathematically as follows. The intensity recorded in the first frame is given by

$$I_1(x,y) = a_1^2(x,y) + a_2^2(x,y) + 2a_1(x,y)a_2(x,y)\cos\phi, \quad \phi = \phi_2 - \phi_1 \qquad (7.59a)$$

where we have taken a speckled reference wave. The output of the detector is assumed to be proportional to the intensity incident on it. The intensity recorded in the second

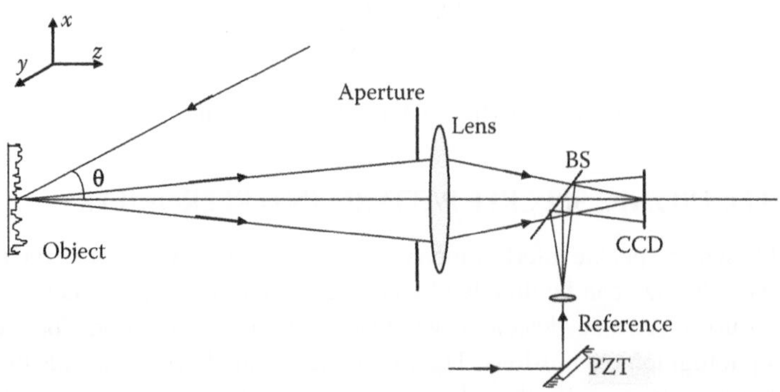

FIGURE 7.22 A configuration for electronic speckle pattern interferometry (ESPI). BS, beam-splitter.

frame is

$$I_2(x,y) = a_1^2(x,y) + a_2^2(x,y) + 2a_1(x,y)a_2(x,y)\cos(\phi + \delta) \tag{7.59b}$$

where the phase difference δ introduced by deformation is given by $\delta = (\mathbf{k}_2 - \mathbf{k}_1) \cdot \mathbf{d}$. The subtracted signal $I_2 - I_1$ will generate a voltage signal ΔV as follows:

$$\begin{aligned} \Delta V \propto I_2 - I_1 &= 2a_1(x,y)a_2(x,y)[\cos(\phi + \delta) - \cos\phi] \\ &= 4a_1(x,y)a_2(x,y)\sin\left(\phi + \frac{\delta}{2}\right)\sin\frac{\delta}{2} \end{aligned} \tag{7.60}$$

The brightness on the monitor will be proportional to the voltage signal ΔV (difference signal) from the detector, and hence

$$B = 4\wp a_1(x,y)a_2(x,y)\sin\left(\phi + \frac{\delta}{2}\right)\sin\frac{\delta}{2} \tag{7.61}$$

where $\wp$ is the constant of proportionality. As δ varies, $\sin(\delta/2)$ will vary between -1 and 1. The negative values of $\sin(\delta/2)$ will appear dark on the monitor, resulting in loss of signal. This loss is avoided by rectifying the signal before it is sent to the monitor. The brightness B is thus given bys

$$B = 4\wp a_1(x,y)a_2(x,y)\left|\sin\left(\phi + \frac{\delta}{2}\right)\sin\frac{\delta}{2}\right| \tag{7.62}$$

The brightness will be zero when $\delta/2 = m\pi$, that is, $\delta = 2m\pi$, with $m = 0, \pm 1, \pm 2, \ldots$. This means that the speckle regions in the speckle pattern that are correlated will appear dark. This is due to the difference operation. Also as a result of this operation, undesirable terms are eliminated. Phase-shifting is easily incorporated by reflecting the reference wave from a PZT-mounted mirror.

7.28.2 In-Plane Displacement Measurement

In-plane displacement can be measured using the arrangement due to Leendertz. The sensitivity can be varied by changing the interbeam angle. Unfortunately, the configurations based on aperturing the lens do not work in ESPI, because the fringe pattern generated by the pair of apertures is not resolved by the CCD camera.

7.28.3 Vibration Analysis

ESPI is an excellent tool for studying vibration modes of an object. It can be used to measure extremely small, moderate, and large vibration amplitudes. The arrangement used is the one suited for out-of-plane displacement measurement. The object is excited acoustically or by directly attaching PZT that is run through a function generator, thereby scanning a large frequency range over which the response of the object can be studied. Since the video rates are very slow compared with the resonance frequencies, the pattern observed on the monitor represents time-average fringes. The intensity distribution is given by

$$\left[J_0\left(\frac{4\pi}{\lambda}A(x,y)\right)\right]^2$$

where A(x, y) is the amplitude of vibration. However, when the reference wave is also modulated at the frequency of object excitation, the intensity distribution in the fringe pattern can be expressed as

$$I(x,y) \propto \left[J_0\left(\frac{4\pi}{\lambda}\left\{[A(x,y)]^2 + a_r^2 - 2A(x,y)a_r\cos(\phi - \phi_r)\right\}^{1/2}\right)\right]^2 \quad (7.63)$$

where a_r and ϕ_r are the amplitude and phase of the reference mirror. Obviously, when the object and reference mirror vibrate in phase, the intensity distribution is proportional to

$$\left[J_0\left(\frac{4\pi}{\lambda}[A(x,y) - a_r]\right)\right]^2$$

The zero-order fringe now occurs where $A(x, y) = a_r$. Therefore, large amplitudes of vibration can be measured. However, if very small vibration amplitudes are to be measured, the frequency of reference-wave modulation is taken slightly different to that of the object vibration, but still within the video frequency. Because of this, the phase of the reference wave varies with time. The intensity distribution is now proportional to

$$\left[J_0\left(\frac{4\pi}{\lambda}\{[A(x,y)]^2 + a_r^2 - 2A(x,y)a_r\cos[\phi - \phi_r(t)]\}^{1/2}\right)\right]^2$$

Since the phase $\phi_r(t)$ varies with time, the argument of the Bessel function varies between $A(x, y) + a_r$ and $A(x, y) - a_r$, and hence the intensity on the monitor will fluctuate. However, if $A(x, y) = 0$, then the argument of the Bessel function remains constant and there is no fluctuation or flicker. Only at those locations where the flicker occurs will the amplitude of vibration be nonzero, thereby allowing very small vibration amplitudes to be detected.

7.28.4 Measurement on Small Objects

EPSI has been used for studying the performance of a variety of objects, ranging in size from large to small. However, there is considerable interest in evaluating the performance of small size objects particularly microelectromechanical systems (MEMS) in real time. MEMS are the result of the integration of mechanical elements, sensors, actuators, and electronics on a common silicon substrate through microfabrication technology. They are used in a number of fields, including telecommunications, computers, aerospace, automobiles, biomedical, and micro-optics. ESPI for the inspection and characterization of MEMS should not alter the integrity or the mechanical behavior of the devices. Since MEMS have an overall size up to few millimeters, a high-spatial-resolution measuring system is required; that is, a long-working-distance microscope with a combination of different magnification objective lenses is incorporated in ESPI. A schematic of a configuration for microscopic ESPI is shown in Figure 7.23a. Instead of a normal camera lens for imaging, a long-working-distance microscope is used for imaging on the CCD array. Phase-shifting is accomplished by a PZT-driven mirror. The MEMS device chosen for study in this example is a pressure transducer. The diaphragm is normally etched out in silicon; the deflection of the diaphragm due to application of pressure is measured using Wheatstone circuitry. However, the deflection profile can be measured using ESPI. ESPI is, in fact, used to calibrate the pressure transducer. Figure 7.23b–d show the results of measurement when pressure is applied to the sensor in between two frames captured by a CCD array camera. Figure 7.23e shows the deflection profile of the pressure sensor.

7.28.5 Shear ESPI Measurement

Again, only the Michelson-interferometer-based shear configurations can be used in ESPI. Other methods of shearing, such as aperture masks with wedges, produce fringes in speckles that are too fine to be resolved by a CCD detector. As mentioned earlier, the fringe pattern carries information about the derivatives of the in-plane and out-of-plane components. This can be seen from the expression

$$\delta \approx \frac{2\pi}{\lambda}\left[\sin\theta\frac{\partial d_x}{\partial x} + (1+\cos\theta)\frac{\partial d_z}{\partial x}\right]\Delta x_0$$

where θ is the angle that the illumination beam makes with the z axis. Obviously, pure partial slope fringes are obtained when $\theta = 0$.

When an in-plane-sensitive configuration is used and shear ESPI is performed, it is possible to obtain both strain and partial slope fringe patterns. Assuming illumination by one beam at a time, we can express the phase difference introduced by deformation in a shear ESPI set-up as

$$\delta_1 \approx \frac{2\pi}{\lambda}\left[\sin\theta\frac{\partial d_x}{\partial x} + (1+\cos\theta)\frac{\partial d_z}{\partial x}\right]\Delta x_0$$

$$\delta_2 \approx \frac{2\pi}{\lambda}\left[-\sin\theta\frac{\partial d_x}{\partial x} + (1+\cos\theta)\frac{\partial d_z}{\partial x}\right]\Delta x_0$$

Obviously, when we subtract these two expressions, we obtain strain fringes, and if we add them, we obtain partial slope fringes.

7.29 CONTOURING IN ESPI

All of the contouring methods discussed in Chapter 6 (Section 6.1.8) can be easily incorporated in ESPI. We give a brief account of these methods and show how they can be adopted in ESPI.

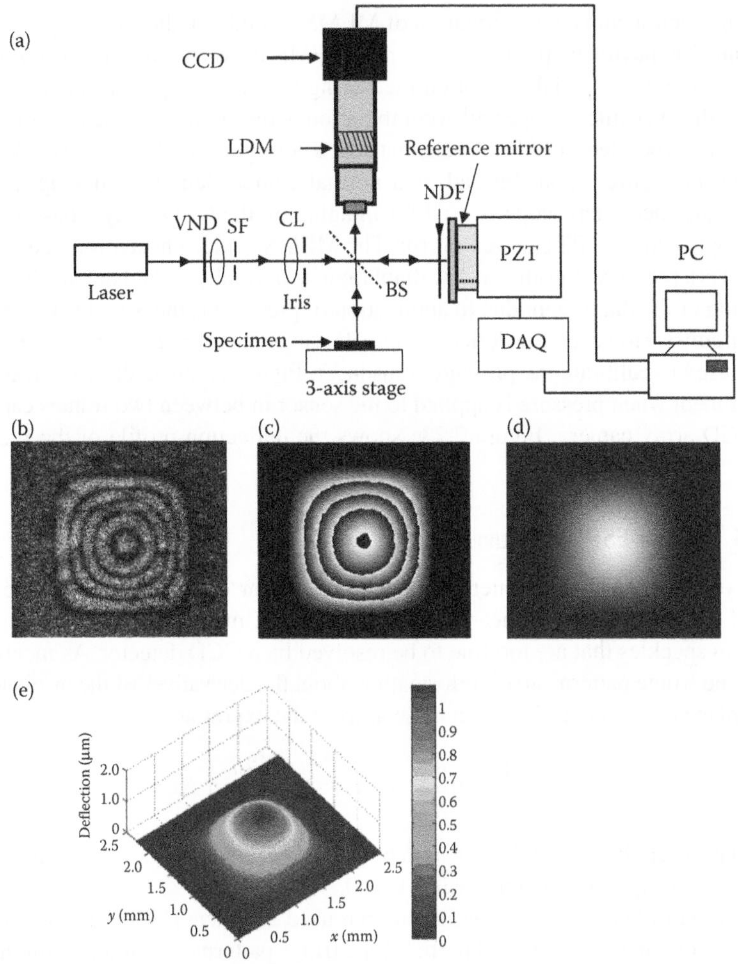

FIGURE 7.23 (a) Schematic of an ESPI system for studying small objects. (b) Correlation fringes. (c) Unwrapped fringes. (d) Wrapped phase. (e) Deflection profile of the sensor. (Courtesy of Dr N. Krishna Mohan, IIT Madras.)

7.29.1 Change of Direction of Illumination

The object is illuminated by a diverging wave from a point source and the usual ESPI set-up is used. The first frame is grabbed and stored. The illumination point source is shifted laterally slightly so as to change the direction of illumination. The second frame captured is now subtracted from the stored frame, and the contour fringes are displayed on the monitor. The contour interval is given by

$$\frac{\lambda L}{2 \sin\theta \Delta s} = \frac{\lambda}{2 \sin\theta \Delta\phi} \tag{7.64}$$

where Δs is the lateral shift of the point source that causes an angular shift of $\Delta\phi$ and L is the distance between the point source and object plane.

7.29.2 Change of Wavelength

The dual-wavelength method requires two sources of slightly different wavelength or a single source that can be tuned. Two frames are grabbed and subtracted, each frame with one of the wavelengths of light. The true depth contours separated by Δz are generated by this method, where the separation Δz is given by

$$\Delta z = \lambda_{\text{eff}}/2 = \frac{\lambda_1\lambda_2}{2\,|\lambda_1 - \lambda_2|} \tag{7.65}$$

where λ_{eff} is the effective (synthetic) wavelength.

7.29.3 Change of Medium Surrounding the Object

Here, the medium surrounding the object is changed between exposures. Subtraction yields true depth contours. In fact, the method is equivalent to the dual-wavelength method, since the wavelength in the medium changes when its refractive index is changed. We can thus arrive at the same result by writing the wavelength in terms of the refractive index and the vacuum wavelength. The contour interval Δz is given by

$$\Delta z = \frac{\lambda}{2\,|n_1 - n_2|} = \frac{\lambda}{2\Delta n} \tag{7.66}$$

Heres, Δn is the change in refractive index when one medium is replaced by the other.

7.29.4 Tilt of the Object

This is a new contouring method, and is applicable to speckle interferometry only. In this method, an in-plane sensitive configuration is used, that is, Leendertz's configuration. The object is rotated by a small amount between exposures. This converts the depth information, due to rotation, into an in-plane displacement to which the set-up is sensitive. The depth contour interval Δz is given by

$$\Delta z = \frac{\lambda}{2\sin\theta\sin\Delta\phi} \approx \frac{\lambda}{2\sin\theta\Delta\phi} \tag{7.67}$$

where 2θ is the interbeam angle of the illumination waves and $\Delta\phi$ is the angle of rotation. Several modifications of this technique have been published.

7.30 SPECIAL TECHNIQUES

7.30.1 Use of Retro-Reflective Paint

One of the features of speckle techniques is that they can be performed with surfaces without any treatment. However, treatment of the surfaces does help in several

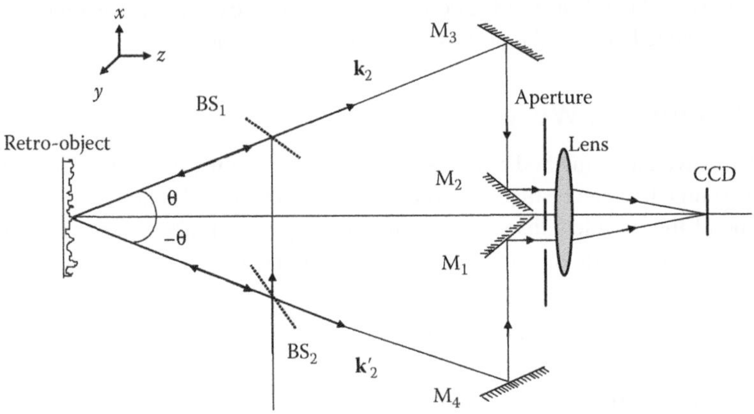

FIGURE 7.24 In-plane displacement measurement with enhanced sensitivity.

cases: for example, coating a dark surface with white paint improves the contrast of fringes. However, when a retro-reflective paint is used, it is possible to essentially enhance the sensitivity of in-plane measurement by a factor of two. The sensitivity remains practically unchanged for out-of-plane displacement measurements. Figure 7.24 shows a schematic of the experimental arrangement to measure the in-plane displacement component.

It can easily be shown that the phase change introduced due to deformation can be written as

$$\delta = \frac{2\pi}{\lambda} 4d_x \sin\theta$$

This shows that the sensitivity is enhanced by a factor of two. It can be seen that this arrangement can also be used for shear ESPI. The shear is introduced by tilting one of the mirrors. The phase difference introduced is given by

$$\delta_2 \approx \frac{2\pi}{\lambda}\left[4d_x \sin\theta + 2\left(\sin\theta\frac{\partial d_x}{\partial x} + \cos\theta\frac{\partial d_z}{\partial x}\right)\Delta x_0\right] \tag{7.68}$$

It can be seen that there is increased sensitivity for the in-plane derivative but almost no change of sensitivity for the out-of-plane displacement derivative.

7.31 SPATIAL PHASE-SHIFTING

It has been pointed out that if the object wave changes during the time interval required for temporal phase shifting, the results of such a measurement are likely to be grossly erroneous. In such a situation, one uses spatial phase-shifting where the information is extracted from a single interferogram. This requires the presence of carrier fringes. Fortunately, these can easily be introduced in Duffy's set-up. The fringe period can also be varied. Moreover, it has been shown that the sensitivity of Duffy's configuration can also be increased. Therefore, this arrangement is best suited for spatial phase-shifting. It has been used to carry out spatial phase-shifting for measuring in-plane displacement, out-of-plane displacement, and shear measurement, as well as for contouring.

BIBLIOGRAPHY

1. J. C. Dainty (Ed.), *Laser Speckle and Related Phenomena*, Springer-Verlag, Berlin, 1975.
2. R. K. Erf (Ed.), *Speckle Metrology*, Academic Press, New York, 1978.
3. M. Françon (Ed.), *Laser Speckle and Applications in Optics*, Academic Press, New York, 1979.
4. R. S. Sirohi (Ed.), *Selected Papers on Speckle Metrology*, Milestone Series MS 35, SPIE, Bellingham, WA, 1991.
5. R. S. Sirohi (Ed.), *Speckle Metrology*, Marcel Dekker, New York, 1993.
6. P. Meinlschmidt, K. D. Hinsch, and R. S. Sirohi, (Eds.), Selected Papers on *Electronic Speckle Pattern Interferometry*, Milestone Series MS 132, SPIE, Bellingham, WA, 1996.
7. P. K. Rastogi (Ed.), *Digital Speckle Pattern Interferometry and Related Techniques*, Wiley-VCH, Weinheim, 2000.
8. W. Steinchen and L. Yang, *Digital Shearography: Theory and Application of Digital Speckle Pattern Shearing Interferometry*, Optical Engineering Press, SPIE, Washington, DC, 2003.

ADDITIONAL READING

1. J. A. Leendertz, Interferometric displacement measurement on scattering surface utilizing speckle effect, *J. Phys. E: Sci. Instrum.*, 3, 214–218, 1970.
2. J. N. Butters and J. A. Leendertz, Speckle pattern and holographic techniques in engineering metrology, *Opt. Laser Technol.*, 3, 26–30, 1971.
3. J. N. Butters and J. A. Leendertz, A double exposure technique for speckle pattern interferometry, *J. Phys.*, 4, 272–279, 1971.
4. J. N. Butters and J. A. Leendertz, Holographic and video techniques applied to engineering measurement, *Trans. Inst. Meas. Control*, 4, 349–354, 1971.
5. A. Macovski, S. D. Ramsey and L. F. Schaefer, Time-lapse interferometry and contouring using television systems, *Appl. Opt.*, 10, 2722–2727, 1971.
6. H. J. Tiziani, Application of speckling for in-plane vibration analysis, *Opt. Acta*, 18, 891–902, 1971.
7. H. J. Tiziani, A study of the use of laser speckle to measure small tilts of optically rough surfaces accurately, *Opt. Commun.*, 5, 271–276, 1972.
8. D. E. Duffy, Moiré gauging of in-plane displacement using double aperture imaging, *Appl. Opt.*, 11, 1778–1781, 1972.
9. E. Archbold and A. E. Ennos, Displacement measurement from double-exposure laser photographs, *Opt. Acta*, 19, 253–271, 1972.
10. J. A. Leendertz and J. N. Butters, An image-shearing speckle-pattern interferometer for measuring bending moments, *J. Phys. E: Sci. Instrum.*, 6, 1107–1110, 1973.
11. Y. Y. Hung, A speckle-shearing interferometer: A tool for measuring derivatives of surface displacements, *Opt. Commun.*, 11, 132–135, 1974.
12. H. M. Pedersen, O. J. Løkberg and B. M. Foerre, Holographic vibration measurement using a TV speckle interferometer with silicon target vidicon, *Opt. Commun.*, 12, 421–426, 1974.

13. K. A. Stetson, Analysis of double-exposure speckle photography with two-beam illumination, *J. Opt. Soc. Am.*, 64, 857–861, 1974.
14. G. Cloud, Practical speckle interferometry for measuring in-plane deformation, *Appl. Opt.*, 14, 878–884, 1975.
15. P. M. Boone, Determination of slope and strain contours by double-exposure shearing interferometry, *Exp. Mech.*, 15, 295–302, 1975.
16. M. Francon, P. Koulev and M. May, Detection of small displacements of diffuse objects illuminated by a speckle pattern, *Opt. Commun.*, 13, 138–141, 1975.
17. R. Jones and J. N. Butters, Some observations on the direct comparison of the geometry of two objects using speckle pattern interferometric contouring, *J. Phys. E: Sci. Instrum.*, 8, 231–234, 1975.
18. C. Forno, White-light speckle photography for measuring deformation, strain, and shape, *Opt. Laser Technol.*, 7, 217–221, 1975.
19. D. A. Gregory, Basic physical principles of defocused speckle photography: A tilt topology inspection technique, *Opt. Laser Technol.*, 8, 201–213, 1976.
20. R. Jones, The design and application of a speckle pattern interferometer for total plane strain field measurement, *Opt. Laser Technol.*, 8, 215–219, 1976.
21. R. P. Khetan and F. P. Chiang, Strain analysis by one-beam laser speckle interferometry. 1: Single aperture method, *Appl. Opt.*, 15, 2205–2215, 1976.
22. F. P. Chiang and R. M. Juang, Vibration analysis of plate and shell by laser speckle interferometry, *Opt. Acta*, 23, 997–1009, 1976.
23. K. Hogmoen and O. J. Løkberg, Detection and measurement of small vibrations using electronic speckle pattern interferometry, *Appl. Opt.*, 16, 1869–1875, 1977.
24. K. P. Miyake and H. Tamura, Observation and explanation of the localization of fringes formed with double-exposure speckle photograph, In *Applications of Holography and Optical Data Processing* (ed. E. Marrom, A. A. Friesem, and E. Wiener), 333–339 Pergamon Press, Oxford, 1977.
25. C. Wykes, De-correlation effects in speckle-pattern interferometry 1. Wavelength change dependent de-correlation with application to contouring and surface roughness measurement, *Opt. Acta*, 24, 517–532, 1977.
26. T. J. Cookson, J. N. Butters, and H. C. Pollard, Pulsed lasers in electronic speckle pattern interferometry, *Opt. Laser Technol.*, 10, 119–124, 1978.
27. J. Brdicko, M. D. Olson, and C. R. Hazell, Theory for surface displacement and strain measurements by laser speckle interferometry, *Opt. Acta*, 25, 963–989, 1978.
28. Y. Y. Hung, I. M. Daniel, and R. E. Rowlands, Full-field optical strain measurement having postrecording sensitivity and direction selectivity, *Exp. Mech.*, 18, 56–60, 1978.
29. O. J. Løkberg, K. Hogmoen, and O. M. Holje, Vibration measurement on the human ear drum in vivo, *Appl. Opt.*, 18, 763–765, 1979.
30. J. Brdicko, M. D. Olson and C. R. Hazell, New aspects of surface displacement and strain analysis by speckle interferometry, *Exp. Mech.*, 19, 160–165, 1979.
31. Y. Y. Hung and A. J. Durelli, Simultaneous measurement of three displacement derivatives using a multiple image-shearing interferometric camera, *J. Strain Anal.*, 14, 81–88, 1979.
32. J. Holoubek and J. Ruzek, A strain analysis technique by scattered-light speckle photography, *Opt. Acta*, 26, 43–54, 1979.
33. G. E. Slettemoen, General analysis of fringe contrast in electronic speckle pattern interferometry, *Opt. Acta*, 26, 313–327, 1979.
34. R. P. Khetan and F. P. Chiang, Strain analysis by one-beam laser speckle interferometry. 2: Multiaperture method, *Appl. Opt.*, 18, 2175–2186, 1979.

35. O. J. Løkberg, Electronic speckle pattern interferometry, *Phys. Technol.*, 11, 16–22, 1980.
36. G. H. Kaufmann, On the numerical processing of speckle photograph fringes, *Opt. Laser Technol.*, 12, 207–209, 1980.
37. G. E. Slettemoen, Electronic speckle pattern interferometric system based on a speckle reference beam, *Appl. Opt.*, 19, 616–623, 1980.
38. S. Nakadate, T. Yatagai and H. Saito, Electronic speckle-pattern interferometry using digital image processing techniques, *Appl. Opt.*, 19, 1879–1883, 1980.
39. S. Nakadate, T. Yatagai and H. Saito, Digital speckle-pattern shearing interferometry, *Appl. Opt.*, 19, 4241–4246, 1980.
40. G. H. Kaufmann, A. E. Ennos, B. Gale and D. J. Pugh, An electro-optical read-out system for analysis of speckle photographs, *J. Phys. E: Sci. Instrum.*, 13, 579–584, 1980.
41. I. Yamaguchi, Speckle displacement and decorrelation in the diffraction and image fields for small object deformation, *Opt. Acta*, 28, 1359–1376, 1981.
42. O. J. Løkberg and G. E. Slettemoen, Interferometric comparison of displacements by electronic speckle pattern interferometry, *Appl. Opt.*, 20, 2630–2634, 1981.
43. I. Yamaguchi, A laser-speckle strain gauge, *J. Phys. E: Sci. Instrum.*, 14, 1270–1273, 1981.
44. G. K. Jaisingh and F. P. Chiang, Contouring by laser speckle, *Appl. Opt.*, 20, 3385–3387, 1981.
45. C. S. Vikram and K. Vedam, Speckle photography of lateral sinusoidal vibrations: Error due to varying halo intensity, *Appl. Opt.*, 20, 3388–3391, 1981.
46. R. Jones and C. Wykes, General parameters for the design and optimization of electronic speckle pattern interferometers, *Opt. Acta*, 28, 949–972, 1982.
47. C. Wykes, Use of electronic speckle pattern interferometry (ESPI) in the measurement of static and dynamic surface displacements, *Opt. Eng.*, 21, 400–406, 1982.
48. R. Krishna Murthy, R. S. Sirohi and M. P. Kothiyal, Detection of defects in plates and diaphragms using split lens speckle-shearing interferometer, *NDT International (UK)*, 15, 329–33, 1982.
49. A. Asundi and F. P. Chiang, Theory and applications of the white light speckle method for strain analysis, *Opt. Eng.*, 21, 570–580, 1982.
50. R. K. Mohanty, C. Joenathan, and R. S. Sirohi, Speckle shear interferometry with double dove prisms, *Opt. Commun.*, 47, 27–30, 1983.
51. T. D. Dudderar, J. A. Gilbert, A. J. Boehnlein and M. E. Schultz, Application of fiber optics to speckle metrology – a feasibility study, *Exp. Mech.*, 23, 289–297, 1983.
52. D. W. Robinson, Automatic fringe analysis with a computer image-processing system, *Appl. Opt.*, 22, 2169–2176, 1983.
53. D. K. Sharma, R. S. Sirohi, and M. P. Kothiyal, Simultaneous measurement of slope and curvature with three aperture speckle interferometer, *Appl. Opt.*, 23, 1542–1546, 1984.
54. C. Shakher and G. V. Rao, Use of holographic optical elements in speckle metrology, *Appl. Opt.*, 23, 4592–4595, 1984.
55. R. Krishna Murthy, R. K. Mohanty, R. S. Sirohi, and M. P. Kothiyal, Radial speckle shearing interferometer and its engineering applications, *Optik*, 67, 85–94, 1984.
56. C. Joenathan, R. K. Mohanty, and R. S. Sirohi, On the methods of multiplexing in speckle shear interferometry, *Optik*, 69, 8–12, 1984.
57. R. K. Mohanty, C. Joenathan, and R. S. Sirohi, Speckle interferometric methods of measuring out of plane displacements, *Opt. Lett.*, 9, 475–477, 1984.

58. R. S. Sirohi, Speckle shear interferometry—A review, *J. Opt. (India)*, 13, 95–113, 1984.
59. B. P. Holownia, Non-destructive testing of overlap shear joints using electronic speckle pattern interferometry, *Opt. Lasers Eng.*, 6, 79–90, 1985.
60. C. Joenathan, R. K. Mohanty, and R. S. Sirohi, Hololens in speckle and speckle shear interferometry, *Appl. Opt.*, 24, 1294–1298, 1985.
61. R. K. Mohanty, C. Joenathan, and R. S. Sirohi, Multiplexing in speckle shear interferometry – An optimal method, *Appl. Opt.*, 24, 2043–2044, 1985.
62. S. Nakadate and H. Saito, Fringe scanning speckle-pattern interferometry, *Appl. Opt.*, 24, 2172–2180, 1985.
63. K. Creath, Phase-shifting speckle interferometry, *Appl. Opt.*, 24, 3053–3058, 1985.
64. R. K. Mohanty, C. Joenathan, and R. S. Sirohi, Speckle and speckle shear interferometers combined for simultaneous determination of out of plane displacement and slope, *Appl. Opt.*, 24, 3106–3109, 1985.
65. O. J. Løkberg, J. T. Malmo, and G. E. Slettemoen, Interferometric measurements of high temperature objects by electronic speckle pattern interferometry, *Appl. Opt.*, 24, 3167–3172, 1985.
66. K. A. Stetson and W. R. Brohinsky, Electrooptic holography and its application to hologram interferometry, *Appl. Opt.*, 24, 3631–3637, 1985.
67. R. Erbeck, Fast image processing with a microcomputer applied to speckle photography, *Appl. Opt.*, 24, 3838–3841, 1985.
68. F. P. Chiang and D. W. Li, Random (speckle) patterns for displacement and strain measurement: Some recent advances, *Opt. Eng.*, 24, 936–943, 1985.
69. K. Creath and G. E. Slettemoen, Vibration-observation techniques for digital speckle-pattern interferometry, *J. Opt. Soc. Am. A*, 2, 1629–1636, 1985.
70. C. Joenathan, R. K. Mohanty, and R. S. Sirohi, Contouring by speckle interferometry, *Opt. Letts.*, 10, 579–581, 1985.
71. R. K. Mohanty, C. Joenathan, and R. S. Sirohi, NDT speckle rotational shear interferometry, *NDT International (UK)*, 18, 203–205, 1985.
72. R. K. Mohanty, C. Joenathan, and R. S. Sirohi, High sensitivity tilt measurement by speckle shear interferometry, *Appl. Opt.*, 25, 1661–1664, 1986.
73. J. R. Tyrer, Critical review of recent developments in electronic speckle pattern interferometry, *Proc. SPIE*, 604, 95–111, 1986.
74. J. M. Huntley, An image processing system for the analysis of speckle photographs, *J. Phys. E: Sci. Instrum.*, 19, 43–49, 1986.
75. C. Joenathan, C. S. Narayanamurthy, and R. S. Sirohi, Radial and rotational slope contours in speckle shear interferometry, *Opt. Commun.*, 56, 309–312, 1986.
76. C. Joenathan and R. S. Sirohi, Elimination of errors in speckle photography, *Appl. Opt.*, 25, 1791–1794, 1986.
77. S. Nakadate, Vibration measurement using phase-shifting speckle pattern interferometry, *Appl. Opt.*, 25, 4162–4167, 1986.
78. A. R. Ganesan, C. Joenathan and R. S. Sirohi, Real-time comparative digital speckle pattern interferometry, *Opt. Commun.*, 64, 501–506, 1987.
79. F. Ansari and G. Ciurpita, Automated fringe measurement in speckle photography, *Appl. Opt.*, 26, 1688–1692, 1987.
80. R. Hoefling and W. Osten, Displacement measurement by image-processed speckle patterns, *J. Mod. Opt.*, 34, 607–617, 1987.
81. O. J. Løkberg and J. T. Malmo, Detection of defects in composite materials by TV holography, *NDT International (UK)*, 21, 223–228, 1988.
82. S. Winther, 3D Strain Measurements using ESPI, *Opt. Lasers Eng.*, 8, 45–57, 1988.

83. M. Owner-Petersen and P. Damgaard Jensen, Computer-aided electronic speckle pattern interferometry (ESPI): Deformation analysis by fringe manipulation, *NDT International (UK)*, 21, 422–426, 1988.
84. C. Joenathan, C. S. Narayanamurthy, and R. S. Sirohi, Localization of fringes in speckle photography that are due to axial motion of the diffuse object, *J. Opt. Soc. Am. A*, 5, 1035–1040, 1988.
85. R. W. T. Preater, Measuring rotating component in-plane strain using conventional pulsed ESPI and optical fibres, *Proc. SPIE*, 952, 245–250, 1988.
86. A. R. Ganesan and R. S. Sirohi, New method of contouring using digital speckle pattern interferometry (DSPI), *Proc. SPIE*, 954, 327–332, 1988.
87. O. J. Løkberg and J. T. Malmo, Long-distance electronic speckle pattern interferometry, *Opt. Eng.*, 27, 150–156, 1988.
88. Boxiang Lu, H. Abendroth, H. Eggers, E. Ziolkowski, and X. Yang, Real time investigation of rotating objects using ESPI system, *Proc. SPIE*, 1026, 218–221, 1988.
89. V. Parthiban, C. Joenathan, and R. S. Sirohi, A simple inverting interferometer with holo elements, *Appl. Opt.*, 27, 1913–1914, 1988.
90. A. R. Ganesan, D. K. Sharma, and M. P. Kothiyal, Universal digital speckle shearing interferometer, *Appl. Opt.*, 27, 4731–4734, 1988.
91. J. Davies and C. Buckberry, Applications of fibre optic TV holography system to the study of large automotive structures, *Proc. SPIE*, 1162, 279–291, 1989.
92. R. J. Pryputniewicz and K. A. Stetson, Measurement of vibration patterns using electro-optic holography, *Proc. SPIE*, 1162, 456–467, 1989.
93. V. M. Chaudhari, A. R. Ganesan, C. Shakher, P. B. Godbole, and R. S. Sirohi, Investigations of in-plane stresses on bolted flange joint using digital speckle pattern interferometry, *Opt. Lasers Eng.*, 11, 257–264, 1989.
94. E. Vikhagen, Vibration measurement using phase shifting TV-holography and digital image processing, *Opt. Commun.*, 69, 214–218, 1989.
95. M. Kujawinska, A. Spik, and D.W. Robinson, Quantitative analysis of transient events by ESPI, *Proc. SPIE*, 1121, 416–423, 1989.
96. W. Arnold and K. D. Hinsch, Parallel optical evaluation of double-exposure records in optical metrology, *Appl. Opt.*, 28, 726–729, 1989.
97. J. M. Huntley, Speckle photography fringe analysis: Assessment of current algorithms, *Appl. Opt.*, 28, 4316–4322, 1989.
98. K. D. Hinsch, Fringe positions in double-exposure speckle photography, *Appl. Opt.*, 28, 5298–5304, 1989.
99. A. J. Moore and J. R. Tyrer, An electronic speckle pattern interferometer for complete in-plane displacement measurement, *Meas. Sci. Technol.*, 1, 1024–1030, 1990.
100. R. P. Tatam, J. C. Davies, C. H. Buckberry, and J. D. C. Jones, Holographic surface contouring using wavelength modulation of laser diodes, *Opt. Laser Technol*, 22, 317–321, 1990.
101. G. Gulker, K. Hinsch, C. Holscher, A. Kramer, and H. Neunaber, Electronic speckle pattern interferometry system for in situ deformation monitoring on buildings, *Opt. Eng.*, 29, 816–820, 1990.
102. E. Vikhagen, Nondestructive testing by use of TV holography and deformation phase gradient calculation, *Appl. Opt.*, 29, 137–144, 1990.
103. M. Owner-Petersen, Decorrelation and fringe visibility: On the limiting behavior of various electronic speckle-pattern correlation interferometers, *J. Opt. Soc. Am. A*, 8, 1082–1089, 1991.
104. M. Owner-Petersen, Digital speckle pattern shearing interferometry: Limitations and prospects, *Appl. Opt.*, 30, 2730–2738, 1991.

105. C. Joenathan and B. M. Khorana, A simple and modified ESPI system, *Optik*, 88, 169–171, 1991.
106. H. Kadono, S. Toyooka, and Y. Iwasaki, Speckle-shearing interferometry using a liquid-crystal cell as a phase modulator, *J. Opt. Soc. Am. A*, 8, 2001–2008, 1991.
107. R. Hoefling, P. Aswendt, W. Totzauer, and W. Jüptner, DSPI: A tool for analysing thermal strains on ceramic and composite materials, *Proc. SPIE*, 1508, 135–142, 1991.
108. S. Takemoto, Geophysical applications of holographic and ESPI techniques, *Proc. SPIE*, 1553, 168–174, 1991.
109. M. M. Ratnam, W. T. Evans, and J. R. Tyrer, Measurement of thermal expansion of a piston using holographic and electronic speckle pattern interferometry, *Opt. Eng.*, 31, 61–69, 1992.
110. S. Ellingsrud and G. O. Rosvold, Analysis of a data-based TV-holography system used to measure small vibration amplitudes, *J. Opt. Soc. Am. A*, 9, 237–251, 1992.
111. R. R. Vera, D. Kerr, and F. M. Santoyo, Electronic speckle contouring, *J. Opt. Soc. Amer. A*, 9, 2000–2008, 1992.
112. R. S. Sirohi and N. Krishna Mohan, Speckle interferometry for deformation measurement, *J. Mod. Opt.*, 39, 1293–1300, 1992.
113. A. R. Ganesan, B. C. Tan, and R. S. Sirohi, Tilt measurement using digital speckle pattern interferometry, *Opt. Laser Technol.*, 24, 257–261, 1992.
114. G. Gulker, O. Haack, K. D. Hinsch, C. Holscher, J. Kuls, and W. Platen, Two-wavelength electronic speckle-pattern interferometry for the analysis of discontinuous deformation fields, *Appl. Opt*, 31, 4519–4521, 1992.
115. R. Spooren, Double-pulse subtraction TV holography, *Opt. Eng.*, 31, 1000–1007, 1992.
116. R. Rodriguez-Vera, D. Kerr, and F. Mendoza-Santoyo, Electronic speckle contouring, *J. Opt. Soc. Am. A*, 9, 2000–2007, 1992.
117. X. Peng, H. Y. Diao, Y. L. Zou, and H. Tiziani, A novel approach to determine decorrelation effect in a dual-beam electronic speckle pattern interferometer, *Optik*, 90, 129–133, 1992.
118. I. Yamaguchi, Z. B. Liu, J. Kato, and J. Y. Liu, Active phase-shifting speckle interferometer, In *Fringe '93: Proceedings of 2nd International Workshop on Automatic Processing of Fringe Pattern* (ed. W. Jüptner and W. Osten), 103–108, Akademie Verlag, Bremen, 1993.
119. P. Aswendt and R. Hoefling, Speckle interferometry for analysing anisotropic thermal expansion—Application to specimens and components, *Composites*, 24, 611–617, 1993.
120. R. Spooren, A. Aksnes Dyrseth, and M. Vaz, Electronic shear interferometry: application of a (double-)pulsed laser, *Appl. Opt.*, 32, 4719–4727, 1993.
121. A. J. P. von Haasteren and H. J. Frankena, Real time displacement measurement using a multi-camera phase stepping interferometer, In *Fringe '93: Proceedings of 2nd International Workshop on Automatic Processing of Fringe Patterns* (ed. W. Jüptner and W. Osten), 417–422, Akademie Verlag, Bremen, 1993.
122. G. Pedrini, B. Pfister, and H. Tiziani, Double pulse-electronic speckle pattern interferometry, *J. Mod. Opt.*, 40, 89–96, 1993.
123. N. Krishna Mohan, H. Saldner, and N.-E. Molin, Electronic speckle pattern interferometry for simultaneous measurement of out-of-plane displacement and slope, *Opt. Lett.*, 18, 1861–1863, 1993.
124. J. Kato, I. Yamaguchi, and Q. Ping, Automatic deformation analysis by a TV speckle interferometer using a laser diode, *Appl. Opt.*, 32, 77–83, 1993.
125. D. Paoletti and G. S. Spagnolo, Automatic digital speckle pattern interferometry contouring in artwork surface inspection, *Opt. Eng.*, 32, 1348–1353, 1993.

126. B. F. Pouet and S. Krishnaswamy, Additive/subtractive decorrelated electronic speckle pattern interferometry, *Opt. Eng.*, 32, 1360–1368, 1993.
127. R. S. Sirohi and N. Krishna Mohan, In-plane displacement measurement configuration with two fold sensitivity, *Appl. Opt. 32*, 6387–6390, 1993.
128. C. Joenathan, B. Franze, and H. J. Tiziani, Oblique incidence and observation electronic speckle-pattern interferometry, *Appl. Opt.*, 33, 7307–7311, 1994.
129. A. R. Ganesan, P. Meinlschmidt, and K. Hinsch, Vibration mode separation using comparative electronic speckle pattern interferometry (ESPI), *Opt. Commun.*, 107, 28–34, 1994.
130. J. Sivaganthan, A. R. Ganesan, B. C. Tan, K. S. Law, and R. S. Sirohi, Study of steel weldments using electronic speckle pattern interferometry, *J. Test. Eval.*, 22, 42–44, 1994.
131. P. R. Sreedhar, N. Krishna Mohan, and R. S. Sirohi, Real time speckle photography with two wave mixing in photorefractive $BaTiO_3$ crystal, *Opt. Eng.*, 33, 1989–1995, 1994.
132. G. Jin, N. Bao, and P. S. Chung, Applications of a novel phase shift method using a computer-controlled polarization mechanism, *Opt. Eng.*, 33, 2733–2737, 1994.
133. O. J. Løkberg, Sound in flight: Measurement of sound fields by use of TV holography, *Appl. Opt.*, 33, 2574–2584, 1994.
134. H. M. Pedersen, O. J. Løkberg, H. Valø, and G. Wang, Detection of non-sinusoidal periodic vibrations using phase modulated TV-holography, *Opt. Commun.*, 104, 271–276, 1994.
135. R. S. Sirohi and N. Krishna Mohan, Role of lens aperturing in speckle metrology, *J. Sci. Ind. Res.*, 54, 67–74, 1995.
136. G. Gulker and K. D. Hinsch, Detection of microstructure changes by electronic speckle pattern interferometry, *Opt. Lasers Eng.*, 26, 1348–1353, 1996.
137. R. S. Sirohi, N. Krishna Mohan, and T. Santhanakrishnan, Optical configuration for measurement in speckle interferometry, *Opt. Lett.*, 21, 1958–1959, 1996.
138. N. Krishna Mohan, T. Santhanakrishnan, P. Senthilkumaran, and R. S. Sirohi, Simultaneous implementation of Leendertz and Duffy methods for in-plane displacement measurement, *Opt. Commun.*, 124, 235–239, 1996.
139. P. K. Rastogi, Measurement of in plane strains using electronic speckle and electronic speckle shearing pattern interferometry, *J. Mod. Opt.*, 43, 1577–1581, 1996.
140. M. Sjödahl, Accuracy in electronic speckle photography, *Appl. Opt.*, 36, 2875–2885, 1997.
141. T. Bothe, J. Burke, and H. Helmers, Spatial phase shifting in electronic speckle pattern interferometry: Minimization of phase reconstruction errors, *Appl. Opt.*, 36, 5310–5316, 1997.
142. R. S. Sirohi, J. Burke, H. Helmers, and K. D. Hinsch, Spatial phase shifting for pure in-plane displacement and displacement-derivative measurements in electronic speckle pattern interferometry (ESPI), *Appl. Opt.*, 36, 5787–5791, 1997.
143. P. K. Rastogi, An electronic pattern speckle shearing interferometer for measurement of surface slope variations of three-dimensional objects, *Opt. Lasers Eng.*, 26, 93–100, 1997.
144. K. S. Kim, J. H. Kim, J. K. Lee, and S. S. Jarng, Measurement of thermal expansion coefficients by electronic speckle pattern interferometry at high temperature, *J. Mater. Sci. Lett.*, 16, 1753–1756, 1997.
145. P. K. Rastogi, Determination of surface strains by speckle shear photography, *Opt. Lasers Eng.*, 29, 103–116, 1998.
146. R. S. Sirohi and N. Krishna Mohan, An in-plane insensitive multiaperture speckle shear interferometer for slope measurement, *Opt. Laser Technol.*, 29, 415–417, 1997.

147. P. K. Rastogi, Speckle shearing photography: A tool for direct measurement of surface strains, *Appl. Opt.*, 37, 1292–1298, 1998.
148. J. N. Petzing and J. R. Tyrer, Recent developments and applications in electronic speckle pattern interferometry, *J. Strain Anal. Eng. Des.*, 33, 153–169, 1998.
149. C. Joenathan, B. Franze, P. Haible, and H. J. Tiziani, Speckle interferometry with temporal phase evaluation for measuring large-object deformations, *Appl. Opt.*, 37, 2608–2614, 1998.
150. J. Zhang, Two-dimensional in-plane electronic speckle pattern interferometer and its application to residual stress determination, *Opt. Eng.*, 37, 2402, 1998.
151. J. Zhang and T. C. Chong, Fiber electronic speckle pattern interferometry and its applications in residual stress measurements, *Appl. Opt.*, 37, 6707–6715, 1998.
152. A. J. Moore, D. P. Hand, J. S. Barton, and J. D. C. Jones, Transient deformation measurement with electronic speckle pattern interferometry and a high speed camera, *Appl. Opt.*, 38, 1159–1162, 1999.
153. J. M. Huntley, G. H. Kaufmann, and D. Kerr, Phase-shifted dynamic speckle pattern interferometry at 1 kHz, *Appl. Opt.*, 38, 6556–6563, 1999.
154. A. Andersson, A. Runnemalm, and M. Sjödahl, Digital speckle-pattern interferometry: Fringe retrieval for large in-plane deformations with digital speckle photography, *Appl. Opt.*, 38, 5408–5412, 1999.
155. B. B. García, A. J. Moore, C. Pérez-López, L. Wang, and T. Tschudi, Transient deformation measurement with electronic speckle pattern interferometry by use of a holographic optical element for spatial phase stepping, *Appl. Opt.*, 38, 5944–5947, 1999.
156. H. A. Aebischer and S. Waldner, A simple and effective method for filtering speckle-interferometric phase fringe patterns, *Opt. Commun.*, 162, 205–210, 1999.
157. F. Chen, G. M. Brown, and M. Song, Overview of three-dimensional shape measurement using optical methods, *Opt. Eng.*, 39, 10–22, 2000.
158. H. Van der Auweraer, H. Steinbichler, C. Haberstok, R. Freymann, and D. Storer, Integration of pulsed-laser ESPI with spatial domain modal analysis: Results from the SALOME project, *Proc. SPIE*, 4072, 313, 2000.
159. B. Kemper, D. Dirksen, W. Avenhaus, A. Merker, and G. Von Bally, Endoscopic double-pulse electronic-speckle-pattern interferometer for techinal and medical intracavity inspection, *Appl. Opt.*, 39, 3899–3905, 2000.
160. H. Van der Auweraer, H. Steinbichler, C. Haberstok, R. Freymann, D. Storer, and V. Linet, Industrial applications of pulsed-laser ESPI vibration analysis, *Proc. SPIE*, 4359, 490–496, 2001.
161. A. Federico and G. H. Kaufmann, Comparative study of wavelet thresholding methods for denoising electronic speckle pattern interferometry fringes, *Opt. Eng.*, 40, 2598–2604, 2001.
162. A. Kishen, V. M. Murukeshan, V. Krishnakumar, and A. Asundi, Analysis on the nature of thermally induced deformation in human dentine by electronic speckle pattern interferometry (ESPI), *J. Dentistry*, 29(8), 531–537, 2001.
163. C. Vial-Edwards, I. Lira, A. Martinez, and M. Münzenmayer, Electronic speckle pattern interferometry analysis of tensile tests of semihard copper sheets, *Exp. Mech.*, 41, 58–62, 2001.
164. D. Karlsson, G. Zacchi, and A. Axelsson, Electronic speckle pattern interferometry: A tool for determining diffusion and partition coefficients for proteins in gels, *Biotechnol. Prog.*, 18, 1423–1430, 2002.
165. H. Van der Auweraer, H. Steinbichler, S. Vanlanduit, C. Haberstok, R. Freymann, D. Storer, and V. Linet, Application of stroboscopic and pulsed-laser electronic speckle

pattern interferometry (ESPI) to modal analysis problems, *Meas. Sci. Technol.*, 13, 451–463, 2002.
166. C. Shakher, R. Kumar, S. K. Singh, and S. A. Kazmi, Application of wavelet filtering for vibration analysis using digital speckle pattern interferometry, *Opt. Eng.*, 41, 176–180, 2002.
167. G. H. Kaufmann and G. E. Galizzi, Phase measurement in temporal speckle pattern interferometry: Comparison between the phase-shifting and the Fourier transform methods, *Appl. Opt.*, 41, 7254–7263, 2002.
168. L. Yang and A. Ettemeyer, Strain measurement by three-dimensional electronic speckle pattern interferometry: Potentials, limitations, and applications, *Opt. Eng.*, 42, 1257–1266, 2003.
169. V. Madjarova, H. Kadono, and S. Toyooka, Dynamic electronic speckle pattern interferometry (DESPI) phase analyses with temporal Hilbert transform, *Opt. Exp.*, 11, 617–623, 2003.
170. A. Martínez, R. Rodríguez-Vera, J. A. Rayas, and H. J. Puga, Fracture detection by grating moiré and in-plane ESPI techniques, *Opt. Lasers Eng.*, 39, 525–536, 2003.
171. N. Krishna Mohan and P. K. Rastogi, Recent developments in digital speckle pattern interferometry, *Opt. Lasers Eng.*, 40, 439–445, 2003.
172. F. Chen, W. D. Luo, M. Dale, A. Petniunas, P. Harwood, and G. M. Brown, High-speed ESPI and related techniques: Overview and its application in the automotive industry, *Opt. Lasers Eng.*, 40, 459–485, 2003.
173. X. Li, K. Wang, and B. Deng, Matched correlation sequence analysis in temporal speckle pattern interferometry, *Opt. Laser Technol.*, 36, 315–322, 2004.
174. R. R. Cordero, A. Martínez, R. Rodríguez-Vera, and P. Roth, Uncertainty evaluation of displacements measured by electronic speckle-pattern interferometry, *Opt. Commun.*, 241, 279–292, 2004.
175. Y. Fu, C. J. Tay, C. Quan, and L. J. Chen, Temporal wavelet analysis for deformation and velocity measurement in speckle interferometry, *Opt. Eng.*, 43, 2780–2787, 2004.
176. T. R. Moore, A simple design for an electronic speckle pattern interferometer, *Am. J. Phys.*, 72, 1380–1384, 2004.
177. Y. Y. Hung and H. P. Ho, Shearography: An optical measurement technique and applications, *Mater. Sci. Eng. R: Rep.*, 49(3), 61–87, 2005.
178. P. D. Ruiz, J. M. Huntley, and R. D. Wildman, Depth-resolved wholefield displacement measurement by wavelength-scanning electronic speckle pattern interferometry, *Appl. Opt.*, 44, 3945–3953, 2005.
179. G. A. Albertazzi, Radial metrology with electronic speckle pattern interferometry, *J. Hologr. Speckle*, 3, 117–124, 2006.
180. H. Joost and K. D. Hinsch, Sound field monitoring by tomographic electronic speckle pattern interferometry, *Opt. Commun.*, 259, 492–498, 2006.
181. E. A. Barbosa and A. C. L. Lino, Multi-wavelength electronic speckle pattern interferometry for surface shape measurement, *Appl. Opt.*, 46, 2624–2631, 2007.

8 Photoelasticity

So far, we have described techniques that are based on interference or diffraction of light, and even on the geometrical theory of light. The phenomena of interference and diffraction are manifestations of wave nature, and hence are not peculiar to light alone. On the other hand, light falls in a very small portion of a very wide electromagnetic spectrum. Electromagnetic waves are transverse waves: the electric and magnetic field vectors are orthogonal and vibrate perpendicular to the direction of propagation in free space or in an isotropic medium. In fact, the electric field vector **E**, magnetic field vector **B**, and propagation vector form an orthogonal triplet. Photo effects, such as vision and recording of images on a photographic emulsion, are attributable to **E**, and hence when dealing with light, we shall be concerned only with this field vector and not with **B**.

Let us consider a plane wave propagating along the z direction, with the electric vector confined to the $(y,\ z)$ plane. The tip of the electric vector describes a line in the $(y,\ z)$ plane as the wave propagates. The $(y,\ z)$ plane is called the plane of vibration and the $(x,\ z)$ plane the plane of polarization. Such a wave is called a plane polarized wave. The light emitted by an incandescent lamp or a fluorescent tube is not plane polarized, since the waves emitted by the source, although plane polarized, are randomly oriented. Such a wave is called unpolarized or natural light. We can obtain polarized light from unpolarized light. This will be discussed in Section 8.4.

8.1 SUPERPOSITION OF TWO-PLANE POLARIZED WAVES

Let us consider two orthogonally polarized plane waves propagating in the z direction: in one wave, the **E** vector is vibrating along the y direction and the other along the x direction. These waves are described as follows:

$$E_y(z;t) = E_{0y}\cos\left(\omega t - kz + \delta_y\right) \tag{8.1}$$

$$E_x(z;t) = E_{0x}\cos(\omega t - kz + \delta_x) \tag{8.2}$$

where δ_y and δ_x are the phases of the waves. These waves satisfy the wave equation. Owing to the superposition principle, the sum of the waves will also satisfy the wave equation. In general, a wave will have both x and y components and can be written as

$$\mathbf{E}(z;t) = \mathbf{i}E_x(z;t) + \mathbf{j}E_y(z;t) \tag{8.3}$$

We wish to find out what track the tip of the electric vector traces when the waves are superposed. To do this, we introduce a variable $\tau = \omega t - kz$ and express the plane waves as

$$E_y(z;t) = E_{0y}\left(\cos\tau\cos\delta_y - \sin\tau\sin\delta_y\right)$$
$$E_x(z;t) = E_{0x}\left(\cos\tau\cos\delta_x - \sin\tau\sin\delta_x\right)$$

From these equations, we obtain

$$\frac{E_y}{E_{0y}}\sin\delta_x - \frac{E_x}{E_{0x}}\sin\delta_y = \cos\tau\sin\left(\delta_x - \delta_y\right) \tag{8.4a}$$

$$\frac{E_y}{E_{0y}}\cos\delta_x - \frac{E_x}{E_{0x}}\cos\delta_y = \sin\tau\sin\left(\delta_x - \delta_y\right) \tag{8.4b}$$

Squaring both the left-hand and right-hand sides of Equations 8.4a,b and summing them, we obtain

$$\left(\frac{E_y}{E_{0y}}\right)^2 + \left(\frac{E_x}{E_{0x}}\right)^2 - 2\frac{E_y}{E_{0y}}\frac{E_x}{E_{0x}}\cos\delta = \sin^2\delta \tag{8.5}$$

where $\delta = \delta_x - \delta_y$. This equation represents an ellipse. The ellipse is inscribed in a rectangle of sides $2E_{0y}$ and $2E_{0x}$ that are parallel to the y and x axes, respectively. Hence, in the general case of the propagation of a monochromatic wave, the tip of its electric vector traces out an ellipse in any z plane. Such a wave is called elliptically polarized. Since it is a propagating wave, the tip of the **E** vector traces out a spiral. The tip of the **E** vector can rotate either clockwise or anticlockwise in the plane. These cases are termed right-hand polarization (the **E** vector rotates clockwise when facing the source of light) and left-hand polarization (the **E** vector rotates anticlockwise when facing the source), respectively.

It can be shown that the direction of rotation is governed by the sign of the phase difference δ. Let us consider a moment of time t_0 when $\omega t_0 - kz + \delta_y = 0$. At this moment, $E_y = E_{0y}$ and $E_x = E_{x0}\cos(\delta_x - \delta_y) = E_{x0}\cos\delta$ and $dE_x/dt = -\omega E_{x0}\sin\delta$. The rate of change of E_x, dE_x/dt, is negative when $0 < \delta < \pi$ and positive when $\pi < \delta < 2\pi$. Obviously, the former case corresponds to the right-hand polarized wave and the latter to the left-hand polarized wave.

There are two special cases of elliptical polarization, which are known as plane or linear polarization and circular polarization. Various polarization states of light are shown in Figure 8.1.

8.2 LINEAR POLARIZATION

When the phase difference between the two waves is a multiple of π, we obtain a linearly polarized wave. When both waves are in phase (i.e., $\delta = 2m\pi$, with $m = 0, 1, 2$), the **E** vector traces a line in the first and third quadrants. When they are in antiphase (i.e., $\delta = (2m+1)\pi$), the **E** vector traces a line in the second and fourth quadrants.

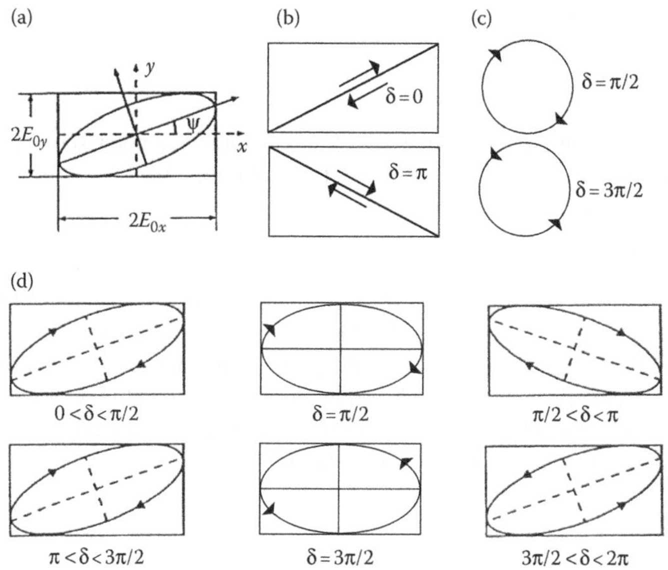

FIGURE 8.1 Polarization states of a sinusoidal electromagnetic wave for different values of δ: (a) elliptical; (b) linear; (c) circular; (d) elliptical.

8.3 CIRCULAR POLARIZATION

When the phase difference between the two waves is an odd multiple of $\pi/2$ [i.e., $\delta = (2m + 1)\pi/2$], the wave is elliptically polarized, but its major and minor axes are now aligned parallel to the x and y axes. However, if the amplitudes of the two waves are equal, then it becomes a circularly polarized wave. It is right-hand circularly polarized if the phase difference is $\pi/2, 5\pi/2, 9\pi/2, \ldots$ and left-hand circularly polarized when $\delta = 3\pi/2, 7\pi/2, 11\pi/2, \ldots$. Therefore, two conditions must be met to obtain circularly polarized light: the phase difference must be an odd multiple of $\pi/2$ and the amplitudes of the waves must be equal.

8.4 PRODUCTION OF POLARIZED LIGHT

Natural light is unpolarized, with the **E** vector taking all possible orientations randomly. We can, however, resolve the **E**-vector orientations into two components: one oscillating in the plane of incidence (p-polarized) and the other orthogonal to this (s-polarized). The amplitudes of these components at any instant are equal. It can also be seen that, in general, it is possible to obtain elliptically polarized wave by introducing a phase difference between the orthogonally polarized components. However, most often a linearly polarized wave is required. Fortunately, there are a number of ways to obtain a linearly polarized wave from natural light. These methods are based on

1. Reflection at a dielectric interface
2. Refraction at a dielectric interface

3. Double refraction
4. Dichroism
5. Scattering

Of these, the first four methods are used for making practical polarizers—the devices that produce linearly polarized light wave from natural light.

8.4.1 Reflection

When a beam of light is incident on the air–dielectric interface at an angle of incidence θ_B, called the Brewster angle, the reflected beam is linearly polarized: the **E** vector oscillates in the plane perpendicular to the plane of incidence. This is also called a p-polarized beam. The transmitted beam is partially polarized. The Brewster angle θ_B is governed by the refractive index n of the dielectric medium. For reflection at an air-dielectric interface, it is given by the relation

$$\tan\theta_B = n$$

It may be noted that reflection at θ_B at an air–dielectric interface fixes the direction of vibration of the electric vector in the reflected beam, and hence is used for calibration of polarizers, among other things.

8.4.2 Refraction

As mentioned earlier, when the light is incident at the angle θ_B, the reflected beam is p-polarized and the transmitted beam is partially polarized. The transmitted beam has less of a p-component, since some of this has been removed by reflection at the angle θ_B. However, if several successive reflections are allowed at a number of plane parallel plates aligned at θ_B, most of the p-polarized light will be removed by reflection, and the transmitted beam is then s-polarized. Such a polarizer is known as a pile-of-plates polarizer. This polarizer is often used with high-power lasers.

8.4.3 Double Refraction

There is a class of crystals in which an incident beam is decomposed into two linearly orthogonally polarized beams inside the crystal. The structure of the crystal supports two orthogonally polarized beams. These are anisotropic crystals. In such crystals, there is a direction along which there is no decomposition. This is known as the optic axis. Some crystals have only one optic axis, and are called uniaxial crystals; others have two optic axes, and are called biaxial crystals. From the point of view of polarizers or other polarization components, uniaxial crystals are of importance, and hence we will discuss them in more detail. Two well-known examples of uniaxial crystals are calcite and quartz.

Let us consider a plate of a uniaxial crystal on which a beam of light is obliquely incident. This beam is decomposed into two orthogonally polarized beams inside the plate. One beam obeys Snell's law of refraction, and is called the ordinary beam

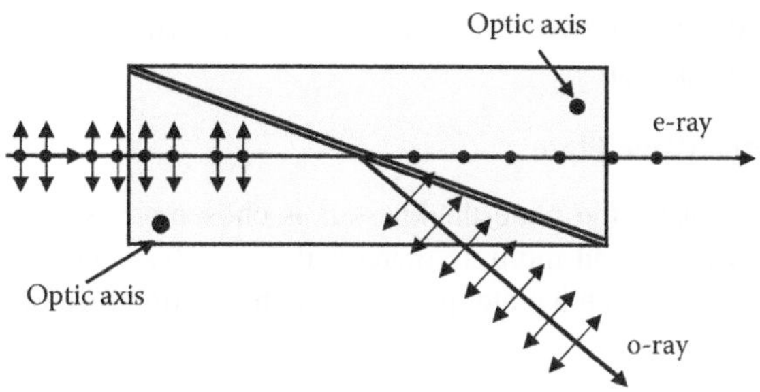

FIGURE 8.2 A Glan–Thompson polarizer.

(o-beam); the other beam does not obey this law, and is called the extra-ordinary beam (e-beam). The refractive index n_o of the o-beam is independent of the direction of propagation, whereas that of the e-beam, n_e, varies with this direction, taking extreme values in a direction orthogonal to the optic axis. If $n_o > n_e$, the crystal is a negative uniaxial crystal: calcite is one such crystal ($n_o = 1.658$ and $n_e = 1.486$ for the yellow sodium wavelength). The quartz crystal is a positive uniaxial crystal, since $n_e > n_o$ ($n_o = 1.544$ and $n_e = 1.553$). The optic axis is a slow axis in calcite, and the axis orthogonal to this is a fast axis. The **E** vector in the o-beam oscillates in a plane that is perpendicular to the principal section of the crystal. The principal section contains the optic axis and the direction of propagation. The **E** vector of the e-beam lies in the principal plane.

Since there are two linearly polarized beams inside the crystal, it is easy to obtain a linearly polarized beam by eliminating one of these beams. Fortunately, owing to the angle-dependent refractive index of the e-beam and the availability of media of refractive index intermediate to n_o and n_e, it is possible to remove the o-beam by total internal reflection in a calcite crystal. One of the early devices based on this principle is the Nicol prism. Its more versatile companion is a Glan–Thompson prism, which is shown in Figure 8.2. It has two halves, cemented by Canada balsam. When an unpolarized beam is incident on the polarizer, the outgoing light beam is linearly polarized. However, a linearly polarized beam will be completely blocked if the transmission axis of the polarizer is orthogonal to the beam. Polarizers obtained from anisotropic crystals are generally small, but have very high extinction ratios. Such polarizers are not generally used for photo-elastic work, which requires large polarizers, since photo-elastic models are usually moderately large.

8.4.3.1 Phase Plates

Besides obtaining polarizers from these crystals, we can also obtain phase plates. These produce a fixed but wavelength-dependent phase difference between the two components. Let us consider a plane parallel plate of a uniaxial crystal with the optic axis lying in the surface of the plate. A linearly polarized beam is incident normally on this plate. This beam is decomposed into two beams, which propagate with different velocities along the slow and fast axes of the plate. Let the refractive indices along these axes be n_o and n_e, respectively. A plate of thickness d will introduce a path difference $|(n_o - n_e)|d$ between the two waves. Therefore, any required path difference can be introduced between two waves by an appropriate choice of plate thickness of a given anisotropic crystal.

8.4.3.2 Quarter-Wave Plate

In a quarter-wave plate, the plate thickness d is chosen so as to introduce a path difference of $\lambda/4$ or an odd multiple thereof, that is, $(2m+1)\lambda/4$, where m is an integer. In other words, such a plate introduces a phase difference of a quarter-wave. Therefore,

$$d = \frac{2m+1}{|n_o - n_e|}\frac{\lambda}{4}$$

For a half-wave plate, the path difference introduced is $\lambda/2$ or $(2m+1)\lambda/2$.

A quarter-wave plate is used to convert a linearly polarized beam into a circularly polarized beam. It is oriented such that the **E** vector of the incident beam makes an angle of 45° with either the fast or the slow axis of the quarter-wave plate. The components in the plate are then of equal amplitude, and the plate introduces a path difference of $\lambda/4$ between these components. The outgoing beam is thus circularly polarized. The handedness of circular polarization can be changed by rotating the plate by 90° about the axis of the optical beam.

8.4.3.3 Half-Wave Plate

A half-wave plate, on the other hand, rotates the plane of a linearly polarized beam. For example, if a beam of linearly polarized light with its azimuth 45° is incident on a half-wave plate, its azimuth is rotated by 90°. In other words, the beam emerges still linearly polarized, but the orientation of the **E** vector is rotated by 90°.

8.4.3.4 Compensators

Phase plates are devices that introduce fixed phase differences. In some applications, it is necessary either to introduce a path difference that could be varied or to compensate for a path difference. This is achieved by compensators. There are two well-known compensators: the Babinet and the Soleil–Babinet compensators. The Babinet compensator consists of two wedge plates with their optic axes orthogonal to each other, as shown in Figure 8.3a. The role of the o- and e-beams changes when the beams pass from one wedge to the other. The path difference introduced by the compensator is given by $|n_o - n_e|[d_2(y) - d_1(y)]$, where $d_2(y)$ and $d_1(y)$ are the thicknesses of the two wedge plates at any position $(0, y)$. Obviously, the path difference varies along the y direction on the wedge plate.

If a constant path difference between the two beams is required, the Soleil–Babinet compensator (Figure 8.3b) is used. This again consists of two elements: one is a plate and the other is a combination of two identical wedge plates forming a plane parallel plate. The optic axes in the plate and the wedge combination are orthogonal. The thickness of the plate formed as a result of the wedge combination is varied by sliding one wedge over the other. Thus, the thickness difference $d_2 - d_1$ remains constant over the whole surface, where d_2 and d_1 are the thicknesses of the plate and the wedge combination, respectively.

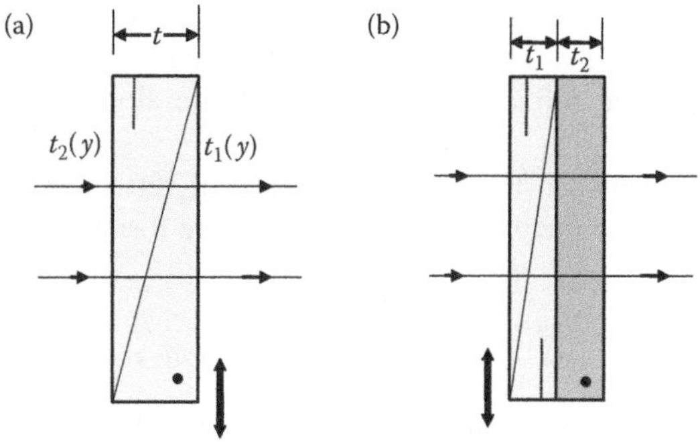

FIGURE 8.3 (a) Babinet compensator. (b) Soleil–Babinet compensator.

8.4.4 Dichroism

There are anisotropic crystals that are characterized by different absorption coefficients with respect to o- and e-beams. For example, a tourmaline crystal strongly absorbs an o-beam. Therefore, we can obtain an e-polarized beam when a beam of natural light passes through a sufficiently thick plate of this crystal. Very large polarizers based on selective absorption are available as sheets, and are known as sheet polarizers or Polaroids. These are the ones often used in photo-elastic work.

8.4.5 Scattering

Light scattered by particles is partially polarized. However, polarizers based on scattering are not used in practice.

8.5 MALUS'S LAW

Consider a linearly polarized light beam incident on a polarizer. The **E** vector of the beam makes an angle θ with the transmission axis of the polarizer. The beam is resolved into two components, one parallel to the transmission axis and the other perpendicular to it. The component perpendicular to the transmission axis is blocked. Therefore, the amplitude of the light transmitted by the polarizer is $E(\theta) = E_0 \cos\theta$. Hence, the intensity of the transmitted light is given by $I(\theta) = I_0 \cos^2\theta$, where I_0 is the intensity of the incident beam. This is a statement of Malus's law. It can be seen that a polarizer could also be used as an attenuator in the beam.

8.6 THE STRESS-OPTIC LAW

The phenomenon of double refraction or optical anisotropy may also occur in certain isotropic materials, such as glass and plastics, when subjected to stress or strain. This condition is temporary, and disappears when the stress is removed. This phenomenon was first observed by Brewster, and forms the basis of photoelasticity. In photoelasticity, models of objects are cast or fabricated from isotropic materials, and are then subjected to stress. The stress produces physical deformations that completely alter the initial isotropic character of the material. We can then characterize the material with three principal refractive indices, which are along the principal axes of the stress.

The relationship between the principal indices of refraction n_i of a temporary birefringent material and the principal stresses σ_i were formulated by Maxwell, and are given by

$$n_1 - n_0 = C_1\sigma_1 + C_2(\sigma_2 + \sigma_3) \tag{8.6a}$$

$$n_2 - n_0 = C_1\sigma_2 + C_2(\sigma_3 + \sigma_1) \tag{8.6b}$$

$$n_3 - n_0 = C_1\sigma_3 + C_2(\sigma_1 + \sigma_2) \tag{8.6c}$$

where n_0 is the refractive index of the unstressed (isotropic) material and C_1, C_2 are constants depending on the material.

For materials under general triaxial stress, the stress-optic law is expressed as

$$n_1 - n_2 = C(\sigma_1 - \sigma_2) \tag{8.7a}$$

$$n_2 - n_3 = C(\sigma_2 - \sigma_3) \tag{8.7b}$$

$$n_1 - n_3 = C(\sigma_1 - \sigma_3) \tag{8.7c}$$

where $C = C_1 - C_2$ is the stress-optic coefficient of the photo-elastic material.

Let us now consider a plate of isotropic material. This could be subjected to either (a) a uniaxial state of stress or (b) a biaxial state of stress. In the first case, $\sigma_2 = \sigma_3 = 0$, and hence $n_2 = n_3$. The stress-optic law takes the very simple form

$$n_1 - n_2 = C\sigma_1 \tag{8.8}$$

The plate behaves like a uniaxial crystal. When the plate is subjected to a biaxial state of stress (i.e., $\sigma_3 = 0$), the stress-optic law takes the form

$$n_1 - n_2 = C(\sigma_1 - \sigma_2) \tag{8.9a}$$

$$n_2 - n_3 = C\sigma_2 \tag{8.9b}$$

$$n_1 - n_3 = C\sigma_1 \tag{8.9c}$$

The plate behaves like a biaxial crystal.

Now let us assume that a beam of linearly polarized light of wavelength λ is incident normally on a plate of photo-elastic material of thickness d. Within the plate, there are two linearly polarized beams, one vibrating in the (x, z) plane and the other in the (y, z) plane. While traversing the plate, these two waves acquire a phase difference, the value of which at the exit surface is given by

$$\delta = \frac{2\pi}{\lambda}|n_1 - n_2|\,d = \frac{2\pi C}{\lambda}(\sigma_1 - \sigma_2)\,d \tag{8.10}$$

The phase change δ depends linearly on the difference of the principal stresses, on the thickness of the plate, and inversely on the wavelength of light used. If the beam strikes the plate at an angle θ, the phase difference is given by

$$\delta = \frac{2\pi C}{\lambda}\left(\sigma_1 - \sigma_2 \cos^2\theta\right) d \sec\theta \tag{8.11}$$

In photo-elastic practice, it is more convenient to write Equation 8.10 in the form

$$\sigma_1 - \sigma_2 = \frac{m f_\sigma}{d} \tag{8.12}$$

where $m = \delta/2\pi$ is the fringe order and $f_\sigma = \lambda/C$ is the material fringe value for a given wavelength of light. This relationship is known as the stress-optic law.

The principal stress difference $\sigma_1 - \sigma_2$, in a two-dimensional model, can be determined by measuring the fringe order m, if the material fringe value f_σ of the material is known or obtained by calibration. The fringe order at each point in the photo-elastic model can be measured by observing the model in a polariscope.

At this juncture, it should be mentioned that a plate of thickness d and refractive index n_0 introduces a phase delay of $k(n_0 - 1)d$; $k = 2\pi/\lambda$. When the plate is stressed, the linearly polarized components travel with different speeds and acquire phase delays $k(n_1 - 1)d_1$ and $k(n_2 - 1)d_1$, where d_1 is the thickness of the stressed plate and is related to the thickness d of the unstressed plate by

$$d_1 = d\left\lfloor 1 - \tfrac{\nu}{E}(\sigma_1 + \sigma_2)\right\rfloor \tag{8.13}$$

where E and ν are the Young's modulus and Poisson's ratio of the material. This change in thickness, $d_1 - d$, is very important in interferometry and also in holophotoelasticity.

8.7 THE STRAIN-OPTIC LAW

The stress-strain relationships for a material exhibiting perfectly linear elastic behavior under a two-dimensional state of stress are:

$$\varepsilon_1 = \frac{1}{E}(\sigma_1 - \nu\sigma_2) \tag{8.14a}$$

$$\varepsilon_2 = \frac{1}{E}(\sigma_2 - \nu\sigma_1) \tag{8.14b}$$

From Equations 8.14a,b, the difference between the principal stresses is

$$\sigma_1 - \sigma_2 = \frac{E}{1+\nu}(\varepsilon_1 - \varepsilon_2) \tag{8.15}$$

Substituting this into the stress-optic law, we obtain

$$\varepsilon_1 - \varepsilon_2 = \frac{m f_\varepsilon}{d} \tag{8.16}$$

where $f_\varepsilon = f_\sigma(1 + \nu)/E$ is the material fringe value in terms of strain. The relationship given in Equation 8.16 is known as the strain-optic law in photoelasticity.

8.8 METHODS OF ANALYSIS

The optical system most often used for stress analysis is a polariscope. It takes a variety of forms, depending on the end use. However, in general, a polariscope consists of a light source, a device to produce polarized light called a polarizer, a model, and a second polarizer called an analyzer. In addition, it may contain a set of lenses, quarter-wave plates, and photographic or recording equipment. We will discuss the optical systems of plane polariscopes and circular polariscopes.

8.8.1 Plane Polariscope

The plane polariscope consists of a light source, a light filter, collimating optics to provide a collimated beam, a polarizer, an analyzer, a lens and photographic equipment as shown in Figure 8.4. The model is placed between the polarizer and the analyzer. The polarizer and the analyzer are crossed, thus producing a dark field.

Let the transmission axis of the polarizer be along the y direction. The amplitude of the wave just behind the polarizer is given by

$$E_y(z;t) = E_{0y}\cos(\omega t - kz)$$

where $k = 2\pi/\lambda$. The field incident on the model is also given by this expression, except that z refers to the plane of the model. Let us also assume that one of the principal stress directions makes an angle α with the transmission direction of the polarizer (i.e., the y axis). The incident field just at the entrance face of the model splits into two components that are orthogonally polarized and vibrate in the planes of σ_1 and σ_2. The amplitudes of these components are $E_{0y}\cos\alpha$ and $E_{0y}\sin\alpha$, respectively.

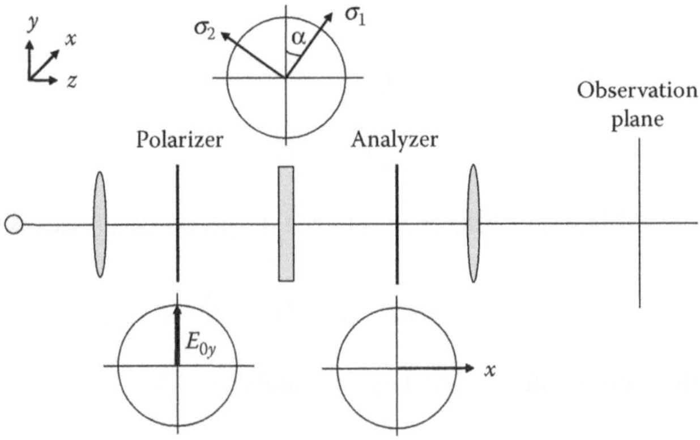

FIGURE 8.4 Schematic of a plane polariscope.

The amplitudes of these components at the exit face of the model are

$$E_1(z;t) = E_{0y} \cos\alpha \cos[\omega t - kz - k(n_1 - 1)d] = E_{0y} \cos\alpha \cos(\omega t - kz + \delta_y) \tag{8.17a}$$

$$E_2(z;t) = E_{0y} \sin\alpha \cos[\omega t - kz - k(n_2 - 1)d] = E_{0y} \sin\alpha \cos(\omega t - kz + \delta_y - \delta) \tag{8.17b}$$

The model introduces a phase difference

$$\delta = \frac{2\pi}{\lambda}(n_1 - n_2)d = \frac{2\pi}{f_\sigma}(\sigma_1 - \sigma_2)d$$

between these components. The analyzer resolves these components further into components along and perpendicular to its direction of transmission, which is along the x direction. The components along the direction of transmission are allowed through, and produce a photo-elastic pattern, while those in the orthogonal direction are blocked. The net transmitted amplitude is

$$\begin{aligned} E_1 \sin\alpha - E_2 \cos\alpha &= \frac{E_{0y}}{2} \sin 2\alpha\left[\cos(\omega t - kz + \delta_y) - \cos(\omega t - kz + \delta_y - \delta)\right] \\ &= E_{0y} \sin 2\alpha \sin\left(\frac{\delta}{2}\right) \sin\left(\omega t - kz + \delta_y - \frac{\delta}{2}\right) \end{aligned} \tag{8.18}$$

This also represents a wave of amplitude $E_{0y} \sin 2\alpha \sin(\delta/2)$ propagating along the z direction. The intensity of this wave is therefore

$$I = E_{0y}^2 \sin^2 2\alpha \sin^2\left(\frac{\delta}{2}\right) = I_0 \sin^2 2\alpha \sin^2\left[\frac{\pi(\sigma_1 - \sigma_2)d}{f_\sigma}\right] \tag{8.19}$$

The intensity of the transmitted beam is governed by α, the orientation of the principal stress direction with respect to the polarizer's transmission axis, and the phase retardation δ. The transmitted intensity will be zero when $\sin 2\alpha \sin(\delta/2) = 0$. In other words, the transmitted intensity is zero when either $\sin 2\alpha = 0$ or $\sin(\delta/2) = 0$. When $\sin 2\alpha = 0$, the angle $\alpha = 0$ or $\pi/2$. In either case, one of the principal stress directions is aligned with the polarizer's transmission axis. Therefore, these dark fringes give the directions of the principal stresses at any point on the model, and are known as isoclinics or isoclinic fringes.

When $\sin(\delta/2) = 0,\ \delta = 2m\pi$, or

$$\sigma_1 - \sigma_2 = \frac{mf_\sigma}{d} \tag{8.20}$$

The transmitted intensity is zero when $\sigma_1 - \sigma_2$ is an integral multiple of f_σ/d. Therefore, the fringes are loci of constant $\sigma_1 - \sigma_2$, and are referred to as isochromatics. The adjacent isochromatics differ by f_σ/d. When white light is used for illumination of the model, these fringes are colored; each color corresponds to a constant value of $\sigma_1 - \sigma_2$, hence the name isochromatics.

It can be seen that both the isoclinics and the isochromatics appear simultaneously in a plane polariscope. It is desirable to separate these fringe patterns. A circular polariscope performs this function and provides only isochromatics.

The isoclinics are the loci of points at which the directions of the principal stresses are parallel to the transmission axes of the polarizer and the analyzer. The isoclinic pattern is independent of the magnitude of the load applied to the model and the material fringe value. When white light is used for illumination, the isoclinics appear dark in contrast to the isochromatics, which, with the exception of the zero-order fringe, are colored. In regions where the directions of the principal stresses do not vary greatly from point to point, the isoclinics appear as wide diffuse bands. The isoclinics do not intersect each other except at an isotropic point, which is a point where the principal stresses are equal in magnitude and sign, that is, $\sigma_1 - \sigma_2 = 0$. Further, at a point on a shear-free boundary where the stress parallel to the boundary has a maximum or a minimum value, the isoclinic intersects the boundary orthogonally.

8.8.2 Circular Polariscope

It can be seen that the isoclinics appear because a linearly polarized light wave is incident on the model. They will disappear if the light incident on the model is circularly polarized. Therefore, a circular polarizer, which is a combination of a linear polarizer and a quarter-wave plate at 45° azimuth, is required. Further, to analyze this light, we also need a circular analyzer. Therefore, a circular polariscope consists of a light source, collimating optics, a polarizer, two quarter-wave plates, an analyzer, and recording optics, as shown in Figure 8.5.

The model is placed between the two quarter-wave plates. Since these plates are designed with slow and fast axes, they can be arranged in two ways, namely with axes parallel or with axes crossed. Similarly, the transmission axes of the polarizer and the analyzer can be parallel or crossed. Therefore, there are four ways of assembling a

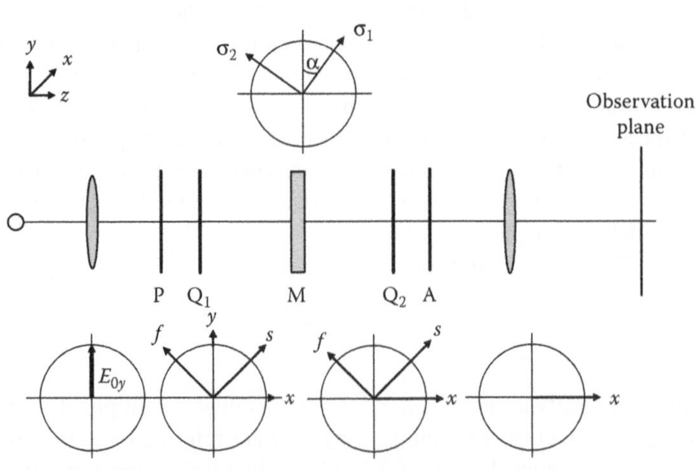

FIGURE 8.5 Schematic of a circular polariscope: P, polarizer; Q_1, Q_2, quarter-wave plates; M, model; A, analyzer; for other symbols, see text.

TABLE 8.1
Four Configurations of a Circular Polariscope

Configuration	Polarizer and Analyzer Axes	Quarter-Wave Plate Axes	Field
1	Parallel	Parallel	Dark
2	Parallel	Crossed	Dark
3	Crossed	Parallel	Bright
4	Crossed	Crossed	Bright

circular polariscope: two of these configurations give a dark field and the remaining two a bright field at the output, as shown in Table 8.1.

We now consider a configuration that has the analyzer and the polarizer crossed and the quarter-wave plates in parallel, resulting in a bright-field output.

Let the transmission axis of the polarizer be along the y direction. The field transmitted by the polarizer is given by

$$E_y(z;t) = E_{0y}\cos(\omega t - kz)$$

This field is split into two components, which propagate along the fast and slow axes of the quarter-wave plate. The field at the exit face of the first quarter-wave plate is

$$E_1(z;t) = \frac{E_{0y}}{\sqrt{2}}\cos\left[\omega t - kz - k(n' - 1)d'\right] = \frac{E_{0y}}{\sqrt{2}}\cos(\omega t - kz + \psi_1) \quad (8.21a)$$

$$E_2(z;t) = \frac{E_{0y}}{\sqrt{2}}\cos\left[\omega t - kz - k(n'' - 1)d'\right]$$

$$= \frac{E_{0y}}{\sqrt{2}}\cos\left(\omega t - kz + \psi_1 - \frac{\pi}{2}\right) = \frac{E_{0y}}{\sqrt{2}}\sin(\omega t - kz + \psi_1) \quad (8.21b)$$

where the phase difference $\pi/2$, introduced by the quarter-wave plate of thickness d', is expressed as

$$\frac{2\pi}{\lambda}(n'' - n')d' = \frac{\pi}{2}$$

and

$$\psi_1 = -\frac{2\pi}{\lambda}(n' - 1)d'$$

This indicates that n' corresponds to the fast axis of the quarter-wave plate. This field is now incident on the model, and hence becomes further decomposed along the directions of the principal stresses. We assume that one of the principal stress directions makes an angle α with the polarizer's transmission axis (i.e., with the

y axis). The field at the entrance face of the model as decomposed along the σ_1 and σ_2 directions is now given by

$$E_{\sigma_1}(z;t) = \frac{E_{0y}}{\sqrt{2}}\cos\left(\frac{\pi}{4}-\alpha\right)\cos(\omega t - kz + \psi_1)$$
$$+\frac{E_{0y}}{\sqrt{2}}\sin\left(\frac{\pi}{4}-\alpha\right)\sin(\omega t - kz + \psi_1)$$
$$= \frac{E_{0y}}{\sqrt{2}}\cos\left(\omega t - kz + \psi_1 - \frac{\pi}{4} + \alpha\right) = \frac{E_{0y}}{\sqrt{2}}\cos\tau \quad (8.22a)$$

$$E_{\sigma_2}(z;t) = -\frac{E_{0y}}{\sqrt{2}}\sin\left(\frac{\pi}{4}-\alpha\right)\cos(\omega t - kz + \psi_1)$$
$$+\frac{E_{0y}}{\sqrt{2}}\cos\left(\frac{\pi}{4}-\alpha\right)\sin(\omega t - kz + \psi_1)$$
$$= \frac{E_{0y}}{\sqrt{2}}\sin\left(\omega t - kz + \psi_1 - \frac{\pi}{4} + \alpha\right) = \frac{E_{0y}}{\sqrt{2}}\sin\tau \quad (8.22b)$$

The field amplitudes at the exit face of the model are given by

$$E_{\sigma_1}(z;t) = \frac{E_{0y}}{\sqrt{2}}\cos(\tau + \psi_2) \quad (8.23a)$$

$$E_{\sigma_2}(z;t) = \frac{E_{0y}}{\sqrt{2}}\sin(\tau + \psi_2 + \delta) \quad (8.23b)$$

where

$$\psi_2 = -\frac{2\pi}{\lambda}(n_1 - 1)d$$

$$\delta = \frac{2\pi}{\lambda}(n_1 - n_2)d$$

We now decompose these fields along the axes of the second quarter-wave plate, which are inclined at 45° and −45° to the y axis. These are given by

$$E_1'(z;t) = \frac{E_{0y}}{\sqrt{2}}\cos\left(\frac{\pi}{4}-\alpha\right)\cos(\tau+\psi_2) - \frac{E_{0y}}{\sqrt{2}}\sin\left(\frac{\pi}{4}-\alpha\right)\sin(\tau+\psi_2+\delta) \quad (8.24a)$$

$$E_2'(z;t) = \frac{E_{0y}}{\sqrt{2}}\sin\left(\frac{\pi}{4}-\alpha\right)\cos(\tau+\psi_2) + \frac{E_{0y}}{\sqrt{2}}\cos\left(\frac{\pi}{4}-\alpha\right)\sin(\tau+\psi_2+\delta) \quad (8.24b)$$

We now assume that both quarter-wave plates are aligned in parallel; that is, their fast and slow axes are parallel. The field at the exit face of the second plate is

$$E_1'(z;t) = \frac{E_{0y}}{\sqrt{2}}\cos\left(\frac{\pi}{4}-\alpha\right)\cos(\tau+\psi_2+\psi_1) - \frac{E_{0y}}{\sqrt{2}}\sin\left(\frac{\pi}{4}-\alpha\right)\sin(\tau+\psi_2+\delta+\psi_1) \quad (8.25a)$$

$$\begin{aligned} E_2'(z;t) &= \frac{E_{0y}}{\sqrt{2}}\sin\left(\frac{\pi}{4}-\alpha\right)\cos\left(\tau+\psi_2+\psi_1-\frac{\pi}{2}\right) + \frac{E_{0y}}{\sqrt{2}}\cos\left(\frac{\pi}{4}-\alpha\right)\sin\left(\tau+\psi_2+\delta+\psi_1-\frac{\pi}{2}\right) \\ &= \frac{E_{0y}}{\sqrt{2}}\sin\left(\frac{\pi}{4}-\alpha\right)\sin(\tau+\psi_2+\psi_1) - \frac{E_{0y}}{\sqrt{2}}\cos\left(\frac{\pi}{4}-\alpha\right)\cos(\tau+\psi_2+\delta+\psi_1) \end{aligned} \quad (8.25b)$$

Since the analyzer is crossed, it takes the components along the x direction. Therefore, the amplitude of the transmitted wave is given by

$$\begin{aligned} E_1'(z;t) = \frac{E_{0y}}{2}\Big[&\cos\left(\frac{\pi}{4}-\alpha\right)\cos(\tau+\psi_2+\psi_1) \\ &- \sin\left(\frac{\pi}{4}-\alpha\right)\sin(\tau+\psi_2+\delta+\psi_1) - \sin\left(\frac{\pi}{4}-\alpha\right)\sin(\tau+\psi_2+\psi_1) \\ &+ \cos\left(\frac{\pi}{4}-\alpha\right)\cos(\tau+\psi_2+\delta+\psi_1)\Big] \end{aligned} \quad (8.26)$$

Equation 8.26 can be rewritten as

$$\begin{aligned} E'(z;t) &= \frac{E_{0y}}{2}\left[\cos\left(\tau+\psi_2+\psi_1+\frac{\pi}{4}-\alpha\right)+\cos\left(\tau+\psi_2+\delta+\psi_1+\frac{\pi}{4}-\alpha\right)\right] \\ &= E_{0y}\cos\left(\frac{\delta}{2}\right)\cos\left(\tau+\psi_2+\frac{\delta}{2}+\psi_1+\frac{\pi}{4}-\alpha\right) \end{aligned} \quad (8.27)$$

Again, this represents a wave with an amplitude $E_{0y}\cos(\delta/2)$. Therefore, the transmitted intensity is given by

$$I = I_0\cos^2(\delta/2) \quad (8.28)$$

When there is no stress distribution, $\delta = 0$, and the transmitted intensity is maximum and uniform. This therefore represents a bright-field configuration. It can be seen that when the quarter-wave plates are crossed, the field exiting from the second plate is

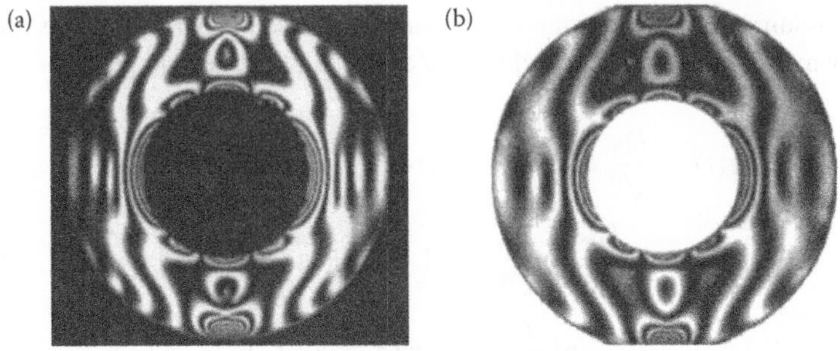

FIGURE 8.6 Isochromatics in (a) a dark field and (b) a bright field.

given by

$$E_1'(z;t) = \frac{E_{0y}}{\sqrt{2}}\cos\left(\frac{\pi}{4}-\alpha\right)\cos\left(\tau+\psi_2+\psi_1-\frac{\pi}{2}\right) - \frac{E_{0y}}{\sqrt{2}}\sin\left(\frac{\pi}{4}-\alpha\right)\sin\left(\tau+\psi_2+\delta+\psi_1-\frac{\pi}{2}\right) \quad (8.29a)$$

$$E_2'(z;t) = \frac{E_{0y}}{\sqrt{2}}\sin\left(\frac{\pi}{4}-\alpha\right)\cos(\tau+\psi_2+\psi_1) + \frac{E_{0y}}{\sqrt{2}}\cos\left(\frac{\pi}{4}-\alpha\right)\sin(\tau+\psi_2+\delta+\psi_1) \quad (8.29b)$$

The intensity transmitted by the polarizer is now given by

$$I = I_0\sin^2(\delta/2) \quad (8.30)$$

This represents a dark field, since the intensity is zero when there is no stress distribution on the model. Equations 8.28 and 8.30 show that the intensity of light emerging from the analyzer in a circular polariscope is a function of the difference of principal stresses $\sigma_1 - \sigma_2$ only. The isoclinics have been eliminated.

In a dark-field configuration, the dark fringes occur wherever $\delta = 2m\pi$ ($m = 0, 1, 2, \ldots$), and they correspond to the integer isochromatic fringe order $m = 0, 1, 2, 3, \ldots$, respectively. An example of this fringe pattern is shown in Figure 8.6a. However, for a bright-field configuration, the dark fringes are obtained when $\delta = (2m+1)\pi$. These corresponds to isochromatic fringes of half order, that is, $m = \frac{1}{2}, \frac{3}{2}, \frac{5}{2}, \ldots$. An example of a light-field fringe pattern is shown in Figure 8.6b.

8.9 EVALUATION PROCEDURE

Directions of principal stresses at any point in the model are determined using a plane polariscope. The polarizer and analyzer are rotated about the optical axis until the isoclinic passes through the point of interest. The inclination of the transmission axis gives the principal stress direction. The principal stress difference $\sigma_1 - \sigma_2$ in the dark field is given by

$$\sigma_1 - \sigma_2 = m\frac{f_\sigma}{d} \tag{8.31a}$$

and that in the bright field by

$$\sigma_1 - \sigma_2 = \left(m + \frac{1}{2}\right)\frac{f_\sigma}{d} \tag{8.31b}$$

Therefore, the material fringe value f_σ must be known before $\sigma_1 - \sigma_2$ can be calculated. f_σ is obtained by calibration. A circular disk of the same photo-elastic material and thickness is used as a model, and is diametrically loaded. The fringe order in the center of the disk is measured, and f_σ is calculated using the formula

$$f_\sigma = \frac{8Fd}{n\pi D} \tag{8.32}$$

where F is the applied force, D is the diameter of the disk, and n is the measured fringe order at the center of the disk. There are other calibration methods that use tensile loading or bending.

The principal stress difference $\sigma_1 - \sigma_2$ in an arbitrary model is then found by using this value of the material fringe value and the order m of the isochromatics. However, the order m is to be counted from $m = 0$, which is normally not known. It can easily be found, if it exists in the model, by using white-light illumination, since $m = 0$ is an achromatic fringe whereas the higher-order fringes are colored. Therefore, a polariscope is usually equipped with both white-light and monochromatic sources: the white-light source for locating the zero-order isochromatic and the monochromatic source for counting the higher-order isochromatics. If the principal stress difference is desired at a point where neither a bright isochromatic nor a dark isochromatic passes, some method of measuring fractional fringe order must be implemented. There are methods to measure fractional fringe orders, which are discussed in the next section.

8.10 MEASUREMENT OF FRACTIONAL FRINGE ORDER

The methods described here assume that the directions of the principal stresses are known. In one method, a Babinet or a Soleil–Babinet compensator is used. The principal axes of the compensator are aligned along the directions of the principal stresses. An additional phase difference can then be introduced to shift the dark isochromatics to the point of interest, and this additional phase shift is read from the compensator. Another method makes use of a quarter-wave plate for compensation, and is known as Tardy's method.

8.10.1 Tardy's Method

This makes use of a plane polariscope in dark-field configuration, where the intensity distribution is given by $I = I_0 \sin^2 2\alpha \sin^2(\delta/2)$. The observation field contains both

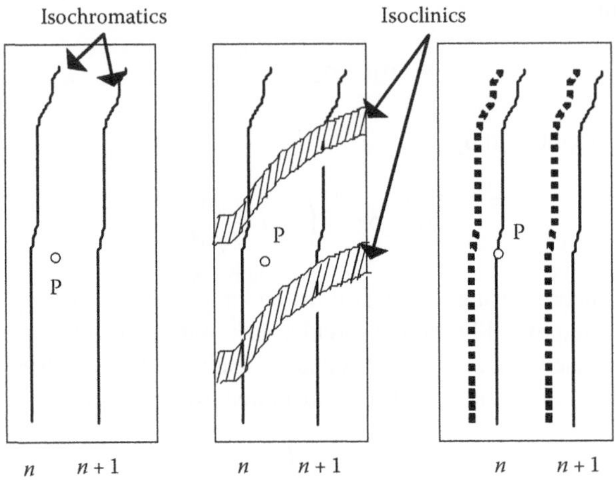

FIGURE 8.7 Tardy's method of compensation.

the isochromatics and the isoclinics. Let us now assume that we wish to measure the fractional isochromatics order at a point P, as shown in Figure 8.7. Since this is a dark-field configuration, the integral isochromatic orders correspond to dark fringes, and in order to work with the dark fringes, we need to make the region at and around the point P bright. For this purpose, the polarizer–analyzer combination is rotated by 45° so that their transmission axes make angles of 45° with the directions of the principal stresses. The intensity distribution is now given by $I = I_0 \sin^2(\delta/2)$. A quarter-wave plate is now inserted between the model and the analyzer in such a way that its principal axes are parallel to the transmission axes of the polarizer and the analyzer. We can now shift the isochromatics by rotation of the analyzer by an angle that is related to the phase shift. In order to understand the working of this principle, we proceed as follows.

We express the amplitude of the wave transmitted by the polarizer as

$$E_{\mathrm{p}} = E_{0\mathrm{p}} \cos(\omega t - kz) \tag{8.33}$$

Since the polarizer–analyzer combination has its transmission axes at 45° to the principal stress directions in the model, this field is resolved along these directions. The components of the field at the exit face of the model are then expressed as

$$E_{\sigma_1} = \frac{E_{0\mathrm{p}}}{\sqrt{2}} \cos[\omega t - kz - k(n_1 - 1)d] = \frac{E_{0\mathrm{p}}}{\sqrt{2}} \cos(\omega t - kz + \psi_1) \tag{8.34a}$$

$$E_{\sigma_2} = \frac{E_{0\mathrm{p}}}{\sqrt{2}} \cos[\omega t - kz - k(n_2 - 1)d] = \frac{E_{0\mathrm{p}}}{\sqrt{2}} \cos(\omega t - kz + \psi_1 + \delta) \tag{8.34b}$$

where $\delta = k(n_1 - n_2)d$. The field transmitted by the analyzer, which is crossed to the polarizer, is

$$E_A = \frac{E_{\sigma_2}}{\sqrt{2}} - \frac{E_{\sigma_1}}{\sqrt{2}} = \frac{E_{0p}}{2}[\cos(\omega t - kz + \psi_1 + \delta) - \cos(\omega t - kz + \psi_1)]$$
$$= \frac{E_{0p}}{2} 2\sin\left(\frac{\delta}{2}\right)\sin\left(\omega t - kz + \psi_1 + \frac{\delta}{2}\right) \quad (8.35)$$

As mentioned earlier, this describes a wave of amplitude $E_{0p}\sin(\delta/2)$, and hence the intensity of the transmitted light is given by $I = I_0 \sin^2(\delta/2)$, as expected. We now introduce a quarter-wave plate after the model and align its axes parallel to the transmission axes of the polarizer and the analyzer. The field components along the fast and slow axes of the quarter-wave plate are

$$E_f = \frac{E_{\sigma_2}}{\sqrt{2}} + \frac{E_{\sigma_1}}{\sqrt{2}} \quad (8.36a)$$

$$E_s = \frac{E_{\sigma_2}}{\sqrt{2}} - \frac{E_{\sigma_1}}{\sqrt{2}} \quad (8.36b)$$

The field components after passage through the quarter-wave plate can be expressed as

$$E_f = \frac{E_{0p}}{2}\left\{\cos\left[\omega t - kz + \psi_1 - k(n_1' - 1)d'\right]\right.$$
$$\left. + \cos\left[\omega t - kz + \psi_1 + \delta - k(n_1' - 1)d'\right]\right\}$$
$$= \frac{E_{0p}}{2}[\cos\tau + \cos(\tau + \delta)] \quad (8.37a)$$

where $\tau = \omega t - kz + \psi_1 - k(n_1' - 1)d'$ and

$$E_s = \frac{E_{0p}}{2}\left\{\cos\left[\omega t - kz + \psi_1 + \delta - k(n_2' - 1)d'\right]\right.$$
$$\left. - \cos\left[\omega t - kz + \psi_1 - k(n_2' - 1)d'\right]\right\}$$
$$= \frac{E_{0p}}{2}\left[\cos\left(\tau + \delta - \frac{\pi}{2}\right) - \cos\left(\tau - \frac{\pi}{2}\right)\right] = \frac{E_{0p}}{2}[\sin(\tau + \delta) - \sin\tau] \quad (8.37b)$$

where

$$k(n_2' - n_1')d' = \frac{\pi}{2}$$

The analyzer will transmit only the E_s component. However, if the analyzer is rotated by an angle χ from this position, then the components of the field along the transmission axis of the analyzer are $E_s\cos\chi + E_f\sin\chi$. After substitution of E_s and E_f, and a little trigonometry, we obtain the amplitude of the wave transmitted through the analyzer as

$$E_A = E_{0p} \sin\left(\chi + \frac{\delta}{2}\right) \cos\left(\tau + \frac{\delta}{2}\right) \tag{8.38}$$

This represents a wave with an amplitude $E_{0p} \sin(\chi + \delta/2)$. Therefore, the intensity of the wave is given by $I = I_0 \sin^2(\chi + \delta/2)$. For the mth isochromatic, $\delta = 2m\pi$, and hence the intensity at this location must be $I = I_0 \sin^2 \chi$, which is evidently zero when $\chi = 0$. Further, the intensity at the same location will also be zero when $\chi = \pi$, but then the mth isochromatic will have moved to the $(m + 1)$th isochromatic. In other words, a rotation of the analyzer by π shifts the isochromatics by one order. Therefore, if the analyzer is rotated by an angle χ_p to shift the isochromatics to the point P, then the fractional order at that point must be χ_p/π.

8.11 PHASE-SHIFTING

Phase-shifting is a technique for the automatic evaluation of phase maps from the intensity data, and has been described in detail in Chapter 4. However, it assumes a special significance in photoelasticity, since both beams participating in interference travel along the same path and the phase of one cannot be changed independently of other, as has been done for the other interferometric methods. We therefore present methods of phase-shifting in photoelasticity.

8.11.1 Isoclinics Computation

For this purpose, a bright-field plane polariscope is used. The transmitted intensity in this configuration is given by

$$\begin{aligned} I &= I_0 - I_0 \sin^2 2\alpha \sin^2\left(\frac{\delta}{2}\right) = I_0\left[1 - \sin^2\left(\frac{\delta}{2}\right)(1 - \cos^2 2\alpha)\right] \\ &= I_0\left[1 - \frac{1}{2}\sin^2\left(\frac{\delta}{2}\right)(1 - \cos 4\alpha)\right] = I_B + V \cos 4\alpha \end{aligned} \tag{8.39}$$

When the whole polariscope is rotated by an angle β, the intensity transmitted can be expressed as

$$I = I_B + V \cos 4(\alpha - \beta) \tag{8.40}$$

Both I_B and V depend on the value of the isochromatic parameter, and consequently on the wavelength of light used. However, the isoclinic parameter does not depend on the wavelength. The phase of the isoclinics is obtained using a four-step algorithm with the intensity data obtained at $\beta_i = (i - 1)\pi/8$, with $i = 1, \ldots, 4$, and the relation

$$\tan 4\alpha = \frac{I_4 - I_2}{I_3 - I_1} \tag{8.41}$$

The phase of the isoclinics is obtained from this relation, except at regions where the modulation V is very small. Since V depend on δ, the low-modulation areas depend on the wavelength. If monochromatic light is used, there may be several areas where the value of δ makes the modulation unusable. This problem, however, can be overcome by using a white-light source, since the low-modulation areas corresponding to a given wavelength will be high-modulation areas for another wavelength. Hence, the modulation is kept high enough for use, except at the zero-order fringe, where the modulation is obviously zero.

8.11.2 Computation of Isochromatics

As can be seen from Tardy's method of compensation, the isochromatic fringe at a point can be shifted by rotation of the analyzer from the crossed position: a rotation of π shifts the isochromatic by one order. This provides a nice method of phase-shifting, but it suffers from drawbacks: the principal axes must be known beforehand and the isochromatics are calculated for a fixed value of the isoclinic parameter. The phase of the isochromatics can be obtained pixel by pixel by taking intensity data at four positions of the analyzer, say, at 0°, 45°, 90°, and 135°.

We now present a method that is free from these shortcomings. It takes intensity data at several orientations of a circular polariscope. We present in Table 8.2 the configurations, along with the corresponding expressions for the transmitted intensity.

From the eight transmitted intensity data, the phase of the isochromatic pattern is computed at each pixel using the relation

$$\tan\delta = \frac{(I_1 - I_2)\cos 2\alpha + (I_5 - I_6)\sin 2\alpha}{\frac{1}{2}\left[(I_4 - I_3) + (I_8 - I_7)\right]} \quad (8.42)$$

TABLE 8.2
Polariscope Configurations and the Corresponding Transmitted Intensities

No.	Polariscope Configuration	Transmitted Intensity
1	$P_{\pi/2}Q_{\pi/4}Q_{\pi/4}A_{\pi/4}$	$I_1 = \frac{I_0}{2}(1 + \cos 2\alpha \sin \delta)$
2	$P_{\pi/2}Q_{\pi/4}Q_{-\pi/4}A_{\pi/4}$	$I_2 = \frac{I_0}{2}(1 - \cos 2\alpha \sin \delta)$
3	$P_{\pi/2}Q_{\pi/4}Q_{-\pi/4}A_0$	$I_3 = \frac{I_0}{2}(1 - \cos \delta)$
4	$P_{\pi/2}Q_{\pi/4}Q_{\pi/4}A_0$	$I_4 = \frac{I_0}{2}(1 + \cos \delta)$
5	$P_{-\pi/4}Q_{\pi/2}Q_{\pi/2}A_0$	$I_5 = \frac{I_0}{2}(1 + \sin 2\alpha \sin \delta)$
6	$P_{-\pi/4}Q_{\pi/2}Q_0A_{\pi/2}$	$I_6 = \frac{I_0}{2}(1 - \sin 2\alpha \sin \delta)$
7	$P_{-\pi/4}Q_{\pi/2}Q_0A_{\pi/4}$	$I_7 = \frac{I_0}{2}(1 - \cos \delta)$
8	$P_{-\pi/4}Q_{\pi/2}Q_{\pi/2}A_{\pi/4}$	$I_8 = \frac{I_0}{2}(1 + \cos \delta)$

It can be seen that I_3 and I_4 are theoretically equal to I_7 and I_8, respectively. However, in practice, owing to polariscope imperfections, they may differ, and hence all four values are used in the algorithm.

The Fourier transform method can also be used for phase evaluation. The carrier fringe pattern is introduced by a birefringent wedge plate of an appropriate wedge angle. Usually, a carrier frequency of 3–5 lines/mm is adequate. The plate is placed close to the model.

8.12 BIREFRINGENT COATING METHOD: REFLECTION POLARISCOPE

The use of a birefringent coating on the surface of an object extends photoelasticity to the measurement of surface strains on opaque objects and eliminates the need to make models. In this method, a thin layer of a birefringent material is bonded onto the surface of the object. Assuming the adhesion to be good, the displacements on the surface of the object on loading are transferred to the coating, which induces birefringence in the latter. The strain-induced birefringence is observed in reflection. In order to obtain good reflected intensity, either the surface of the object is polished to make it naturally reflecting or some reflective particles are added to the cement that bonds the birefringent coating to the surface of the object. Figure 8.8 shows a schematic of a reflection polariscope used with birefringent coatings.

Such an instrument can be used either as a plane polariscope or a circular polariscope. The isochromatics obtained with a circular polariscope give the difference between the principal stresses in the coating, that is,

$$(\sigma_1 - \sigma_2)_c = m\frac{f_{\sigma c}}{2d} \tag{8.43}$$

where d is the thickness and $f_{\sigma c}$ the fringe value of the coating. Since the light travels through nearly the same region twice, the effective thickness is $2d$. The principal strains are related to the principal stresses through Hooke's law. We thus obtain the

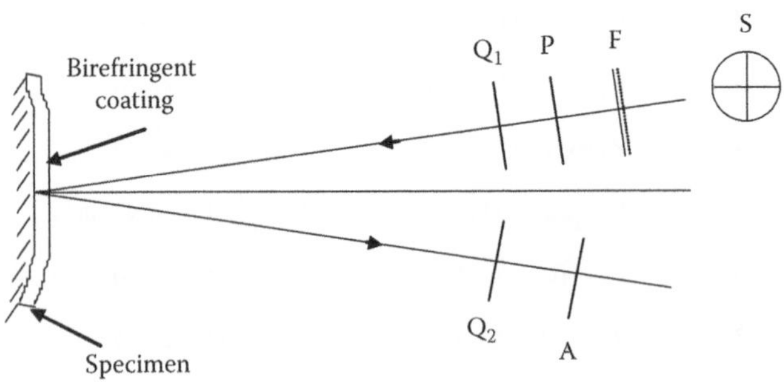

FIGURE 8.8 Schematic of a reflection polariscope.

difference of the principal strains as

$$\varepsilon_1 - \varepsilon_2 = \frac{1+\nu_c}{E_c}(\sigma_1 - \sigma_2)_c \tag{8.44}$$

where E_c and ν_c are the elastic constants of the birefringent coating material. Similarly, we can express the difference of principal strains at the surface of the object as

$$\varepsilon_1 - \varepsilon_2 = \frac{1+\nu_o}{E_o}(\sigma_1 - \sigma_2)_o \tag{8.45}$$

Assuming that the strains in the coating and at the surface of the object are same, we have

$$(\sigma_1 - \sigma_2)_o = \frac{E_o}{E_c}\frac{1+\nu_c}{1+\nu_o}(\sigma_1 - \sigma_2)_c \tag{8.46}$$

Separation of stresses in the coating is accomplished by the oblique incidence method. Hence, the principal strains in the coating are calculated. Having obtained these, the principal stresses at the surface of the object are obtained from the following equations:

$$\sigma_{1o} = \frac{E_o}{1-\nu_o^2}(\varepsilon_1 + \nu_o\varepsilon_2) \tag{8.47a}$$

$$\sigma_{2o} = \frac{E_o}{1-\nu_o^2}(\varepsilon_2 + \nu_o\varepsilon_1) \tag{8.47b}$$

The analysis is based on the assumption that the strains in the coating and at the surface of the object are the same.

8.13 HOLOPHOTOELASTICITY

Separation of stresses requires that σ_1, σ_2, or $\sigma_1 + \sigma_2$ be known in addition to $\sigma_1 - \sigma_2$ obtained from photoelasticity. The sum of the principal stresses is obtained interferometrically, for example by using a Mach–Zehnder interferometer. On the other hand, holophotoelasticity provides fringe patterns belonging to $\sigma_1 - \sigma_2$ and $\sigma_1 + \sigma_2$ simultaneously, thereby effecting easy separation of stresses. However, the method requires coherent light for illumination. Here, we use holography to record the waves transmitted through the model and later reconstruct this record to extract the information. We can indeed use the technique in two ways. In one method, we obtain only isochromatics, and hence the method is equivalent to a circular polariscope; it also provides the flexibility of leisurely evaluation of the fringe pattern. In the other method, both the isochromatic and isopachic fringe patterns are obtained. This method, which requires two exposures, is termed double-exposure holophotoelasticity, while the first method is a single-exposure method.

8.13.1 Single-Exposure Holophotoelasticity

The experimental arrangement is shown in Figure 8.9. The model is already stressed, and hence is birefringent. The light from a laser is generally polarized with its **E** vector vibrating in the vertical plane when the beam propagates in the horizontal plane. The beam is expanded and collimated. In the case where the laser output is randomly polarized, a polarizer is used, followed by a quarter-wave plate oriented at 45°. In brief, the model is illuminated by a circularly polarized wave. The reference wave is also circularly polarized and of the same handedness, so that both components are interferometrically recorded.

The components of the wave just after the model can be expressed as

$$E_{\sigma_1}(z;t) = \frac{E_{0y}}{\sqrt{2}}\cos(\tau + \psi_2) \tag{8.48a}$$

$$E_{\sigma_2}(z;t) = \frac{E_{0y}}{\sqrt{2}}\sin(\tau + \psi_2 + \delta) \tag{8.48b}$$

Consistent with the treatment presented in Chapter 6, we write these components as

$$E_{\sigma_1}(z;t) = \mathrm{Re}\left\{\frac{E_{0y}}{\sqrt{2}}e^{i(\tau_1+\psi_2)}\right\} \tag{8.49a}$$

$$E_{\sigma_2}(z;t) = \mathrm{Re}\left\{\frac{E_{0y}}{\sqrt{2}}e^{i(\tau_1+\psi_2+\pi/2+\delta)}\right\} \tag{8.49b}$$

with

$$\psi_2 = -\frac{2\pi}{\lambda}(n_1 - 1)d_1, \quad \delta = \frac{2\pi}{\lambda}(n_1 - n_2)d_1$$

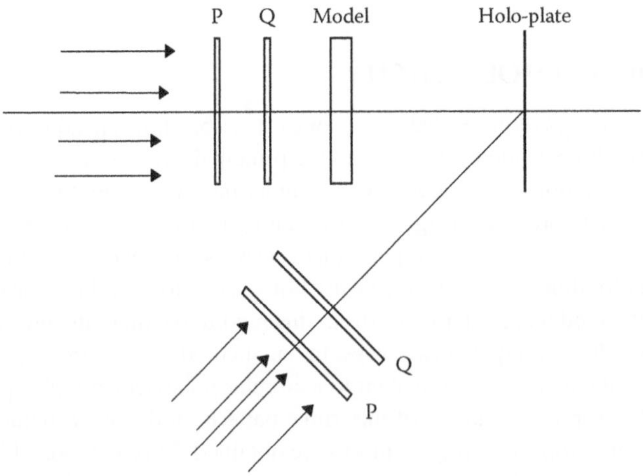

FIGURE 8.9 An experimental arrangement for single-exposure holophotoelasticity.

and τ_1 does not have any time dependence—which has been ignored since a monochromatic wave is used for illumination—and d_1 is the thickness of the stressed model. As usual, Re{ } denotes the real part. Similarly, the reference wave components are written as

$$E_{r_1} = \text{Re}\left\{a_r e^{i\phi_r}\right\} \tag{8.50a}$$

$$E_{r_2} = \text{Re}\left\{a_r e^{i(\phi_r+\pi/2)}\right\} \tag{8.50b}$$

Since these components are orthogonally polarized, they will interfere with the respective components—essentially, we record two holograms. The recorded intensity is given by

$$I = \left|E_{\sigma_1} + E_{r_1}\right|^2 + \left|E_{\sigma_2} + E_{r_2}\right|^2 \tag{8.51}$$

This record, on processing, is a hologram. Assuming linear recording and illumination with a reference beam that releases two beams, these beams interfere and generate an intensity distribution of the type

$$I = I_0' \left|e^{i(\tau_1+\psi_2)} + e^{i(\tau_1+\psi_2+\delta)}\right|^2 = I_0\,(1+\cos\delta) \tag{8.52}$$

This is the intensity distribution as obtained in a bright-field circular polariscope. It may be noted that a quarter-wave plate–analyzer combination is not placed behind the model during recording. The state of polarization in the reference wave serves the function of this assembly. If the state of polarization in the reference wave is orthogonal to that in the object wave from the model, that is, the reference wave is of opposite handedness, then the isochromatics pattern corresponding to a dark-field circular polariscope will be obtained.

8.13.2 Double-Exposure Holophotoelasticity

The experimental arrangement is similar to that shown in Figure 8.9. The model is illuminated by a circularly polarized wave, and another circularly polarized wave of the same handedness is used as a reference wave. The first exposure is made with the model unstressed and the second exposure with the model stressed. During the first exposure, the model is isotropic. However, to be consistent with our earlier treatment, we write the amplitudes of the object and reference waves recorded during the first exposure as

$$E_1(z;t) = \text{Re}\left\{\frac{E_{0y}}{\sqrt{2}} e^{i(\tau_1+\psi_0)}\right\} \tag{8.53a}$$

$$E_2(z;t) = \text{Re}\left\{\frac{E_{0y}}{\sqrt{2}} e^{i(\tau_1+\psi_0+\pi/2)}\right\} \tag{8.53b}$$

with

$$\psi_0 = -\frac{2\pi}{\lambda}(n_0 - 1)d$$

and

$$E_{r_1} = \text{Re}\{a_r e^{i\phi_r}\} \tag{8.50a}$$

$$E_{r_2} = \text{Re}\left\{a_r e^{i(\phi_r+\pi/2)}\right\} \tag{8.50b}$$

In the second exposure, we record two waves from the stressed model. These waves are represented as

$$E_{\sigma_1}(z;t) = \text{Re}\left\{\frac{E_{0y}}{\sqrt{2}} e^{i(\tau_1+\psi_2)}\right\} \tag{8.54a}$$

$$E_{\sigma_2}(z;t) = \text{Re}\left\{\frac{E_{0y}}{\sqrt{2}} e^{i(\tau_1+\psi_2+\pi/2+\delta)}\right\} \tag{8.54b}$$

with

$$\psi_2 = -\frac{2\pi}{\lambda}(n_1 - 1)d_1, \quad \delta = \frac{2\pi}{\lambda}(n_1 - n_2)d_1.$$

The reference waves are the same as those used in the first exposure. As explained earlier, we record two holograms in the second exposure. The total intensity recorded can be written as

$$I = \left|E_1 + E_{r_1}\right|^2 + \left|E_2 + E_{r_2}\right|^2 + \left|E_{\sigma_1} + E_{r_1}\right|^2 + \left|E_{\sigma_2} + E_{r_2}\right|^2 \tag{8.55}$$

The amplitudes of the waves of interest, on reconstruction of the double-exposure hologram, are proportional to

$$2e^{i(\tau_1+\psi_0)} + e^{i(\tau_1+\psi_2)} + e^{i(\tau_1+\psi_2+\delta)} \tag{8.56}$$

These three waves interfere to generate a system of isochromatic $(\sigma_1 - \sigma_2)$ and isopachic $(\sigma_1 + \sigma_2)$ fringe patterns. The intensity distribution in the interferogram is given by

$$I = I_0\left\{1 + 2\cos\left(\frac{2\psi_2 + \delta - 2\psi_0}{2}\right)\cos\frac{\delta}{2} + \cos^2\frac{\delta}{2}\right\} \tag{8.57}$$

Before proceeding further, we need to know what $2\psi_2 + \delta - 2\psi_0$ represents. Substituting for ψ_2, δ, and ψ_0, we obtain

$$\begin{aligned}
2\psi_2 + \delta - 2\psi_0 &= -\frac{2\pi}{\lambda}[2(n_1 - 1)d_1 - (n_1 - n_2)d_1 - 2(n_0 - 1)d] \\
&= -\frac{2\pi}{\lambda}[(n_1 + n_2)d_1 - 2n_0 d - 2\Delta d] \\
&= -\frac{2\pi}{\lambda}[(n_1 - n_0)d + (n_2 - n_0)d + (n_1 + n_2)\Delta d - 2\Delta d]
\end{aligned} \tag{8.58}$$

Assuming the birefringence to be small, so that $n_1 + n_2$ can be replaced by $2n_0$, substituting for $n_1 - n_0$ and $n_2 - n_0$, and using Equation 8.13, we obtain

$$\begin{aligned}2\psi_2 + \delta - 2\psi_0 &= -\frac{2\pi}{\lambda}\left[(C_1 + C_2)(\sigma_1 + \sigma_2)d - 2(n_0 - 1)\frac{\nu}{E}(\sigma_1 + \sigma_2)d\right] \\ &= -\frac{2\pi}{\lambda}\left\{\left[(C_1 + C_2) - 2(n_0 - 1)\frac{\nu}{E}\right](\sigma_1 + \sigma_2)d\right\} \\ &= -\frac{2\pi}{\lambda}\left[(C_1' + C_2')(\sigma_1 + \sigma_2)d\right] = \frac{2\pi}{\lambda}C'(\sigma_1 + \sigma_2)d \qquad (8.59)\end{aligned}$$

It is thus seen that the argument of the cosine in the second term in Equation 8.57 depends only on the sum of the principal stresses, and hence generates an isopachic fringe pattern. We can rewrite Equation 8.57 as

$$I = I_0\left\{1 + 2\cos\left[\frac{\pi}{\lambda}C'(\sigma_1 + \sigma_2)d\right]\cos\left[\frac{\pi}{\lambda}C(\sigma_1 - \sigma_2)d\right] + \cos^2\left[\frac{\pi}{\lambda}C(\sigma_1 - \sigma_2)d\right]\right\} \qquad (8.60)$$

It can be seen that the second term in Equation 8.60 contains information about the isopachics, and the second and third terms contain information about the isochromatics. Figure 8.10 shows an interferogram depicting both types of fringes. We will now examine Equation 8.60 and study the formation of isochromatics and isopachics.

Since we are using a bright-field configuration, the dark isochromatics will occur when

$$\frac{\pi}{\lambda}C(\sigma_1 - \sigma_2)d = (2n+1)\frac{\pi}{2}, \qquad \text{with } n \text{ an integer} \qquad (8.61)$$

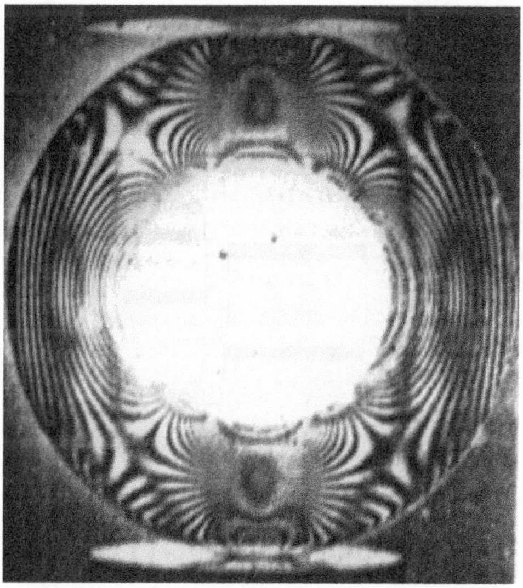

FIGURE 8.10 Interferogram showing both the isochromatics (broad fringes) and isopachics. Note the phase shift of π when the isopachics cross an isochromatic.

However, the intensity of the dark isochromatics is not zero but I_0. The bright isochromatics occur when $(\pi/\lambda)C(\sigma_1 - \sigma_2)d = n\pi$, and the intensity in the bright isochromatics is given by

$$I = 2I_0\left\{1 + (-1)^n \cos\left[\frac{\pi}{\lambda}C'(\sigma_1 + \sigma_2)d\right]\right\} \tag{8.62}$$

The intensity of the bright isochromatics is modulated by isopachics. Let us first consider a bright isochromatics of even order, whose intensity is given by

$$I = 2I_0\left\{1 + \cos\left[\frac{\pi}{\lambda}C'(\sigma_1 + \sigma_2)d\right]\right\} \tag{8.63}$$

The intensity in the bright isochromatics will be zero whenever

$$\frac{\pi}{\lambda}C'(\sigma_1 + \sigma_2)d = (2K+1)\pi, \qquad \text{with } K = 0, 1, 2, 3 \tag{8.64}$$

The integer K gives the order of the isopachic. When this condition is satisfied, the isochromatics will have zero intensity. Therefore, the isopachics modulate the bright isochromatics. Let us now see what happens to the next-order bright isochromatic. Obviously, the intensity distribution in this isochromatic will be

$$I = 2I_0\left\{1 - \cos\left[\frac{\pi}{\lambda}C'(\sigma_1 + \sigma_2)d\right]\right\} \tag{8.65}$$

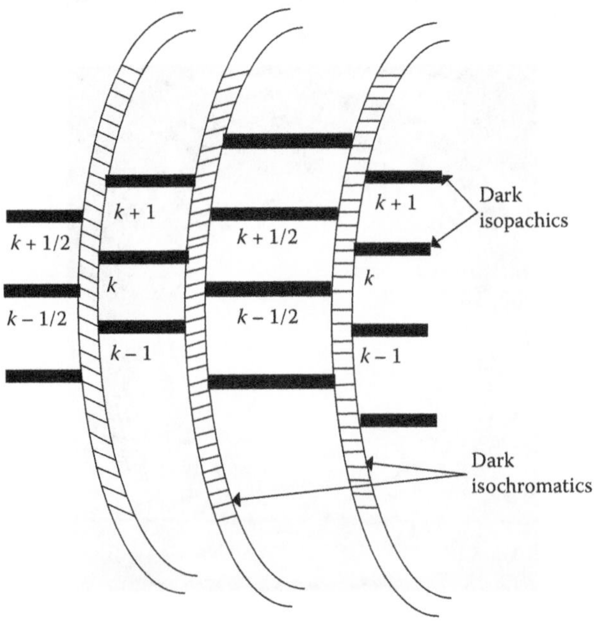

FIGURE 8.11 Simplified combined isochromatic and isopachic pattern.

Substituting the condition for the Kth dark isopachic in this equation, gives a maximum intensity of $4I_0$. This shows that the Kth isopachic has changed by one half-order in going over from one bright isochromatic to the next. This interpretation is simple, and remains valid when the two families of fringes are nearly perpendicular, as shown in Figure 8.11. In the other extreme case, where the isochromatics and isopachics are parallel to each other, this analysis breaks down. It is therefore advisable to use some method to separate out these two fringe patterns. The influence of birefringence can be eliminated by passing the beam twice through the model and a Faraday rotator, thereby eliminating the isochromatic pattern. For the model, one can also use materials such as poly-methyl methacrylate (PMMA), which has almost no or very little birefringence. For such a model, only the isopachic pattern will be observed. Holophotoelasticity can also be performed in real time, which offers certain advantages.

8.14 THREE-DIMENSIONAL PHOTOELASTICITY

Photo-elastic methods, thus far described, cannot be used for investigations of objects under a three-dimensional state of stress. When a polarized wave propagates through such an object, assumed to be transparent, it integrates the polarization changes over the distance of travel. The integrated optical effect is so complex that it is impossible to analyze it or relate it to the stresses that produced it. There are, however, several methods for such investigations. Of these, we discuss two, namely the frozen-stress method and the scattered-light method. The former is restricted in its application to static cases of loading by external forces.

8.14.1 The Frozen-Stress Method

The frozen-stress method is possibly the most powerful method of experimental stress analysis. It takes advantage of the multiphase nature of plastics used as model materials. The procedure for stress freezing consists of heating the model to a temperature slightly above the critical temperature and then cooling it slowly to room temperature, typically at less than 2°C/h under the desired loading condition. The load may be applied to the model either before or after reaching the critical temperature. Extreme care is taken to ensure that the model is subjected to correct loading, since spurious stresses due to bending and gravitational load may be induced as a result of the low rigidity of the model material at the critical temperature.

After the model has been cooled to room temperature, the elastic deformation responsible for the optical anisotropy is permanently locked. The model is now cut into thin slices for examination under the polariscope. The optical anisotropy is normally not disturbed during slicing if the operation is carried out at high speeds and under coolant conditions. Of two methods of data collection and interpretation from these slices—sub-slicing and oblique incidence—the latter is the more practical.

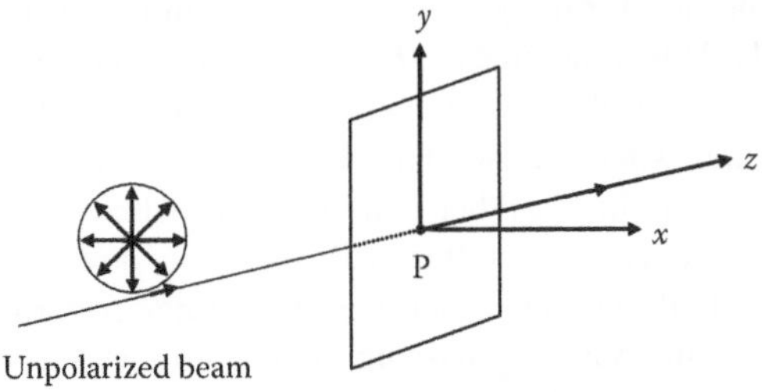

FIGURE 8.12 Scattering of unpolarized beam by a scatterer at P.

8.14.2 Scattered-Light Photoelasticity

When a beam of light passes through a medium containing fine particles dispersed in the volume, part of the beam is scattered. The intensity of the scattered light, when the particles are much smaller than the wavelength of light, varies as ω^4, where ω is the circular frequency of the light waves. This phenomenon was investigated by Rayleigh in detail and is called Rayleigh scattering. The most beautiful observations of red sunset and blue sky are due to scattering from gaseous molecules in the atmosphere. Further, the light from the blue sky is partially linearly polarized. In some observation directions, the scattered light is linearly polarized.

Consider a scattering center located at a point P, as shown in Figure 8.12. Let the incident light be unpolarized light, which can be resolved into two orthogonal linearly polarized components with random phases. The incident component vibrating in the (y, z) plane, when absorbed, will set the particle (rather the electrons in the particle) vibrating along the y direction. The re-radiated wave will have zero amplitude along the y direction. On the other hand, if the particle is oscillating along the x direction, the re-radiated wave will have zero amplitude in that direction. Thus, when the observation direction lies along the y direction in the (x, y) plane passing through the point P, the scattered wave will be plane polarized. The particle acts as a polarizer.

Let us now assume that the incident wave is linearly polarized, with the **E** vector vibrating in the (y, z) plane. The electrons in the particle will oscillate along the y direction. The re-radiated wave will have zero amplitude when observed along the y axis. The particle thus acts as an analyzer. This picture is equivalent to placing a polarizer and an analyzer anywhere in the model. Therefore, stress information can be obtained without freezing the stress and slicing the model. The scattered-light method therefore provides a nondestructive means of optical slicing in three dimensions.

8.15 EXAMINATION OF THE STRESSED MODEL IN SCATTERED LIGHT

8.15.1 Unpolarized Incident Light

Let us consider a stressed model in the path of a narrow beam of unpolarized light. We assume that there are a large number of scatterers in the model. Let us now consider the light scattered by a scatterer at point P inside the model when the observation direction is perpendicular to the incident beam. The scattered light is resolved into components along the directions of principal stresses σ_2 and σ_3, as shown in Figure 8.13. In traversing a distance PQ in the model, these two orthogonally polarized components acquire a phase difference. If an analyzer is placed in the observation path, the transmitted intensity will depend on the phase difference acquired. Since the incident beam is unpolarized, there is no influence of the transverse distance AP in the model. Assume that the transmitted intensity is zero for a certain location P of the scatterer; this occurs when the phase difference is a multiple of 2π. As the beam is moved to illuminate another scatterer at point P′ in the same plane along the line of sight, the transmitted intensity will undergo cyclic variation between minima and maxima, depending on the additional phase acquired when traversing the distance PP′. It is, however, assumed that the directions of the principal stresses σ_2 and σ_3 do not change over the distance PP′.

Let m_1 and m_2 be the fringe orders when the light scattered from scatterers at points P and P′ is analyzed. Then

$$x_1(\sigma_2 - \sigma_3) = m_1 f \tag{8.66a}$$

$$x_2(\sigma_2 - \sigma_3) = m_2 f \tag{8.66b}$$

where x_1 = PQ and x_2 = P′Q. Therefore, we obtain

$$\sigma_2 - \sigma_3 = \frac{dm}{dx} f \tag{8.67}$$

The principal stress difference at any point along the observation direction is proportional to the gradient of the fringe order.

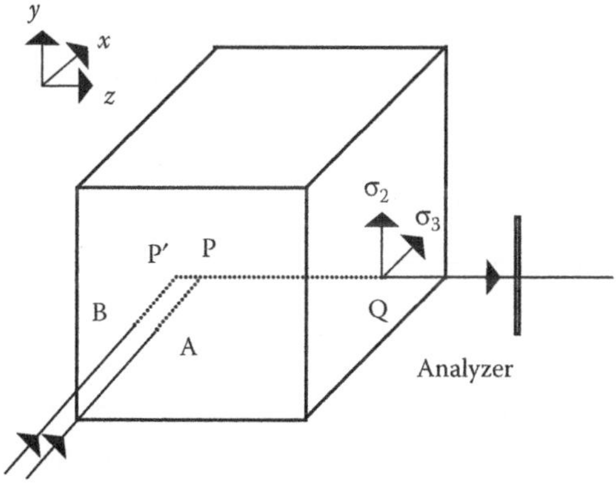

FIGURE 8.13 Stressed model: illumination by an unpolarized beam and observation through an analyzer.

8.15.2 Linearly Polarized Incident Beam

We now consider another situation of a linearly polarized beam incident on the model, as shown in Figure 8.14. We assume for the sake of simplicity that the principal stress directions are along the x and y axes. The transmission axis of the polarizer makes an angle α with the x axis. The incident wave of amplitude E_0 is resolved along the x and y directions. These linearly polarized components travel with different velocities in the model, and hence pick up a phase difference δ. Thus, at any plane normal to the direction of propagation, the state of polarization of the wave in general will be elliptical. It can be expressed as

$$\frac{E_x^2}{E_0^2\cos^2\alpha} + \frac{E_y^2}{E_0^2\sin^2\alpha} - \frac{2E_xE_y}{E_0^2\cos\alpha\sin\alpha}\cos\delta = \sin^2\delta \tag{8.68}$$

where E_x and E_y are the components along the x and y directions, respectively. The major axis of the ellipse makes an angle ψ with the x axis, where

$$\tan 2\psi = \tan 2\alpha\ \cos\delta \tag{8.69}$$

When $\delta = 2p\pi, p = 0, \pm 1, \pm 2, \pm 3, \pm 4, \ldots$, the state of polarization of the wave is linear, with orientation $\psi = \pm\alpha$. For positive values of the integer p, the state of polarization of the wave at any plane is the same as that of the incident wave. In general, a scatterer at point P in any plane is excited by an elliptically polarized wave. In scattered-light photoelasticity, we are looking in the model normal to the direction of propagation; that is, the observation is confined to the plane of the elliptically polarized light. If the observation is made in the direction along the major axis of the ellipse, the amplitude of the re-radiated wave received by the observer will be proportional to the magnitude of the minor axis of the ellipse, and hence a minimum. On the other hand, if the observation direction coincides with the minor axis, the intensity will be a maximum.

As the beam propagates in the stressed model, the ellipse just described continues to rotate as the phase difference changes. Therefore, the observation made in the scattered light normal to the direction of the incident beam will show a variation in intensity along the length of the model in the direction of the incident beam. There is no

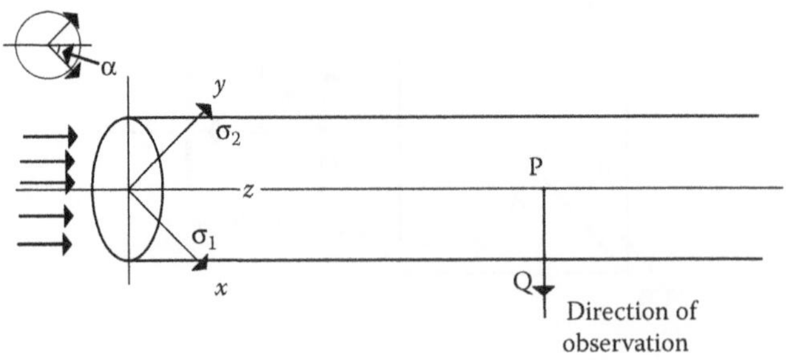

FIGURE 8.14 Stressed model: illumination by a linearly polarized beam.

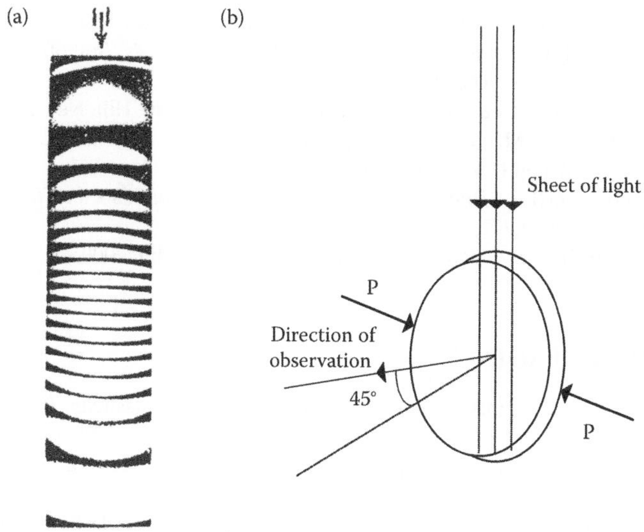

FIGURE 8.15 (a) Scattered-light pattern of a disk under radial compression. (b) Schematic showing the directions of illumination and observation. (From L. S. Srinath, *Scattered Light Photoelasticity*, Tata McGraw-Hill, New Delhi, 1983. With permission.)

influence of birefringence in the model on traverse through distance PQ. Figure 8.15 shows a scattered-light stress pattern of a disk under radial compression. The directions of the incident beam and that of the scattered light are also shown in Figure 8.15.

Owing to the weak intensity of the scattered light, an intense incident beam is used. This beam is generated by a high-pressure mercury lamp with suitable collimating optics. The laser is an attractive alternative, since the beam can then be used without optics. The model is placed in a tank containing an index-matching liquid to avoid refraction and polarization changes at the model surfaces. To facilitate proper adjustments, the model in the tank is mounted on a stage capable of translational and rotational movement.

BIBLIOGRAPHY

1. M. Frocht, *Photoelasticity*, Wiley, New York, 1941.
2. H. T. Jessop and F. C. Harris, *Photoelasticity: Principles and Methods*, Dover, New York, 1949.
3. E. G. Coker and L. N. G. Filon, *A Treatise on Photoelasticity*, Cambridge University Press, New York, 1957.
4. A. J. Durelli and W. F. Riley, *Introduction to Photomechanics*, Prentice-Hall, Englewood Cliffs, NJ, 1965.
5. A. S. Holister, *Experimental Stress Analysis*, Cambridge University Press, New York, 1967.
6. A. Kuske and G. Robertson, *Photoelastic Stress Analysis*, Wiley, New York, 1974.
7. J. W. Dally and W. F. Riley, *Experimental Stress Analysis*, McGraw-Hill, New York, 1978.

8. H. Aben, *Integrated Photoelasticity*, McGraw-Hill, New York, 1979.
9. A. Lagarde (Ed.), *Optical Methods in Mechanics of Solids*, Sijthoff en Noordhoff, Alphen aan den Rijn, The Netherlands, 1981.
10. L. S. Srinath, *Scattered Light Photoelasticity*, Tata McGraw-Hill, New Delhi, 1983.
11. R. S. Sirohi and M. P. Kothiyal, *Optical Components, Systems, and Measurement Techniques*, Marcel Dekker, New York, 1991.
12. A. S. Kobayashi (Ed.), *Handbook on Experimental Mechanics*, Society for Experimental Mechanics, Bethel, CT, 1993.
13. K. Ramesh, *Digital Photoelasticity*, Springer-Verlag, Berlin, 2000.

ADDITIONAL READING

1. D. C. Drucker, The method of oblique incidence in photoelasticity, *Proc. SESA, VIII*, 1, 51–66, 1950.
2. K. Ito, New model material for photoelasticity and photoplasticity, *Exp. Mech.*, 2, 373–376, 1962.
3. A. S. Redner, New oblique-incidence method for direct photoelastic measurements of principal strains, *Exp. Mech.*, 3, 67–72, 1963.
4. A. Robert and E. Guillemet, New scattered light method in three dimensional photoelasticity, *Br. J. Appl. Phy.*, 15, 567–578, 1964.
5. M. Nishida and H. Saito, A new interferometric method of two dimensional stress analysis, *Proc. SESA*, 21, 366–376, 1964.
6. R. O'Regan, New method of determining strain on the surface of a body with photoelastic coating, *Exp. Mech.*, 5, 241–246, 1965.
7. J. W. Dally and E. R. Erisman, An analytic separation method for photoelasticity, *Exp. Mech.*, 6, 493–499, 1966.
8. A. J. Robert, New methods in photoelasticity, *Exp. Mech.*, 7, 224–232, 1967.
9. M. E. Fourney, Application of holography to photoelasticity, *Exp. Mech.*, 8, 33–38, 1968.
10. D. Hovanesian, V. Broic, and R. L. Powell, A new experimental stress-optic method: Stress-holo-interferometry, *Exp. Mech.*, 8, 362–368, 1968.
11. H. H. M. Chau, Holographic interferometer for isopachic stress analysis, *Rev. Sci. Instrum.*, 39, 1789–1792, 1968.
12. W. Wetzels, Holographie als Hilfsmittel zur Isopachenbestimmung, *Optik*, 27, 271–272, 1968.
13. E. von Hopp and G. Wutzke, The application of holography in plane photoelasticity, *Materialprufung*, 11, 409–415, 1969.
14. L. S. Srinath, Analysis of scattered-light methods in photoelasticity, *Exp. Mech.*, 9, 463–468, 1969.
15. E. von Hopp and G. Wutzke, Holographic determination of the principal stresses in plane models, *Materialprufung*, 12, 13–22, 1970.
16. M. E. Fourney and K. V. Mate, Further applications of holography to photoelasticity, *Exp. Mech.*, 10, 177–186, 1970.
17. R. C. Sampson, A stress-optic law for photoelastic analysis of orthotropic composites, *Exp. Mech.*, 10, 210–215, 1970.
18. D. Post, Photoelastic fringe multiplication for ten fold increase in sensitivity, *Exp. Mech.*, 10, 305–312, 1970.
19. D. C. Holloway and R. H. Johnson, Advancements in holographic photoelasticity, *Exp. Mech.*, 11, 57–63, 1971.

20. B. Chatelain, Holographic photoelasticity: Independent observation of the isochromatic and isopachic fringes for a single model subjected to only one process, *Opt. Laser Technol.*, 5, 201–204, 1973.
21. S. Redner, New automatic polariscope system, *Exp. Mech.*, 14, 486–491, 1974.
22. J. Ebbeni, J. Coenen, and H. Hermanne, New analysis of holophotoelastic patterns and their application, *J. Strain Anal.*, 11, 11–17, 1976.
23. J. S. Parks and R. J. Sanford, On the role of material and optical properties in complete photoelastic analysis, *Exp. Mech.*, 16, 441–447, 1976.
24. H. Uozato and R. Nagata, Holographic photoelasticity by using dual hologram method, *Jpn J. Appl. Phys.*, 16, 95–100, 1977.
25. M. Arcan, Z. Hashin, and A. Voloshin, A method to produce uniform plane stress states with applications, *Exp. Mech.*, 18, 141–145, 1978.
26. J. F. Doyle and H. T. Denyluk, Integrated photoelasticity for axisymmetric problems, *Exp. Mech.*, 18, 215–220, 1978.
27. J. W. Dally and R. J. Sanford, Classification of stress intensity factors from isochromatic fringe patterns, *Exp. Mech.*, 18, 441–448, 1978.
28. P. S. Theocaris and E. E. Goloutos, A unified interpretation of interferometric and holographic fringe patterns in photoelasticity, *J. Strain Anal.*, 13, 95, 1978.
29. D. Post, Photoelasticity, *Exp. Mech.*, 19, 176–192, 1979.
30. S. B. Mazurikiewicz and J. T. Pindera, Integrated photoelastic method—Application to photoelastic isodynes, *Exp. Mech.*, 19, 225–234, 1979.
31. R. K. Mueller and L. R. Saackel, Complete automatic analysis of photoelastic fringes, *Exp. Mech.*, 19, 245–251, 1979.
32. J. W. Dally, Dynamic photoelastic studies on fracture, *Exp. Mech.*, 19, 349–361, 1979.
33. Y. Seguchi, Y. Tomita and M. Wanatabe, Computer-aided fringe-pattern analyser—A case of photoelastic fringe, *Exp. Mech.*, 19, 362–370, 1979.
34. R. J. Sanford, Application of the least squares method to photoelastic analysis, *Exp. Mech.*, 20, 192–197, 1980.
35. W. F. Swinson, J. L. Turner, and W. F. Ranson, Designing with scattered light photoelasticity, *Exp. Mech.*, 20, 397–402, 1980.
36. J. W. Dally, An introduction to dynamic photoelasticity, *Exp. Mech.*, 20, 409–416, 1980.
37. J. Cernosek, Three dimensional photoelasticity by stress freezing, *Exp. Mech.*, 20, 417–426, 1980.
38. R. J. Sanford, Photoelastic holography—A modern tool for stress analysis, *Exp. Mech.*, 20, 427–436, 1980.
39. H. A. Gomide and C. P. Burger, Three dimensional strain distribution in upset rings by photo-plastic simulation, *Exp. Mech.*, 21, 361–370, 1981.
40. D. G. Berghaus, Simplification for scattered-light photoelasticity when using the unpolarised incident beam, *Exp. Mech.*, 22, 394–400, 1982.
41. C. P. Burger and H. A. Gomide, Three-dimensional strain in rolled slabs by photo-plastic simulation, *Exp. Mech.*, 22, 441–447, 1982.
42. R. Prabhakaran, Photo-orthotropic-elasticity: A new technique for stress analysis of composites, *Opt. Eng.*, 21, 679–688, 1982.
43. C. W. Smith, W. H. Peters, and G. C. Kirby, Crack-tip measurements in photoelastic models, *Exp. Mech.*, 22, 448–453, 1982.
44. C. P. Burger and A. S. Voloshin, Half-fringe photoelasticity: A new instrument for wholefield stress analysis, *ISA Trans.*, 22, 85–95, 1982.
45. R. B. Agarwal and L. W. Teufel, Epon 828 epoxy: A new photoelastic-model material, *Exp. Mech.*, 23, 30–35, 1983.

46. K. A. Jacob, V. Dayal, and B. Ranganayakamma, On stress analysis of anisotropic composites through transmission optical patterns: Isochromatics and isopachics, *Exp. Mech.*, 23, 49–54, 1983.
47. J. Komorowski and J. Stupnicki, Strain measurement with asymmetric oblique-incidence polariscope for birefringent coatings, *Exp. Mech.*, 23, 171–176, 1983.
48. F. Zhang, S. M. Zhao and B. Chen, A digital image processing system for photoelastic stress analysis, *Proc. SPIE*, 814, 806–809, 1987.
49. T. Y. Chen and C. E. Taylor, Computerised fringe analysis in photomechanics, *Exp. Mech.*, 29, 323–329, 1989.
50. I. V. Zhavoronok, V. V. Nemchinov, S. A. Litvin, A. M. Skanavi, U. V. Pavlov, and V. S. Evsenev, Automatization of measurement and processing of experimental data in photoelasticity, *Proc. SPIE, 1554A*, 371–379, 1991.
51. J. T. Pindera, Scattered light optical isodynes basis for 3-D isodyne stress analysis, *Proc. SPIE*, 1554A, 458–471, 1991.
52. C. Quan, P. J. Bryanston-Cross and T. R. Judge, Photoelasticity stress analysis using carrier fringes and FFT techniques, *Opt. Lasers Eng.*, 18, 79–108, 1993.
53. A. Asundi and M. R. Sajan, Digital dynamic photoelasticity, *Opt. Lasers Eng.*, 20, 135–140, 1994.
54. M. Wolna, Polymer materials in practical uses of photoelasticity, *Opt. Eng.*, 34, 3427–3432, 1995.
55. S. J. Haake, E. A. Patterson, and Z. F. Wang, 2D and 3D separation of stresses using automated photoelasticity, *Exp. Mech.*, 36, 269–276, 1996.
56. K. Ramesh and S. S. Deshmukh, Three fringes photoelasticity: Use of colour image processing hardware to automate ordering of isochromatics, *Strain*, 32, 79–86, 1996.
57. T. Kihara, A measurement method of scattered light photoelasticity using unpolarized light, *Exp. Mech.*, 37, 39–44, 1997.
58. J.-C. Dupré and A Lagarde, Photoelastic analysis of a three-dimensional specimen by optical slicing and digital image processing, *Exp. Mech.*, 37, 393–397, 1997.
59. A. D. Nurse, Full-field automated photoelasticity by use of a three-wavelength approach to phase-stepping, *Appl. Opt.*, 36, 5781–5786, 1997.
60. G. Petrucci, Full field automatic evaluation of isoclinic parameter in white light, *Exp. Mech.*, 37, 420–426, 1997.
61. J. A. Quiroga and A. González-Cona, Phase measuring algorithms for extraction of information of photoelastic fringe patterns, In *Fringe'97, Automatic Processing of Fringe Patterns* (ed. W. Jüptner and W. Osten), 77–83, Akademie Verlag, Berlin, 1997.
62. J. A. Quiroga and A. Gonzalez-Cona, Phase measuring algorithm for extraction of isochromatics of photoelastic fringe patterns, *Appl. Opt.*, 36, 8397–8402, 1997.
63. W. Ji and E. A. Patterson, Simulation of errors in automated photoelasticity, *Exp. Mech.*, 38, 132–139, 1998.
64. M. J. Ekman and A. D. Nurse, Absolute determination of the isochromatic parameter by load-stepping photoelasticity, *Exp. Mech.*, 38, 189–195, 1998.
65. E. A. Patterson and Z. W. Wang, Simultaneous observation of phase-stepped images for automated photoelasticity, *J. Strain Anal. Eng. Des.*, 33, 1–15, 1998.
66. A. Ajovalasit, B. Barone, and G. Petrucci, A method for reducing the influence of quarter-wave plate errors in phase stepping photoelasticity, *J. Strain Anal. Eng. Des.*, 33, 207–216, 1998.
67. K. Ramesh and S. K. Mangal, Data acquisition techniques in digital photoelasticity: A review, *Opt. Lasers Eng.*, 30, 53–75, 1998.
68. A. Asundi, L. Tong, and C. G. Boay, Dynamic phase-shifting photoelasticity, *Appl. Opt.*, 40, 3654–3658, 2001.

69. D. K. Tamrakar and K. Ramesh, Simulation of error in digital photoelasticity by Jones calculus, *Strain*, 37, 105–112, 2001.
70. S. Barone, G. Burriesci, and G. Petrucci, Computer aided photoelasticity by an optimum phase stepping method, *Exp. Mech.*, 42, 132–139, 2002.
71. K. Ramesh and G. Lewis, Digital photoelasticity: Advanced techniques and applications, *Appl. Mech. Rev.*, 55, B69–B71, 2002.
72. G. Calvert, J. Lesniak, and M. Honlet, Applications of modern automated photoelasticity to industrial problems, *Insight*, 44(4), 1–4, 2002.
73. T. Kihara, An arctangent unwrapping technique of photoelasticity using linearly polarized light at three wavelengths, *Strain*, 39, 65–71, 2003.
74. I. A. Jones and P. Wang, Complete fringe order determination in digital photoelasticity using fringe combination matching, *Strain*, 39, 121–130, 2003.
75. J. W. Hobbs, R. J. Greene, and E. A. Patterson, A novel instrument for transient photoelasticity, *Exp. Mech.*, 43, 403–409, 2003.
76. V. Sai Prasad, K. R. Madhu, and K. Ramesh, Towards effective phase unwrapping in digital photoelasticity, *Opt. Lasers Eng.*, 42(4), 421–436, 2004.
77. P. Pinit and E. Umezaki, Full-field determination of principal-stress directions using photoelasticity with plane polarized RGB lights, *Opt. Rev.*, 12, 228–232, 2005.
78. S. Yoneyama and H. Kikuta, Phase-stepping photoelasticity by use of retarders with arbitrary retardation, *Exp. Mech.*, 46, 289–296, 2006.
79. E. Patterson, P. Brailly, and M. Taroni, High frequency quantitative photoelasticity applied to jet engine components, *Exp. Mech.*, 46, 661–668, 2006.
80. K. Ashokan and K. Ramesh, A novel approach for ambiguity removal in isochromatic phase map in digital photoelasticity, *Meas. Sci. Technol.*, 17, 2891–2896, 2006.
81. A. Ajovalasit, G. Petrucci, and M. Scafidi, Phase shifting photoelasticity in white light, *Opt. Lasers Eng.*, 45, 596–611, 2007.
82. P. Pinit and E. Umezaki, Digitally wholefield analysis of isoclinic parameter in photoelasticity by four-step color phase-stepping technique, *Opt. Lasers Eng.*, 45, 795–807, 2007.

9 The Moiré Phenomenon

9.1 INTRODUCTION

When two periodic patterns are superposed, a moiré pattern is formed. This is in fact a very common phenomenon. Moiré patterns are formed by periodic structures of lines or stripes. In general, the superposed patterns should have opaque and transparent regions. Although moiré patterns can be observed with superposition of periodic objects of nearly the same periodicity, the most commonly used objects are linear gratings with equal opaque and transparent regions, often called Ronchi gratings. Sometimes, circular gratings with equal opaque and transparent regions are used. The moiré phenomenon is a mechanical effect, and the formation of moiré fringes is best explained by mechanical superposition of the gratings. However, when the period of these periodic structures is very fine, diffraction effects play a very significant role.

The fringes formed in holographic interferometry (HI) can be considered as a moiré pattern between the primary grating structures belonging to the initial and final states of the object. These primary grating structures are a result of interference between the object wave and the reference wave. Two-wavelength interferometric fringes are also moiré fringes. In general, the moiré pattern can be regarded as a mathematical solution to the interference of two periodic functions. Holo-diagram, a tool developed by Abramson to deal with several issues in HI, is a beautiful device to study fringe formation and fringe control using the moiré phenomenon.

We can explain the moiré phenomenon either using the indicial equation or by the superposition of two sinusoidal gratings. We will follow both approaches, beginning with the indicial equation.

9.2 THE MOIRÉ FRINGE PATTERN BETWEEN TWO LINEAR GRATINGS

Let us consider a line grating with lines running parallel to the y axis and having period b (Figure 9.1a). This grating is described as follows:

$$x = mb \tag{9.1}$$

where the various lines in the grating are identified by the index m, which takes values $0, \pm 1, \pm 2, \pm 3, \ldots$. A second grating of period a is inclined at an angle θ with the

y axis as shown in Figure 9.1b. The lines in this grating are represented by

$$y = x \cot\theta - na/\sin\theta \tag{9.2}$$

where a/sinθ is the intercept with the y axis and the index n takes values 0, ±1, ±2, ±3, . . . , and hence identifies various lines in the grating. On superposing the two line gratings, the moiré fringe pattern formed is governed by the indicial equation

$$m \pm n = p \tag{9.3}$$

where p is another integer that takes values 0, ±1, ±2, ±3, ... The plus sign in the indicial equation generates a sum moiré pattern, which usually has high frequency, and the minus sign generates difference moiré fringes, which are the moiré patterns that are most frequently used and observed. We will be using the difference moiré pattern unless mentioned otherwise. We obtain the equation of the moiré fringes by eliminating m and n from Equations 9.1 through 9.3 as

$$y = \frac{x(b\cos\theta - a)}{b\sin\theta} + \frac{pa}{\sin\theta} \tag{9.4}$$

This is shown in Figure 9.1c. Equation 9.4 can be written in a more familiar form like Equation 9.2 as

$$y = x\cot\phi + \frac{pd}{\sin\phi} \tag{9.5}$$

This implies that the moiré pattern is a grating of period d that is inclined at an angle ϕ with the y axis, where

$$d = \frac{ab}{\left(a^2 + b^2 - 2ab\cos\theta\right)^{1/2}} \tag{9.6a}$$

$$\sin\phi = \sin\theta \frac{b}{\left(a^2 + b^2 - 2ab\cos\theta\right)^{1/2}} = \frac{d}{a}\sin\theta \tag{9.6b}$$

It is interesting to study moiré patterns for the following two situations:

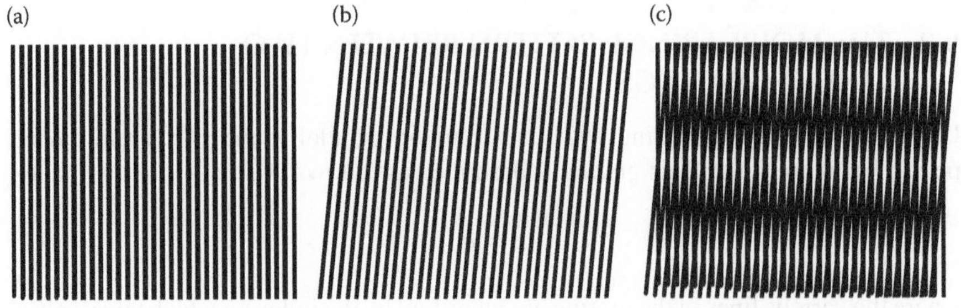

FIGURE 9.1 (a) Vertical grating. (b) Inclined grating. (c) Moiré pattern as a result of superposition.

9.2.1 $a \neq b$ BUT $\theta = 0$

This is a well-known situation of pitch mismatch: the moiré fringes run parallel to the grating lines. The moiré fringe spacing is given by $d = ab/|a - b|$. Here $a - b$ represents the pitch mismatch. If the gratings are of nearly the same pitch, $a \approx b$, then $d = a^2/|a - b|$. Figures 9.2a and 9.2b show gratings of pitches a and b, respectively, with grating elements parallel to the y axis. The moiré fringes due to pitch mismatch are shown in Figure 9.2c. Physically, the moiré spacing is the distance over which pitch mismatch accumulates to the pitch of the grating itself. When the gratings are of equal period, the moiré spacing is infinite. This arrangement is therefore called the infinite fringe mode.

9.2.2 $a = b$ BUT $\theta \neq 0$

This is referred to as an angular mismatch between two identical gratings. This results in moiré pattern formation with a period $d = a/[2\sin(\theta/2)]$ and orientation with the y axis given by $\phi = \pi/2 + \theta/2$. In fact, the moiré fringes run parallel to the bisector of the larger enclosed angle between the gratings.

Moiré fringe formation is easily appreciated when we work in the Fourier domain. In the Fourier domain, a sinusoidal grating of finite size generates three spectra (spots): the spots lie on a line that passes through the origin and is perpendicular to the grating elements. This is due to the fact that the spectrum of a real grating (intensity grating) is centrosymmetric. We need therefore consider only one half of the spectrum. The spots lie on a line that is perpendicular to the grating elements (lines), and the distance between two consecutive spots is proportional to the frequency of the grating. The second grating also generates a spectrum, which is rotated by an angle equal to the angle between the two gratings. When two gratings are superposed, the difference Δr between the two spots (Figure 9.3) generates a grating. If Δr lies within the visibility circle (a spectrum lying in this circle will generate a grating that can be seen by the unaided eye), a moiré pattern is formed: the pitch of the moiré fringes is inversely proportional to the length Δr and the orientation is normal to it. Obviously, when two gratings of equal period are superposed with an angular mismatch, a moiré fringe pattern is formed with fringes running parallel to the bisector of the angle between these gratings and with spacing inversely proportional to the angular mismatch.

(a) (b) (c)

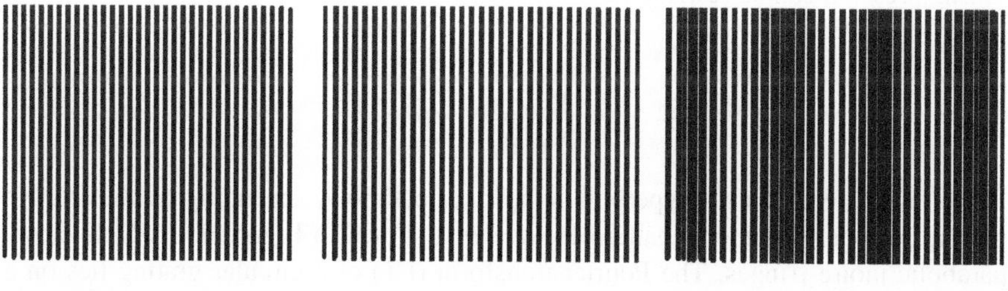

FIGURE 9.2 (a) Vertical grating. (b) Another vertical grating with a different pitch. (c) Moiré pattern as a result of pitch mismatch.

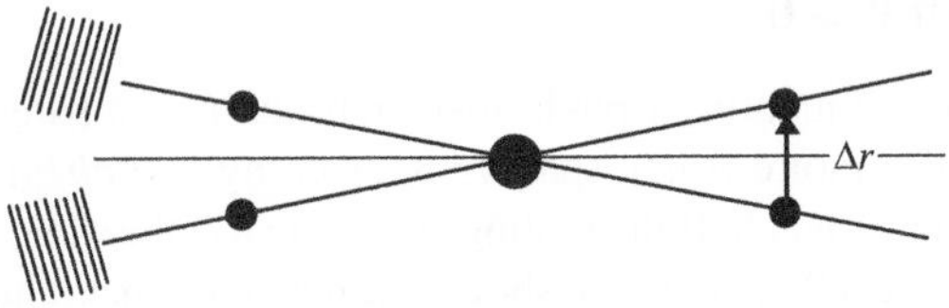

FIGURE 9.3 Spectra of two inclined sinusoidal gratings.

9.3 THE MOIRÉ FRINGE PATTERN BETWEEN A LINEAR GRATING AND A CIRCULAR GRATING

As mentioned earlier, moiré fringes are produced by the superposition of periodic structures. The superposition of a linear grating with a circular grating is of some academic interest: we consider moiré formation by two such gratings. We take a linear grating of period b with its elements running parallel to the y axis. It is represented, as in Equation 9.1, by

$$x = mb \quad \text{for } m = 0, \pm 1, \pm 2, \pm 3, \ldots$$

A circular grating of period a is centered at the origin of the coordinate system, and hence can be represented as

$$x^2 + y^2 = a^2 n^2 \quad \text{for } n = 0, \pm 1, \pm 2, \pm 3, \ldots \tag{9.7}$$

The indicial equation is $m \pm n = p$. Therefore, the moiré pattern for $m - n = p$ is given by

$$\frac{x}{b} - \frac{\left(x^2 + y^2\right)^{1/2}}{a} = p$$

or

$$\frac{x^2 + y^2}{a^2} = p^2 + \frac{x^2}{b^2} - \frac{2xp}{b} \tag{9.8}$$

This expression can be rewritten as

$$x^2\left(\frac{1}{a^2} - \frac{1}{b^2}\right) + \frac{y^2}{b^2} + 2\frac{x}{b}p - p^2 = 0 \tag{9.9}$$

Equation 9.9 represents a hyperbola, ellipse, or parabola, depending on the relative grating periods. The moiré pattern for $a = b$ is shown in Figure 9.4, which shows parabolic moiré fringes. The Fourier transform (FT) of a circular grating lies on a circle; hence an overlap of a linear grating with a circular grating produces a moiré that has a very wide FT spectrum.

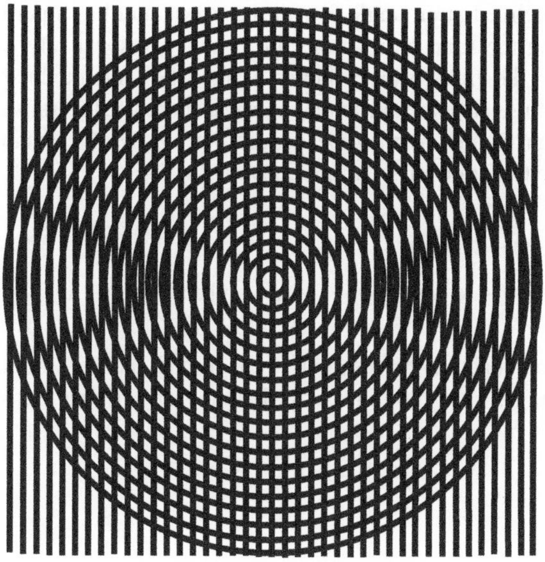

FIGURE 9.4 Moiré pattern between a linear grating and a circular grating.

9.4 MOIRÉ BETWEEN SINUSOIDAL GRATINGS

So far, we have looked into the process of moiré formation using line gratings. Such gratings are seldom used in practice. One normally uses gratings that are binary, since they are very easy to produce. However, when high-frequency gratings are used, they are generally fabricated interferometrically, and hence have a sinusoidal profile. Even binary gratings can be Fourier-decomposed into sinusoidal components, however, so it is instructive to see how the moiré pattern of sinusoidal gratings is formed.

Sinusoidal gratings can be recorded either on the same film or on two separate films, which are then overlapped, as was done with line gratings. We examine these two situations separately. Let us consider a grating defined by a transmittance function $t_1(x)$:

$$t_1(x) = t_0\left(1 - M\cos\frac{2\pi}{b}x\right) = t_0(1 - M\cos 2\pi\mu_0 x) \tag{9.10}$$

where t_0 is the bias transmission, M is the modulation, b is the period, and μ_0 is the spatial frequency of the grating. The grating elements run parallel to the y axis. When $M = 1$, the grating has unit contrast and its transmission function lies between 0 and $2t_0$. Let us now take another sinusoidal grating that is inclined with respect to the first grating. Its transmission function $t_2(x, y)$ is expressed as

$$t_2(x,y) = t_0\left[1 - M\cos\frac{2\pi}{a}(x\cos\theta - y\sin\theta)\right] = t_0\big[1 - M\cos 2\pi(\mu x - \nu y)\big] \tag{9.11}$$

The grating is inclined at an angle θ with the y axis and has a period a. Its spatial frequencies along the $x$and y directions are μ and ν, respectively, such that $a^{-2} = \mu^2 + \nu^2$. Further, the gratings are assumed to have the same modulation.

When the gratings are exposed on the same transparency, the transmission function of the positive transparency, with proper processing, can be taken as being proportional to the sum of the two transmission functions:

$$\begin{aligned} t(x,y) &\propto t_1(x) + t_2(x,y) \\ &= 2t_0\{1 - M\cos\pi[(\mu+\mu_0)x - \nu y]\cos\pi[(\mu-\mu_0)x - \nu y]\} \end{aligned} \tag{9.12}$$

This transmission function corresponds to a grating that is modulated by a low-frequency grating (i.e., the moiré pattern). Bright moiré fringes are formed when

$$\cos\pi[(\mu-\mu_0)x - \nu y] = -1$$

or

$$(\mu-\mu_0)x - \nu y = 2m + 1 \tag{9.13}$$

where m is an integer. Similarly, dark moiré fringes are formed when

$$\cos\pi[(\mu-\mu_0)x - \nu y] = 1, \quad \text{or} \quad (\mu-\mu_0)x - \nu y = 2m$$

The moiré fringes are inclined at an angle ϕ with the y axis such that

$$\cot\phi = \frac{\mu-\mu_0}{\nu} = \frac{\cos\theta - a/b}{\sin\theta} \tag{9.14}$$

When $a = b$, we have

$$\cot\phi = \frac{\cos\theta - 1}{\sin\theta} \tag{9.15}$$

The period of the moiré pattern is also obtained as

$$d = \left[\frac{1}{(\mu-\mu_0)^2} + \frac{1}{\nu^2}\right]^{1/2} = \frac{ab}{\left(a^2 + b^2 - 2ab\cos\theta\right)^{1/2}} \tag{9.16}$$

These are the same formulae that were obtained for line gratings.

When the gratings are recorded on separate films and the moiré pattern due to their overlap is observed, the transmission function is obtained by multiplication of their respective transmission functions: $t(x,y) = t_1(x)t_2(x,y)$. The moiré pattern is then obtained following the procedure explained earlier.

9.5 MOIRÉ BETWEEN REFERENCE AND DEFORMED GRATINGS

When the moiré phenomenon is used for metrology, one of the gratings is mounted on the object that is subjected to deformation. Therefore, one then observes the moiré pattern from the deformed grating and the reference grating (undeformed grating). One can also obtain a moiré pattern between two deformed gratings (i.e., when two deformed states of the object are compared). It is thus instructive to study moiré formation from deformed and undeformed gratings and learn how to extract information about the deformation from the moiré pattern.

The deformation is represented by a function $f(x, y)$, which is assumed to be slowly varying. The deformed grating can be expressed as

$$x + f(x, y) = mb \quad \text{for } m = 0, \pm 1, \pm 2, \pm 3, \pm 4, \ldots \tag{9.17a}$$

This grating is superposed on a reference grating represented by

$$x = nb \quad \text{for } n = 0, \pm 1, \pm 2, \pm 3, \pm 4, \ldots \tag{9.17b}$$

This results in a moiré pattern represented by

$$f(x, y) = pb \quad \text{for } p = 0, \pm 1, \pm 2, \pm 3, \pm 4, \ldots \tag{9.18}$$

The moiré fringes represent a contour map of $f(x, y)$ with period b. Here, the elements in both the gratings run parallel to the y axis with the deformed grating exhibiting slow variation. We can also obtain moiré between the deformed grating and a reference grating that is oriented at an angle θ with the y axis. That is, the gratings are expressed as

$$x + f(x, y) = mb$$

and

$$y = x \cot\theta - \frac{nb}{\sin\theta} \tag{9.19}$$

The moiré pattern, when the grating is inclined by a small angle such that $\cos\theta \approx 1$ and $\sin\theta \approx \theta$ is given by

$$y + \frac{f(x, y)}{\theta} = p\frac{b}{\theta} \tag{9.20}$$

This describes a moiré grating with the period and distortion function magnified by a factor $1/\theta$.

When two distorted gratings are superposed, the moiré pattern gives the difference between the two distortion functions. This difference can also be magnified when the finite-fringe mode of moiré formation is used.

The derivative of the distortion function is obtained by observing the moiré pattern from the deformed grating and its displaced replica. For example, let us consider the deformed grating represented as in Equation 9.17a:

$$x + f(x, y) = mb$$

Its replica has been displaced along the x direction by Δx, and hence is represented by

$$x + \Delta x + f(x + \Delta x, y) = nb \tag{9.21}$$

When these gratings are superposed, the moiré pattern is given by

$$\Delta x + f(x + \Delta x, y) - f(x, y) = pb \tag{9.22}$$

In the limit when the lateral shift Δx is small, we obtain

$$\Delta x + \frac{\partial f(x, y)}{\partial x}\Delta x = pb \tag{9.23}$$

The first term is a constant, and just represents a shift of the moiré pattern. The moiré pattern thus displays the partial x derivative of the distortion function $f(x, y)$. As mentioned earlier, the moiré effect can be magnified by $1/\theta$ using the finite-fringe mode.

Pitch mismatch between the gratings can also be used to magnify the effect of distortion. As an example, we consider a distorted grating and a reference grating given by

$$x + f(x, y) = mb$$

and

$$x = na \tag{9.24}$$

The moiré pattern is given by

$$x + \frac{a}{|a - b|}f(x, y) = \frac{ab}{|a - b|}p \tag{9.25}$$

The period of the moiré pattern is $ab/|a - b|$ and the distortion function has been magnified by $a/|a - b|$.

9.6 MOIRÉ PATTERN WITH DEFORMED SINUSOIDAL GRATING

The transmission function of a deformed sinusoidal grating is given by

$$t_1(x, y) = A_0 + A_1 \cos\left\{\frac{2\pi}{b}\left[x - f(x, y)\right]\right\} \tag{9.26}$$

where A_0 and A_1 are constants specifying the bias transmission and the modulation of the grating and $f(x, y)$ represents the distortion of the grating. The reference grating, oriented at an angle θ, is represented by

$$t_2(x, y) = B_0 + B_1 \cos\left[\frac{2\pi}{a}(x\cos\theta - y\sin\theta)\right] \tag{9.27}$$

where B_0 and B_1 are constants.

9.6.1 Multiplicative Moiré Pattern

Moiré fringes are formed when the angle θ is small, the deformation $f(x, y)$ varies slowly in space, and the periods of the two gratings are nearly equal; in general, $a = Nb$, where N is an integer. The transmission function for multiplicative moiré is the product of the transmission functions of the gratings:

$$t(x, y) = t_1(x, y)\, t_2(x, y) = A_0 B_0 + A_1 B_0 \cos\left\{\frac{2\pi}{b}\left[x - f(x, y)\right]\right\}$$
$$+ A_0 B_1 \cos\left[\frac{2\pi}{a}(x \cos\theta - y \sin\theta)\right]$$
$$+ A_1 B_1 \cos\frac{2\pi}{b}\left[x - f(x, y)\right] \cos\left[\frac{2\pi}{a}(x \cos\theta - y \sin\theta)\right] \tag{9.28}$$

For simplicity, the contrasts of the gratings are assumed to be same (i.e., $A_0 = B_0$ and $A_1 = B_1$). Then,

$$t(x, y) = A_0^2 + A_1 A_0 \cos\left\{\frac{2\pi}{b}\left[x - f(x, y)\right]\right\} + A_0 A_1 \cos\left[\frac{2\pi}{a}(x \cos\theta - y \sin\theta)\right]$$
$$+ \frac{A_1^2}{2}\left\{\cos\left\{2\pi\left[x\left(\frac{1}{b} + \frac{\cos\theta}{a}\right) - y\frac{\sin\theta}{a} - \frac{f(x, y)}{b}\right]\right\}\right.$$
$$\left. + \cos\left\{2\pi\left[x\left(\frac{1}{b} - \frac{\cos\theta}{a}\right) + y\frac{\sin\theta}{a} - \frac{f(x, y)}{b}\right]\right\}\right\} \tag{9.29}$$

In Equation 9.29, the first term is a DC term; the second, third, and fourth terms are the carriers; and the fifth term represents the moiré pattern. Moiré fringes are formed wherever

$$x\left(\frac{1}{b} - \frac{\cos\theta}{a}\right) + y\frac{\sin\theta}{a} - \frac{f(x, y)}{b} = p \tag{9.30}$$

The moiré fringes are deformed straight lines as a result of $f(x, y)$, the local deformation being $f(x, y)/\theta$ when θ is small.

9.6.2 Additive Moiré Pattern

An additive moiré pattern is obtained when the transmission functions of the individual gratings are added. Therefore, the transmission function, assuming gratings of the same modulation, is given by

$$t(x, y) = t_1(x, y) + t_2(x, y) = 2A_0 + A_1 \cos\left\{\frac{2\pi}{b}\left[x - f(x, y)\right]\right\}$$
$$+ A_1 \cos\left[\frac{2\pi}{a}(x \cos\theta - y \sin\theta)\right] \tag{9.31}$$

This can be written as

$$t(x,y) = 2A_0 + 2A_1 \cos\left\{\pi\left[x\left(\frac{1}{b}+\frac{\cos\theta}{a}\right) - \frac{\sin\theta}{a} - \frac{f(x,y)}{b}\right]\right\}$$
$$\times \cos\left\{\pi\left[x\left(\frac{1}{b}-\frac{\cos\theta}{a}\right) + \frac{\sin\theta}{a} - \frac{f(x,y)}{b}\right]\right\} \quad (9.32)$$

The second cosine term represents the moiré pattern, which modulates the carrier grating. Owing to the cosine variation of the moiré term, the phase of the carrier changes intrinsically at the crossover point. The visibility of the moiré pattern is generally poor, and imaging optics are required to resolve the carrier grating.

9.7 CONTRAST IMPROVEMENT OF THE ADDITIVE MOIRÉ PATTERN

It has been mentioned that the contrast of the additive moiré pattern is poor. In order to observe the fringe pattern, it is necessary either to resolve the carrier grating or to use nonlinear recording. Alternately, one could use an FT processor to filter out the required information. The transparency $t(x,y)$ is placed at the input of the FT processor, and the spectrum, which consists of the zero order and two first orders, is observed at the filter plane. Either of the first orders is filtered out and used for imaging. The intensity distribution at the output plane is proportional to

$$A_1^2 \cos^2\left\{\pi\left[x\left(\frac{1}{b}-\frac{\cos\theta}{a}\right) + \frac{\sin\theta}{a} - \frac{f(x,y)}{b}\right]\right\}$$
$$= \frac{1}{2}A_1^2\left\{1 + \cos\left\{2\pi\left[x\left(\frac{1}{b}-\frac{\cos\theta}{a}\right) + \frac{\sin\theta}{a} - \frac{f(x,y)}{b}\right]\right\}\right\} \quad (9.33)$$

This represents a unit-contrast moiré pattern.

9.8 MOIRÉ PHENOMENON FOR MEASUREMENT

The moiré phenomenon has been used extensively for the measurement of length and angle. It has also been applied to the study of deformation of objects and also for shape measurement. We will now investigate how to employ this tool in the area of experimental solid mechanics.

9.9 MEASUREMENT OF IN-PLANE DISPLACEMENT

9.9.1 Reference and Measurement Gratings of Equal Pitch and Aligned Parallel to Each Other

A measurement grating is bonded to the object, and the reference grating is aligned parallel to it. In-plane displacement in the direction of the grating vector (the direction normal to the grating elements) causes the period of the bonded grating to change from b to a. The moiré fringes formed will have a period $d = ab/|a - b|$. The normal strain ε as measured by the moiré method is $|a - b|/a$. Therefore, $\varepsilon = b/d$: the ratio of the grating period to that of the moiré pattern. At this juncture, it should be noted that the moiré method measures the Lagrangian (engineering) strain. However, if the deformation is small, the Lagrangian and Eulerian strains are practically equal.

The shear strain is obtained likewise: the measurement grating (the grating bonded on the object), as a result of shear, is inclined at an angle θ with the reference grating, resulting in the formation of a moiré pattern. Moiré fringes are formed wherever

$$x(1 - \cos\theta) + y\sin\theta = pb \tag{9.34}$$

The period d of the moiré fringes for very small rotation is b/θ. In fact, the shear strain, when the rotation is small and also when strains are small, is equal to θ. Thus, the shear strain γ is given by

$$\gamma = b/d \tag{9.35}$$

The shear strain is also obtained as the ratio of the grating pitch to that of the moiré. This is valid for the homogeneous normal strain and simple shear strain.

9.9.2 Two-Dimensional In-Plane Displacement Measurement

The analysis of the moiré pattern becomes quite easy when it is recognized that the moiré fringes are the contours of the constant displacement component—the so-called isothetics. The zero-order moiré fringe runs through regions where the periods of the reference and measurement gratings are equal (i.e., the displacement component is zero). Similarly, the Nth-order moiré fringe runs through the regions where the displacement component is N times the period of the grating. If the reference grating has its grating vector along the x direction, the moiré fringes represent loci of constant u displacement (i.e., $u = Na$). If the v component of the displacement is to be measured, both the reference grating and the measurement grating are aligned to have their grating vectors along the y direction. The moiré fringes are now the loci of constant v component, and hence $v = N'a$, where N' is the moiré fringe order. To obtain both u and v components simultaneously, a cross grating may be used. The u and v components are isolated by optical filtering.

Let us consider an example where the moiré pattern is obtained when both the reference and measurement gratings are aligned with their grating vectors along the x direction. The moiré pattern then represents the u displacement component. Since strain is a derivative of the displacement, absolute displacement measurement is not necessary. Hence, it is not necessary to assign the absolute fringe order to the moiré fringes; instead, the orders are assigned arbitrarily. However, it should be kept in mind that the increasing orders represent tensile strain and the decreasing orders compressive strain. For the sake of analysis, the moiré fringes are assigned orders 8, 7, 6, 5, 4 in the moiré pattern shown in Figure 9.5. We now wish to obtain the strain at a point A. Through A, a line parallel to the x axis is drawn, and a displacement curve (the moiré fringe order versus the position on the object) is drawn as shown in Figure 9.5: this curve represents the u displacement on the line over the object. We

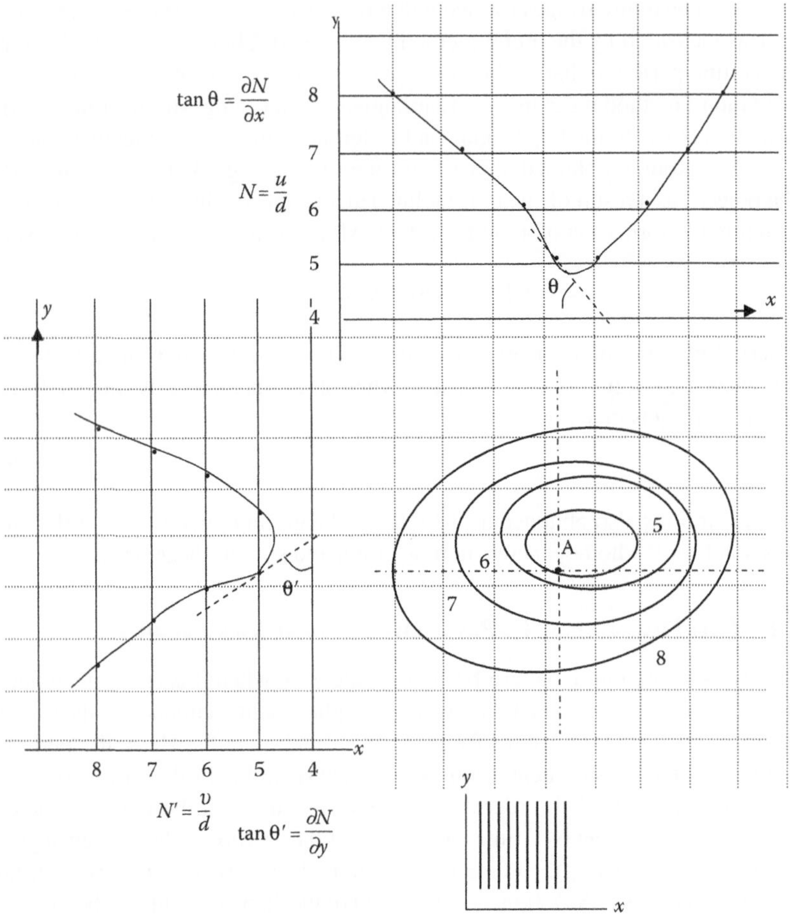

FIGURE 9.5 In-plane displacement and normal strain measurement from the moiré pattern.

now draw the tangent at the point A. The slope of the tangent at this point gives the normal strain along the x direction. That is,

$$\tan\theta = \frac{\partial N}{\partial x} = \frac{1}{a}\frac{\partial u}{\partial x}$$

or

$$\frac{\partial u}{\partial x} = \varepsilon_x = a\frac{\partial N}{\partial x} \tag{9.36}$$

In fact, the strain field along this line on the object can be obtained from this curve. If another line parallel to the y axis but passing through A is drawn and the displacement curve plotted as shown in Figure 9.5, we can obtain

$$\frac{\partial u}{\partial y} = a\frac{\partial N}{\partial y}$$

at point A. Similarly, we can obtain

$$\frac{\partial v}{\partial y} = \varepsilon_y = a\frac{\partial N'}{\partial y}$$

from the v family of displacement fringes obtained by conducting the experiment with the gratings aligned with their grating vectors parallel to the y axis. Also, we obtain the quantity

$$a\frac{\partial N'}{\partial x}$$

The shear strain γ_{xy} is then calculated from

$$\gamma_{xy} = a\left(\frac{\partial N}{\partial y} + \frac{\partial N'}{\partial x}\right) \tag{9.37}$$

Thus, both the normal strains and the shear strain can be obtained from the moiré phenomenon. However, it is very clear from this analysis that the following two factors are very important: first, correct assignment of the fringe orders to the moiré fringes; second, accurate generation of the displacement curve. Assignment of the fringe orders is not trivial, and requires considerable effort and knowledge. The rules of topography of continuous surfaces govern the order of fringes. Adjacent fringes differ by plus or minus one fringe order, except in zones of local fringe maxima or minima, where they can have equal fringe orders. Local maxima and minima are usually distinguished by closed loops or intersecting fringes; in topography, such contours represent hills and saddle-like features, respectively. Fringes of unequal orders cannot intersect. The fringe order at any point is unique and independent of the path of the fringe count used to reach that point. A fringe can be assigned any number, since absolute displacement information is not required for strain measurement. Relative displacement can be determined using an arbitrary datum.

For generating the displacement curve accurately, one needs a large number of data points, and so methods have been found to increase the number of moiré fringes, and hence the data points, for the same loading conditions. This can be accomplished by (i) pitch mismatch, (ii) angular mismatch, or (iii) their combination.

The strains can also be obtained by shearing: the moiré patterns representing the displacement fringes are sheared to obtain the fringes corresponding to the strain.

9.9.3 High-Sensitivity In-Plane Displacement Measurement

The sensitivity of in-plane displacement measurement depends on the period of the grating. In moiré work, low-frequency gratings are usually employed, and the analysis is based on geometrical optics. However, with fine-period gratings, although

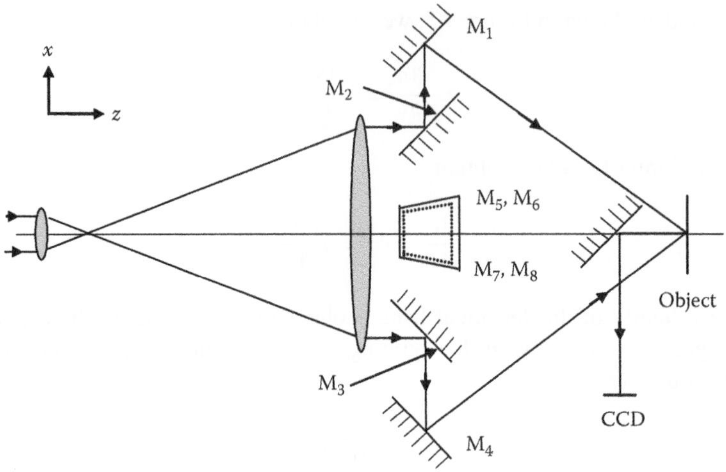

FIGURE 9.6 High-sensitivity in-plane measurement set-up: mirrors M_5, M_6, M_7, and M_8 generate two beams in the (y, z) plane.

the sensitivity is increased, diffraction effects are prominent, and consequently quasi-monochromatic, spatially coherent light needs to be used. Indeed, with laser radiation, in-plane displacement measurements with a very high degree of accuracy can be made. A cross grating, say 1200 lines/mm in either direction, recorded holographically and aluminum-coated for reflection, is bonded onto the surface under test. The use of high-density gratings puts a limitation on the size of the object that can be examined. The grating is illuminated by four collimated beams, as shown in Figure 9.6. Two beams lie in the (x, z) plane and two in the (y, z) plane, and make angles such that first-order diffracted beams propagate along the z axis. These beams, on interference, generate fringe patterns characteristic of the u and v families of the displacement.

To understand the working of the technique, let us consider a one-dimensional grating with its grating vector along the x direction. The grating is recorded holographically by interference between two plane waves, one propagating along the z axis and the other making an angle θ with this but lying in the (x, z) plane. The period of the grating is b, or its spatial frequency $\mu = (\sin\theta)/\lambda$ (i.e., $b \sin\theta = \lambda$). This grating is bonded onto the surface of the object. When the grating is illuminated normally by a collimated beam, the first-order diffracted beams make angles of θ and $-\theta$ with the normal to the grating. Let us assume that the object is loaded, resulting in distortions in the grating period. Let the modified spatial frequency be $\mu(x)$. The grating is illuminated symmetrically at angles of θ and $-\theta$ by two plane waves that can be represented by $Re^{2\pi i\mu x}$ and $Re^{-2\pi i\mu x}$, where R is the amplitude of the wave. These plane waves will be diffracted by the grating, and the diffracted field can be expressed as

$$R\left(e^{2\pi i\mu x} + e^{-2\pi i\mu x}\right)\frac{1}{2}\{1 + \cos[2\pi\mu(x)x]\} \tag{9.38}$$

Collecting terms of interest (i.e., those terms representing waves that propagate along the z axis), we obtain

$$\frac{1}{4}R\left(e^{2\pi i[\mu-\mu(x)]x}+e^{-2\pi i[\mu-\mu(x)]x}\right) \tag{9.39}$$

The intensity distribution in the image is

$$I(x)=\frac{1}{4}R^2\cos^2\{2\pi[\mu-\mu(x)]x\} \tag{9.40}$$

This represents a moiré pattern. The moiré fringe width is

$$d=\frac{1}{2}\frac{aa(x)}{|a-a(x)|}=\frac{1}{2}Na \tag{9.41}$$

where $a(x)$ is the period of the deformed grating and N is the order of the moiré fringe. It should be noted that the sensitivity of the method is twice as large as would be obtained with the same grating using the conventional method. This is due to the fact that the deformed grating in $+1$ and -1 diffraction orders is being compared. This increase in sensitivity by a factor of two has been explained by Post as being due to moiré formation between a grating on the test surface and a virtual grating of twice the frequency formed by interference between the two beams, thus providing a multiplicative factor of two. Use of higher diffraction orders results in increased fringe multiplication.

An arrangement to measure u and v components of the displacement simultaneously uses a high-frequency cross grating bonded onto the surface of the object. The grating is illuminated simultaneously and symmetrically along the x and y directions to generate four beams travelling axially. The moiré fringes representing u and v displacement components are then obtained by interference of these four beams. Figure 9.7 shows the u and v families of fringes for an electronic component that was heated to show the influence of heat (generated internally during operation) on the packaging. A cross grating with 1200 lines/mm is transferred to the face of the electronic component and He–Ne laser light is used for illumination.

9.10 MEASUREMENT OF OUT-OF-PLANE COMPONENT AND CONTOURING

The Moiré technique is well suited for the measurement of in-plane displacement components: the sensitivity is controlled by the period of the grating. Further, the techniques used for moiré formation are either pitch mismatch or angular mismatch. Therefore, moiré formation when measuring out-of-plane displacement will also be governed by these techniques. Consequently, the moiré method is not as sensitive for out-of-plane measurement as for in-plane measurement. Out-of-plane displacement and surface topography can be measured by the shadow moiré method or the projection moiré method.

We will discuss these methods in detail.

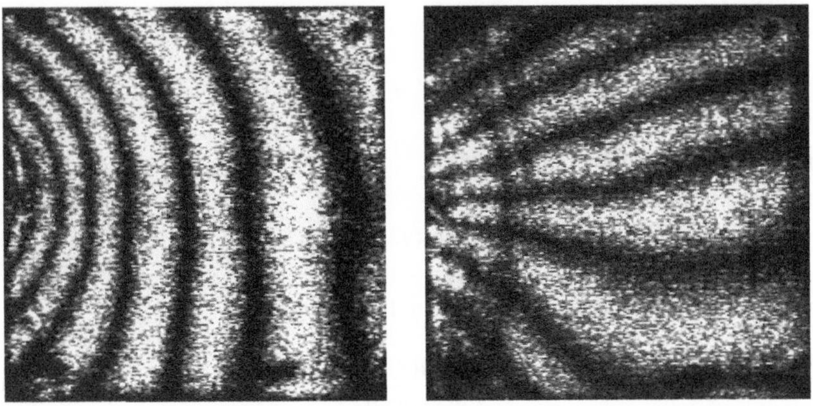

FIGURE 9.7 Interferograms showing (a) u-displacement,and (b) v-displacement fringes at the face of an electronic component when energized. (Courtesy of Fu Yu, National University of Singapore.)

9.10.1 The Shadow Moiré Method

As the name suggests, the moiré pattern is formed between the grating and its shadow on the object. The shadow grating will be distorted by the object topography, and hence moiré fringes from the distorted and reference gratings are observed.

9.10.1.1 Parallel Illumination and Parallel Observation

Figure 9.8 shows an arrangement for shadow moiré that relies on parallel illumination and observation; a reference grating is placed on the object. Without loss of generality, we may assume that the point A on the object surface is in contact with the grating. The grating is illuminated by a collimated beam incident at an angle α with the normal to the grating surface (i.e., the z axis). It is viewed from infinity at an angle β. It is obvious that the grating elements contained in a distance AB occupy a distance AD on the object surface. The elements on AD will form a moiré pattern with the grating elements contained in distance AC. Let us assume that AB and AC have p and q grating elements, respectively. Therefore, AB = pa, and AC = $qa = pb$. From geometry, BC = AC − AB = $(q - p)a = Na$ for $N = 0, \pm 1, \pm 2, \ldots$; N is the order of the moiré fringes.

Therefore, we may write

$$z(x, y)(\tan\alpha + \tan\beta) = Na$$

or

$$z(x, y) = \frac{Na}{\tan\alpha + \tan\beta} \tag{9.42}$$

where $z(x, y)$ is the depth as measured from the grating. This is the governing equation of this method. It can be seen that the moiré fringes are contours of equal depth measured from the grating. If the viewing is along the normal to the grating (i.e., $\beta = 0$),

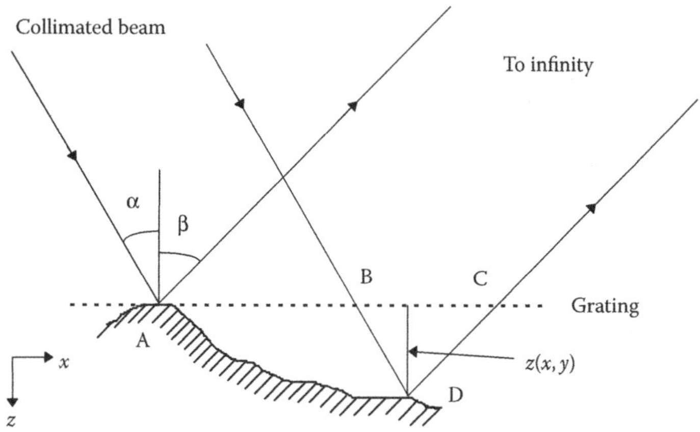

FIGURE 9.8 Schematic for shadow moiré.

then $z(x, y) = Na/\tan\alpha$. Alternatively, the grating may be illuminated normally and viewed obliquely. Then $z(x, y) = Na/\tan\beta$. The assumption that the viewing is done with a collimated beam is not always valid. However, when the object under study is small and the camera is placed sufficiently far away, this requirement is nearly met, although the method is thus not suited for large objects.

9.10.1.2 Spherical-Wave Illumination and Camera at Finite Distance

The assumption made earlier that both the source and the camera are at infinity limits the application of the method to the study of small objects. However, when divergent illumination is used, larger objects can be studied. In general, the source and the camera may be located at different distances from the reference grating. However, a special case where the source and the camera are at equal distances from the grating is of considerable practical importance, and hence is discussed here in detail.

Let the point source S and the camera be at a distance L from the grating surface and let their separation be P, as shown in Figure 9.9. The object is illuminated by a divergent wave from a point source. As before, the number of grating elements p contained in AB on the grating are projected onto AD on the object surface. These elements interact with the q elements in AC, thus producing a moiré pattern.

If we assume that an Nth-order moiré fringe is observed at the point D, then

$$\mathrm{BC} = \mathrm{AC} - \mathrm{AB} = (q - p)a = Na \tag{9.43}$$

But $\mathrm{BC} = z(x, y)(\tan\alpha' + \tan\beta')$, where $z(x, y)$ is the depth of the point D from the grating. Therefore, we obtain

$$z(x, y) = \frac{Na}{\tan\alpha' + \tan\beta'} \tag{9.44}$$

Here, α' and β' vary over the surface of the grating or over the surface of the object. From Figure 9.9, we have $\tan\alpha' = x/[L + z(x, y)]$ and $\tan\beta' = (P - x)/[L + z(x, y)]$.

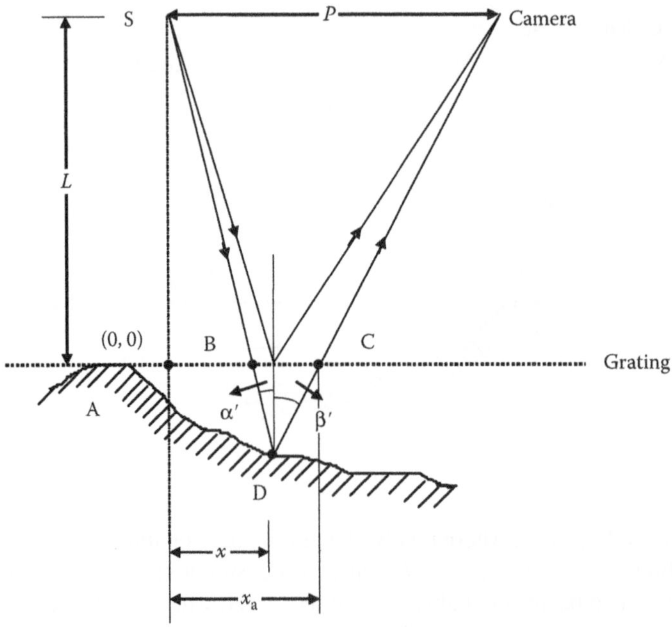

FIGURE 9.9 Shadow moiré method with point source illumination and point receiving.

Substituting these expressions into Equation 9.44, we obtain

$$z(x,y) = \frac{Na}{\dfrac{x}{L+z} + \dfrac{P-x}{L+z}} = Na\frac{L+z(x,y)}{P}$$

Rearranging, we obtain

$$z(x,y) = \frac{NaL}{P-Na} = \frac{Na}{\dfrac{P}{L} - \dfrac{Na}{L}} \tag{9.45}$$

The ratio P/L is called the base-to-height ratio. This is an extremely simple formula. In fact, it is this simplicity that makes this technique attractive over the one in which the source and the camera are placed at different distances from the grating. The distance $\Delta z(x,y)$ between adjacent moiré fringes (i.e., $\Delta N = 1$) is

$$\Delta z(x,y) = \frac{La}{P}\left(1+\frac{z}{L}\right)^2 \tag{9.46}$$

It can be seen that the fringe spacing is not constant, but increases with depth. Since $z(x,y)/L \ll 1$, then the fringe spacing is constant and is given by $\Delta z(x,y) = La/P$. The moiré fringes then represent true depth contours.

Although the method is applicable to the study of large objects, it is essential to correct for perspective error. This error arises because the actual coordinates (x,y) of a point appear as (x_a, y_a). From Figure 9.9, the actual coordinates (x, y) and apparent coordinates (x_a, y_a) are related by

$$\frac{x_a - x}{z(x, y)} = \frac{P - x_a}{L}$$

or

$$x = x_a - \frac{z}{L}(P - x_a) = x_a\left(1 + \frac{z}{L}\right) - \frac{zP}{L} \tag{9.47a}$$

Similarly,

$$y = y_a\left(1 + \frac{z}{L}\right) - \frac{zP}{L} \tag{9.47b}$$

Using this arrangement, the shadow moiré method can be applied to the study (obviously limited by the size of the grating) of very large structures.

Since the fringe spacing is not constant over the object depth, the sensitivity of the method may not be adequate to map out surfaces that are strongly curved or very steep. Therefore, the use of a composite grating has been suggested. The composite grating consists of two parallel superposed gratings with two discretely different periods. This thus provides two sensitivities.

One of the problems of the shadow moiré method is that the moiré fringes do not localize on the surface of the object. Also, the contrast of the fringes is not constant. Fringes of maximum contrast are obtained when the surface under test is nearly flat. The moiré fringes become fuzzy when objects of steep curvature are examined. As a precaution, objects with steep curvature should be placed as close to the grating as possible, and the depth of field of the photographic objective should be large enough to focus the grating and its shadow simultaneously.

9.10.2 Automatic Shape Determination

When the distance between the grating and the object surface changes, the moiré fringes shift. From the direction of the shift—which can differ locally—information about the local surface gradient can be obtained. In one approach to automatic surface determination, a grating placed on the object is illuminated obliquely, and a moiré pattern is recorded as shown in Figure 9.10. A number of moiré patterns are recorded by shifting the grating axially. The phase of the moiré fringes is calculated using well-known phase-shift techniques. The object topography is then calculated using the appropriate expression.

9.10.3 Projection Moiré

There are several differences between the shadow moiré and projection moiré techniques. In projection moiré, a higher-frequency grating is used, and the grating is projected on the object surface and is re-imaged on a reference grating that is identical to the projection grating and aligned parallel to it. The moiré fringes are formed at the plane of the reference grating. The projection moiré method also provides a

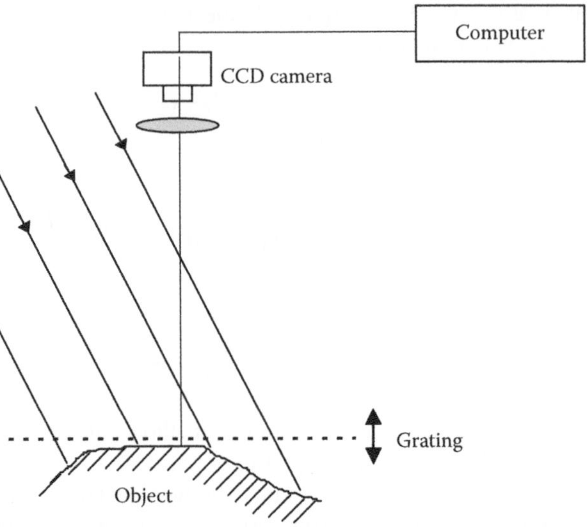

FIGURE 9.10 Automatic shape determination.

certain flexibility that is not possible in the shadow moiré method. We will examine the projection moiré method in the following two cases:

- The optical axes of the projection and imaging (observation) systems are parallel.
- The optical axes are inclined with each other.

Referring to Figure 9.11, a grating G_1 of period b is imaged on a reference plane, where its period is Mb, M being the magnification of the projection system. If the object surface is plane and located on this reference plane, the projected grating will have a constant period. This grating is re-imaged on the reference grating G_2. Assuming the projection and imaging systems to be identical, the pitch of the imaged grating will be equal to that of G_2 and the grating elements will lie parallel to each other owing to their initial alignment. Hence, no moiré pattern will be formed. However, if the surface is curved, the projection grating on the surface will have a varying period, and hence a moiré pattern is formed. We can examine this in exactly the same manner as applied to shadow moiré; that is, the grating of period Mb at the reference plane is illuminated by a spherical wave from the exit pupil of the projection lens. As described in the shadow moiré method, we can write

$$z(x,y) = \frac{NMb}{\tan\alpha' + \tan\beta'} = NMb\frac{L + z(x,y)}{P} \tag{9.48}$$

Equation 9.48, on rearranging, becomes

$$z(x,y) = \frac{NMbL}{P - NMb} \tag{9.49}$$

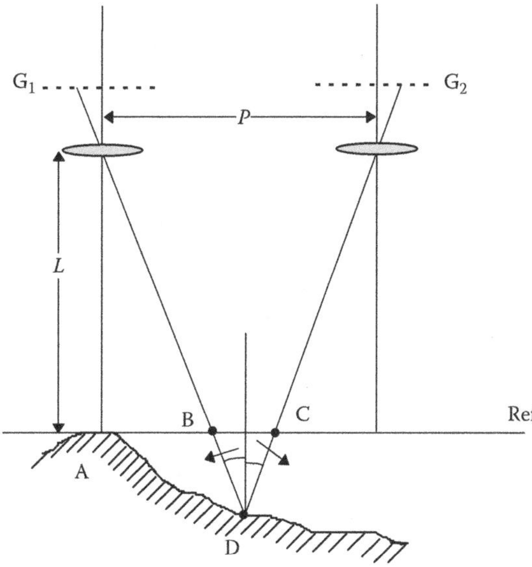

FIGURE 9.11 Projection moiré method.

Writing the magnification of the imaging system as $M = (L - f)/f$, where f is the focal length, we obtain

$$z(x, y) = \frac{NbL(L - f)}{fP - Nb(L - f)} \tag{9.50}$$

Here, $z(x, y)$ corresponds to the depth of the object at the point where the Nth-order moiré fringe is formed. It can be seen that the moiré fringe spacing is not constant. However, when $Mb \ll P$, the moiré fringes are equally spaced.

The disadvantage of this method is that only very small objects can be studied. However, large objects also can be studied when the optical axes of the projection system and the imaging system are inclined, as shown in Figure 9.12. The grating can be imaged on the reference plane where the Scheimpflug condition is satisfied. However, the magnification is not constant, and consequently the period of the grating on the reference plane varies. This problem can be solved by using a special grating whose period remains constant on projection at the reference plane.

The technique can be used for the comparison of two objects if, instead of the reference grating, a photographic record of the projected grating is made first with one object and then with the object replaced by another object. The records can be made on the same film (additive moiré) or on two separate films (multiplicative moiré). Likewise, one can study out-of-plane deformation of the object by recording gratings corresponding to two states of the object.

As should be obvious, the grating plays a very important role: the sensitivity depends on the grating period. Therefore, several gratings may be required to maintain accuracy over a steeply curved surface. Further, for phase-shifting, the grating needs to be shifted. The use of programmed LCD panels for generating gratings of variable

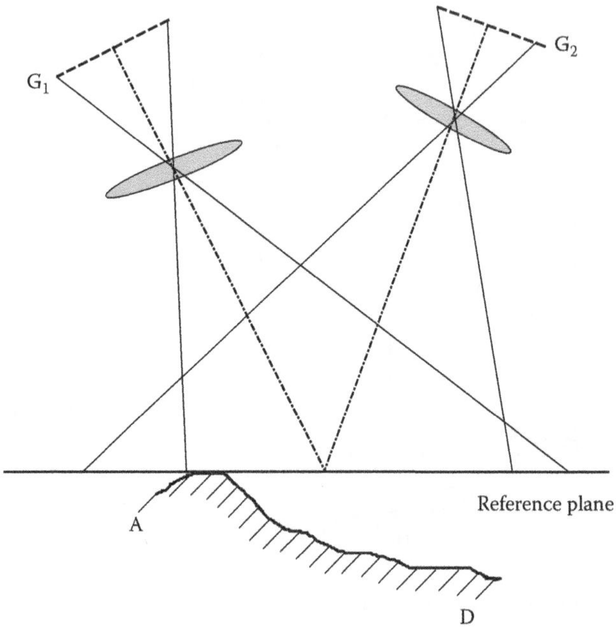

FIGURE 9.12 Projection moiré method with oblique incidence and observation.

period, with a sinusoidal or binary transmission function, and arbitrarily inclined, makes automatic shape determination considerably more convenient.

9.10.4 Light-Line Projection with TDI Mode of Operation of CCD Camera

If depth determination of an object is to be carried out over the whole 360°, then several measurements are needed. These measurements are made by changing the viewing and projection directions. An alternative approach is to unwrap the object surface. This is particularly straightforward if the object has axial symmetry. The object is placed on a turntable with its axis coinciding with the axis of rotation. A beam of light in the form of a line is projected on the surface, as shown in Figure 9.13a. An image is made on a CCD camera; the image of the projected line will be smeared out owing to rotational motion. On the other hand, if the laser is operated in the pulse mode and the camera in TDI (time-delay and integration) mode, then the object surface will be unwrapped and it will be covered with equidistant lines. The line width will be governed by the width of the projected line and the line separation by the pulse rate. If the object departs from axial symmetry, the pitch of the grating will not be uniform, but will vary, depending on the magnitude of departure from the nominal cylindrical surface.

Let us consider a section of the object, as shown in Figure 9.13b. It is illuminated at an angle θ with the optical axis. The dashed line is the nominal circular section and the solid line represents the surface of the object at that section. The point B on the

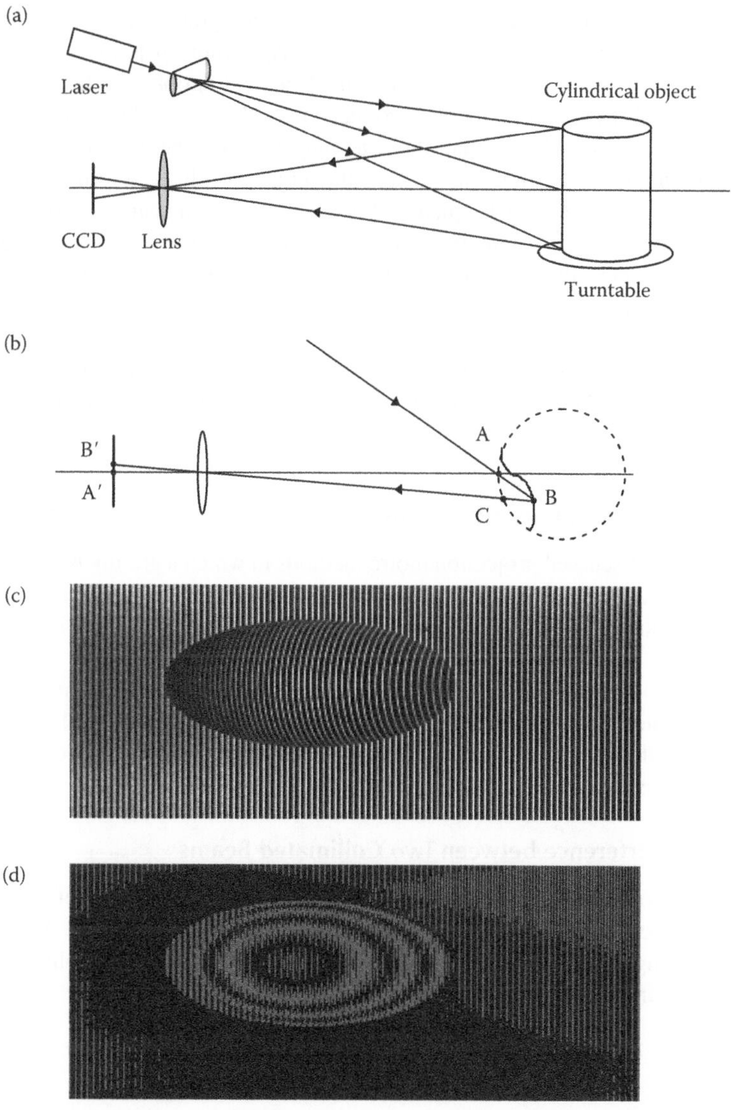

FIGURE 9.13 TDI mode of a CCD camera: (a) line scan of an object; (b) calculation of the shift of the line due to out-of-plane displacement; (c) unwrapped image of a cylindrical object with a dent; (d) moiré pattern showing the dented region. (Courtesy of Dr. M. R. Sajan.)

surface is imaged at B′, which would also be the image of the point C if the object were not deformed. Further, if the object were not deformed, the point A would be imaged on the axis as A′ instead of the point B illuminated now. Therefore, the image of the line on the object will be distorted. From the figure, we have $AC = BC \tan\theta$. The out-of-plane displacement d is BC. Thus $d = AC \cot\theta$. Figure 9.13c shows a distorted image on the unwrapped object. This distortion is seen vividly when a moiré is formed with an internally generated electronic grating, as shown in Figure 9.13d. The moiré condition, as before, is AC $= Np$, where p is the grating pitch. Therefore, $d = Np \cot\theta$. The out-of-plane displacement is obtained from the moiré pattern. Phase-shifting can easily be introduced by a shift of the laser diode.

If the pulse width is larger than the scanning time for one column of detectors, the bright line will be smeared. Hence, the pulse width is determined according to the selected scanning frequency. For example, for a scanning frequency of 50,000 lines/min, that is, a scanning time $T = 1.2$ ms/line, the suggested pulse width should be less than 10% of T, that is, 0.12 ms. Figure 9.13c was obtained with a TDI camera with 192 pixels per column (line) and has 165 lines. In the TDI mode, the charge of each pixel on a line is added to itself as it moves from one end of the detector to the other. There is thus a time delay from the moment a line of pixels is first illuminated to the instant when it is read into the buffer, and the charge of the pixel is integrated over 165 lines.

9.10.5 Coherent Projection Method

So far, we have discussed projection moiré methods in which a grating is projected on the object surface, and its image is formed by another optical system on an identical grating. The illumination is incoherent and the method requires two physical gratings. However, it is possible to create a grating structure on the object by interference between two coherent waves. We shall consider two distinct situations: one where two plane (collimated) waves interfere, generating equidistant equiphasal plane surfaces in space, and the other where interference between two spherical waves is used to generate interference surfaces (hyperboloidal surfaces).

9.10.5.1 Interference between Two Collimated Beams

There are several ways of generating two inclined collimated beams. Let us consider two plane waves, as shown in Figure 9.14, which enclose a small angle $\Delta\theta$ between them. Assuming the amplitudes of these waves to be equal, the total amplitude $u(x, y)$ at the (x, z) plane can be written as

$$u(x, y) = e^{(2\pi/\lambda)i[x\sin(\theta - \Delta\theta/2) + z\cos(\theta - \Delta\theta/2)]} + e^{(2\pi/\lambda)i[x\sin(\theta + \Delta\theta/2) + z\cos(\theta + \Delta\theta/2)]} \tag{9.51}$$

where θ represents the mean direction. When such an amplitude distribution is recorded, the intensity distribution in the record is

$$\begin{aligned} I &= I_0\left\{1 + \cos\left[\frac{2\pi}{\lambda}\left(2\sin\frac{\Delta\theta}{2}\right)(x\cos\theta + z\sin\theta)\right]\right\} \\ &= I_0\left\{1 + \cos\left[\frac{2\pi}{a}(x\cos\theta + z\sin\theta)\right]\right\} \end{aligned} \tag{9.52}$$

where $a = \lambda/[2\sin(\Delta\theta/2)]$. Thus, a grating of period a is formed on the object surface. The period of the grating along the x direction is $a/\cos\theta$ and that along the z

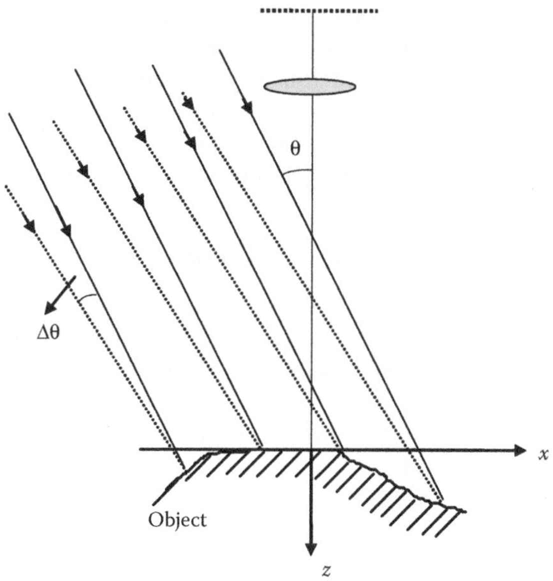

FIGURE 9.14 Coherent projection method.

direction is $a/\sin\theta$. This grating is photographed normally. If the object surface is a plane lying in the (x, y) plane, then the grating period is $a_x = a/\cos\theta$, and the period in the photographic record is Ma_x, where M is the magnification.

In order to visualize the moiré formation, we note that, in collimated illumination, the fringe width along the x direction remains unchanged, irrespective of the depth of the object. However, there is an in-plane shift that varies linearly with depth. So, if the object depth changes by Δz, the x shift is $\Delta z \tan\theta$. If this shift is equal to the grating period a_x, then a moiré fringe is formed. The moiré fringe interval is thus governed by $a_x = \Delta z \tan\theta = a/\cos\theta$. This leads to $\Delta z = a/\sin\theta$. In other words, a moiré fringe is formed wherever the object depth changes by $a/\sin\theta$. This is thus a neat method of measuring out-of-plane displacement.

We again have the possibilities of either additive moiré or multiplicative moiré. The first exposure is made with the object in its initial state and the second exposure when it is deformed. The moiré pattern gives the out-of-plane displacement. Alternatively, a reference record can be made of the grating on a reference plane, and the second record made with dual illumination of the object surface. This way, the surface topography can be obtained. The requirement is that the grating period be resolved. Therefore, the interbeam angle should be only a few degrees: this will give a grating structure of about 100 lines/mm, which can be resolved by a good imaging system. Obviously, the size of the object that can be studied depends on the size of the collimated beams, which is limited by the collimator optics.

9.10.5.2 Interference between Two Spherical Waves

In order to cover large objects, interference between diverging waves is utilized. One simple method employs a single-mode 50 : 50 fiber-optic beam-splitter (coupler).

The laser beam is coupled to one port of the coupler, and the beam is equally divided into two output ports. The separation between the output fibers can be adjusted to achieve the desired fringe width. Unfortunately, the surfaces of constant intensity in the interference pattern are not plane surfaces but hyperboloids. The period of the grating on the reference plane thus varies, but can be calculated given the coordinates of the point sources. This method can therefore be used with large objects.

9.10.6 Measurement of Vibration Amplitudes

Both the shadow moiré and projection moiré methods can be used to measure the amplitude of vibration. We use the method described earlier in which two plane waves are superposed on the object surface. A record of this will be a grating on the surface of the object, with intensity distribution given by Equation 9.52. Let us now assume that the object surface executes simple harmonic motion of amplitude $w(x, y)$ at frequency ω. We can express this as

$$z = z_0 + w(x, y) \sin \omega t \tag{9.53}$$

where z_0 represents the static position of the surface. Owing to surface motion, the phases of the two interfering beams at the surface will change with time; consequently, the instantaneous intensity distribution in the record will be

$$I(x, t) = I_0 \left\{ 1 + \cos \left[\frac{2\pi}{a} [x \cos \theta + (z_0 + w(x, y) \sin \omega t) \sin \theta] \right] \right\} \tag{9.54}$$

This intensity distribution is integrated over a period T much longer than the period of vibration $2\pi/\omega$; that is,

$$I(x) = \frac{1}{T} \int_0^T I(x, t) dt = I_0 \left\{ 1 + J_0 \left(\frac{2\pi}{a} w(x, y) \sin \theta \right) \cos \left[\frac{2\pi}{a} (x \cos \theta + z_0 \sin \theta) \right] \right\} \tag{9.55}$$

where $J_0(x)$ is the Bessel function of zero order. This again represents a grating with modulation given by the Bessel function. Information about the vibration amplitude $w(x, y)$ can be extracted from this record by optical filtering. We assume that the record has an amplitude transmittance proportional to the intensity distribution. The record is placed at the input plane of a FT processor, and one of the first orders is allowed for image formation. The intensity distribution at the output plane is proportional to

$$\left[J_0 \left(\frac{2\pi}{a} w(x, y) \sin \theta \right) \right]^2 \tag{9.56}$$

The intensity distribution exhibits maxima and minima. The amplitude of vibration is obtained as explained for time-average HI in Chapter 6.

9.10.7 Reflection Moiré Method

Unlike the shadow moiré and projection moiré methods, the reflection moiré method yields information about slope and curvature. In the problem of flexure of thin plates, the second derivatives of the deflection are related to the bending moments and twist. The deflection data obtained from shadow moiré or projection moiré need to be differentiated twice in order to obtain curvature, resulting in inaccuracies. Therefore, it is necessary to obtain slope or curvature data so that either only a single differentiation need be performed or differentiation is eliminated altogether. The reflection moiré method serves this purpose. The only disadvantage is that the specimen has to be specular.

Figure 9.15 shows a schematic of the experimental set-up for reflection moiré, also known as Ligtenberg's arrangement. The object is an edge-clamped plate. The surface of the plate is mirror-like so that a virtual image of the reference grating is formed. Consider a ray from a point D on the grating that, after reflection from point P on the object, meets a point I at the image plane, as shown in Figure 9.15. In other words, point I is the image of point D as formed by reflection on the plate surface. When the plate is deformed, point P moves to point P′. If the local slope is ϕ, then point I now receives the ray from point E: the image of point E is formed at I again. In reality, an image of the grating as seen on reflection from the plate is recorded. This image itself may contain distortions due to nonflatness of the plate surface. It can be shown that, owing to self-healing, these distortions do not influence moiré formation. After the plate is deformed (loaded), the image is again recorded on the same photographic film/plate. Owing to loading, the grating image will be further distorted, thus forming a moiré pattern.

Following the arguments presented earlier, a grating AE, after loading, is superposed on another grating AD, leading to moiré formation due to pitch mismatch. Let there be p grating elements in AE and q elements in AD. If the Nth-order moiré fringe appears at point I, then

$$\mathrm{ED} = \mathrm{AE} - \mathrm{AD} = pb - qb = (p - q)b = Nb \tag{9.57}$$

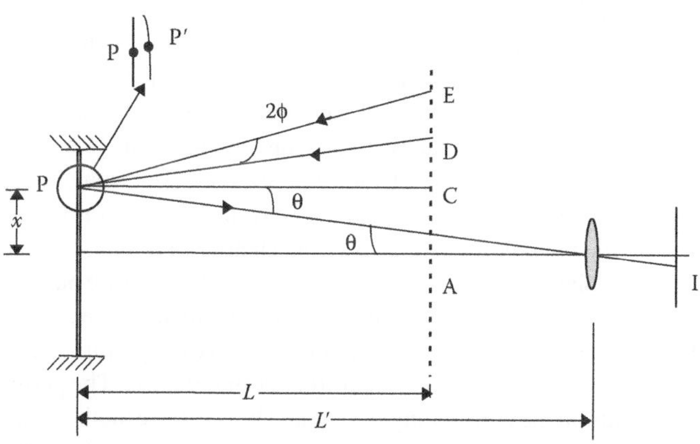

FIGURE 9.15 Reflection moiré: Ligtenberg's arrangement.

But

$$\mathrm{ED} = \mathrm{EC} - \mathrm{CD} = L\tan(\theta + 2\phi) - L\tan\theta$$
$$= L\frac{\tan\theta + \tan 2\phi}{1 - \tan\theta\tan 2\phi} - L\tan\theta$$

Assuming ϕ to be very small, which is usually the case, we write $\tan 2\phi \approx 2\phi$. Therefore,

$$\mathrm{ED} = L\frac{2\phi(1 + \tan^2\theta)}{1 - 2\phi\tan\theta} = Nb \tag{9.58}$$

Further, $2\phi\tan\theta \ll 1$, and, writing the partial x slope $\phi = \partial w/\partial x$, we have

$$\frac{\partial w}{\partial x} = \frac{Nb}{2L(1 + \tan^2\theta)} = \frac{Nb}{2L\left(1 + \dfrac{x^2}{L'^2}\right)} \tag{9.59}$$

The slope depends on the value of x. In order to eliminate the dependence of $\partial w/\partial x$ on x, a curved grating is used. However, Ligtenberg's method utilizing curved gratings also suffers from several disadvantages: the large dimensions of the curved surface gratings, the necessity of using gratings of low spatial frequency to obtain a moiré pattern of good contrast, and the limitation to static loading problems and to models of relatively large flexures. The disadvantages of Ligtenberg's method are largely removed by the Rieder–Ritter arrangement. A cross-grating is used so that $\partial w/\partial x$ and $\partial w/\partial y$ are recorded simultaneously. These can be separated by optical filtering.

Further, as with other moiré methods, the sensitivity is dependent on the grating period. It is therefore desirable to have an arrangement in which the grating period can be varied easily. Figure 9.16 shows an arrangement where a grating is projected onto a ground-glass screen. The projected grating is imaged using the Rieder–Ritter arrangement. The pitch of the grating can be varied by changing the magnification during projection of the grating.

Several other arrangements have been suggested to obtain slope and curvature. A diffraction grating is placed at or near the focal plane of the imaging lens. The diffraction grating produces sheared fields. Double exposure, before and after plate loading, is performed. In another interesting arrangement, an inclined plane parallel plate is placed near the focal plane of the imaging lens. The image is recorded in reflection: two reflected beams, which are sheared, produce a grating in the image plane. Loading of the plate deforms this grating. Therefore, double-exposure recording, before and after plate loading, results in the formation of a moiré pattern that is due to the slope variation. When two plates are used in tandem and properly aligned such that three sheared beams participate in interference, curvature fringes are obtained.

These methods are coherent methods requiring laser radiation. The permissible size of the object depends on the collimating optics, and hence these methods are limited to small to moderate size objects. As mentioned earlier, the surface of the plate (object) need not be optically flat. The departure from flatness causes some distortions in the

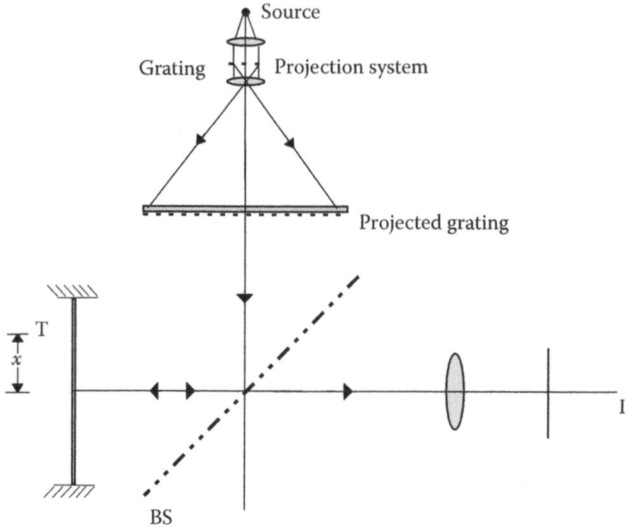

FIGURE 9.16 Reflection moiré: grating with variable pitch.

grating. However, the moiré phenomenon has a self-healing property, and thereby these distortions cancel out, and the moiré fringes are due alone to the deformation of the plate caused by loading.

9.11 SLOPE DETERMINATION FOR DYNAMIC EVENTS

So far, we have described methods that are applicable only to static problems. Pulse illumination is frequently used for the examination of dynamic processes. Grating images corresponding to successive states of the object are recorded. However, it is necessary to record the reference grating each time as well, in order to generate the moiré pattern. This makes these methods rather troublesome to apply to the study of dynamic events.

Several methods have been developed whereby the images of the reference grating and the deformed grating from the surface under investigation are recorded simultaneously. In one such arrangement (Figure 9.17), the beam splitter provides two paths for the imaging. One image is formed by reflection from the object surface T and the other image is formed by reflection from a mirror surface M. Therefore, both images of the grating can be recorded in a single exposure (pulse illumination), and hence an additive moiré pattern can be observed for the various states of the object. This arrangement, however, does not exploit the healing property of the moiré phenomenon. Since, in general, the object surface in its initial state and the mirror surface are not identical, the grating formed by reflection from the object surface will carry some distortions, which will show up as moiré fringes even for the unloaded state of the object. In other words, an initial fringe-free field is not obtained. This problem is solved in the arrangement shown in Figure 9.18. The grating G_1 is a replica of the image formed by reflection from the object surface. Therefore, the images of gratings

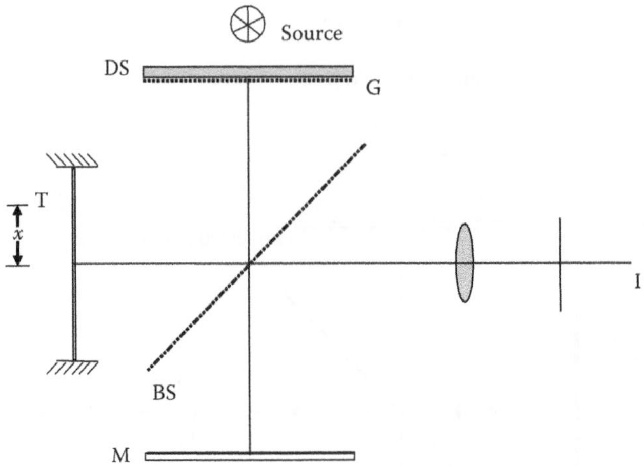

FIGURE 9.17 Arrangement for slope measurement.

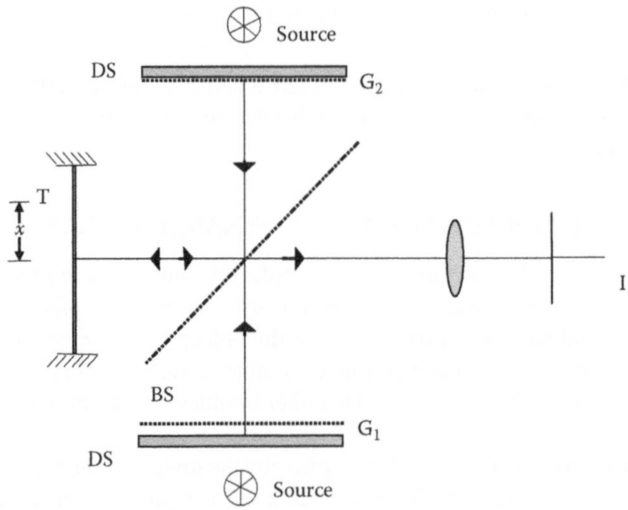

FIGURE 9.18 Modified arrangement for slope measurement.

G_1 and G_2 will be identical, and an initial fringe-free field is obtained. However, for each object surface, a new grating G_1 has to be produced and aligned.

9.12 CURVATURE DETERMINATION FOR DYNAMIC EVENTS

The arrangements shown in Figures 9.17 and 9.18 can be modified to provide for additional shear by incorporating a Mach–Zehnder interferometer. Figure 9.19 shows this arrangement. The images are sheared by the tilt given to one of the mirrors.

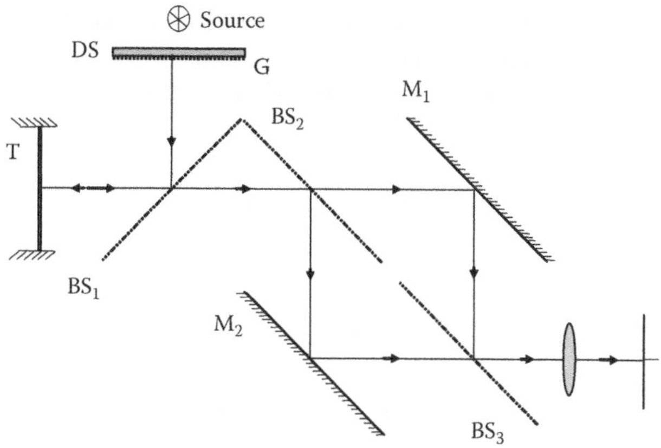

FIGURE 9.19 Arrangement for curvature measurement.

9.13 SURFACE TOPOGRAPHY WITH REFLECTION MOIRÉ METHOD

The methods described for slope and curvature measurements exploit the imaging of a reference grating when the separation between the grating and the object surface is large. However, when the reference grating is placed very close to the object surface, the moiré pattern is formed between the reference grating and its image. Under certain conditions, the moiré fringes are loci of constant depth. To explain the formation of the moiré pattern, we consider the arrangement shown in Figure 9.20.

Grating G_1 (which essentially is an image of G_2) is viewed through grating G_2. Let the gratings G_1 and G_2 be described by

$$x = mb$$

$$x = na$$

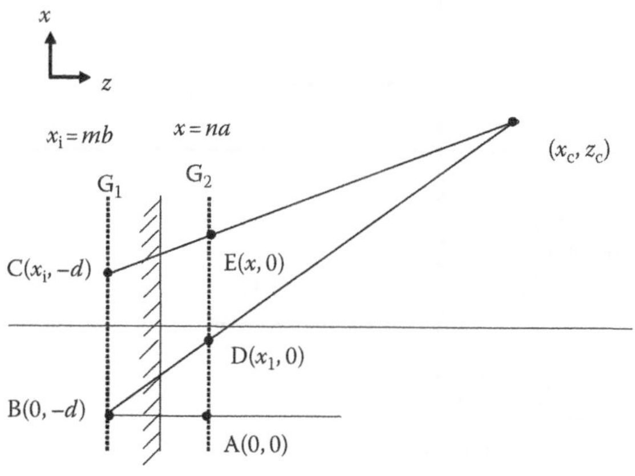

FIGURE 9.20 Surface topography via reflection moiré.

respectively, where m and n are integers and the gratings have different periods b and a respectively. The camera is located at the position (x_c, z_c). The x coordinate of any point projected from grating G_1 to grating G_2 is obtained from

$$\frac{x_c - x}{z_c} = \frac{x_c - mb}{z_c + d}$$

or

$$x = \frac{bz_c}{z_c + d}m + \frac{d}{z_c + d}x_c \tag{9.60}$$

Using the indicial equation $m - n = p$, we obtain an equation for the moiré pattern:

$$x = \frac{abz_c}{(a-b)z_c + ad}p + \frac{ad}{(a-b)z_c + ad}x_c \tag{9.61}$$

In general, d is a function of x and y: $d(x, y)$. Let us now consider another experimental arrangement (Figure 9.21): a grating is placed on a reflecting surface, and an image of the grating is formed by reflection. Owing to imaging, both the period and the separation of the image grating will vary spatially. The moiré pattern is observed as a result of interaction of the image grating with the reference grating.

We now assume that the reflecting surface departs very little from the plane surface and that the reference grating is in contact with the surface. Then, it is safe to assume that the periods of both the reference and the image gratings are equal: $a = b$. The

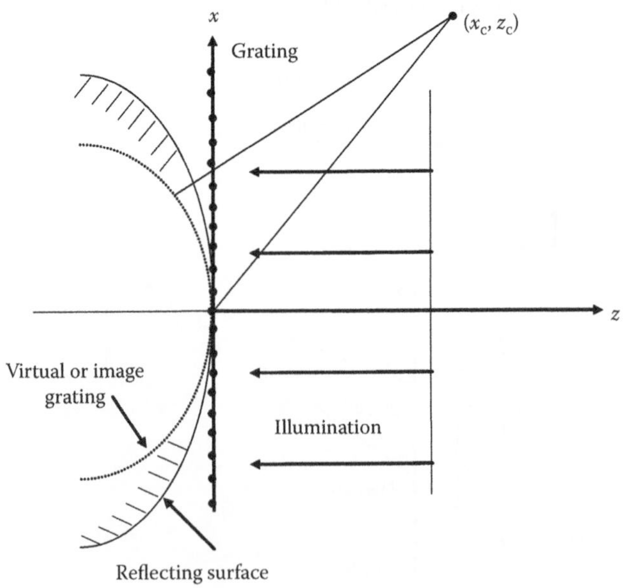

FIGURE 9.21 Reflection moiré for contouring.

moiré pattern is then described by

$$x = \frac{az_c}{d(x,y)}p + x_c \tag{9.62}$$

Since $d(x, y)$ is the separation between the two gratings, the depth $z(x, y)$ of the surface from the reference grating will be half of it: $z(x, y) = d(x, y)/2$. Therefore, we write

$$z(x,y) = \frac{az_c}{2(x - x_c)}p \tag{9.63}$$

Further, if the area to be examined is very small and the illumination source is far away obliquely, then

$$z(x,y) = -\frac{az_c}{2x_c}p \tag{9.64}$$

The moiré fringes now represent true depth contours.

We now apply this technique to study moiré pattern formation when the grating is placed on a spherical surface of radius curvature R. Let us assume that the reference grating is placed at $z = 0$. The equation of the spherical surface is

$$x^2 + y^2 + z^2 + 2zR = 0 \tag{9.65}$$

The distance $z(x, y)$ between the grating and the surface for large R is

$$z = -\frac{x^2 + y^2}{2R}$$

Substituting for $z(x, y)$, the equation of the moiré pattern is now given by

$$x^2 + y^2 = \frac{az_cR}{x_c}p \tag{9.66}$$

The moiré fringes are circles of radii $\sqrt{(az_cR/x_c)p}$. The radii vary as $\sqrt{p}$, as in a Fresnel zone plate. Figures 9.22a and 9.22b show photographs of such moiré patterns taken by placing gratings of 200 lines/inch and 500 lines/inch, respectively, on a concave mirror surface.

9.14 TALBOT PHENOMENON

When a periodic object is illuminated by a coherent monochromatic beam, its image is formed at specific planes called the self-image planes or Talbot planes. The effect was first observed by Talbot in 1836 and its theory was worked out by Rayleigh in 1881. Self-imaging is due to diffraction and can be observed with periodic objects that satisfy the Montgomery condition. A linear (1D) grating is one such object. For a 1D grating of spatial frequency μ, illuminated by a collimated beam of wavelength λ, the self-image planes are equidistant and are located at distances $N/\mu^2\lambda$ from the

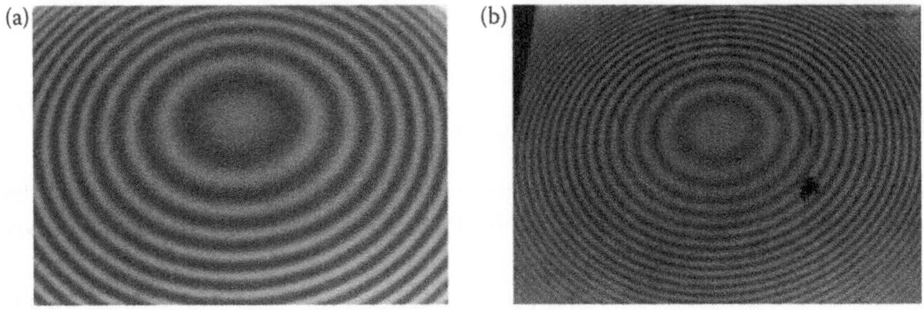

FIGURE 9.22 Contour maps of a spherical surface: (a) with a grating of 200 lines/inch; (b) with a grating of 500 lines/inch.

object, where $N = 0, 1, 2, 3, \ldots$ gives the order of the Talbot planes. In other words, the transverse periodicity of the object manifests itself as longitudinal periodicity. The imaging is called self-imaging because no imaging devices are used. A two-dimensional (cross) grating with the same spatial frequency μ in both directions also self-images at the planes located at $N/\mu^2\lambda$ from the grating.

9.14.1 Talbot Effect in Collimated Illumination

To explain Talbot imaging, let us consider a 1D grating whose transmittance is given by

$$t(x) = \frac{1}{2}(1 + \cos 2\pi\mu x)$$

This grating is placed at the plane $z = 0$ and is illuminated by a collimated beam of amplitude A, as shown in Figure 9.23. The amplitude of the wave just behind the grating ($z = 0$ plane) will be given by

$$u(x, z = 0) = \frac{1}{2}A(1 + \cos 2\pi\mu x) \tag{9.67}$$

Using the Fresnel diffraction approach, the amplitude at any plane z is obtained as

$$u(x_1, z) = \frac{A}{2}e^{ikz}\left(1 + e^{-i\pi\pi^2\lambda z}\cos 2\pi\mu x_1\right) \tag{9.68}$$

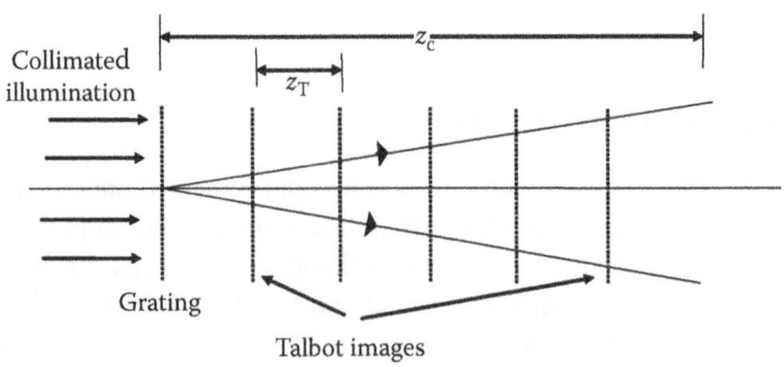

FIGURE 9.23 Formation of Talbot images in collimated illumination.

The amplitude distribution at any plane z will be identical to the grating transmittance function, except for a constant multiplicative phase factor, if $e^{-i\pi\pi^2\lambda z} = 1$. This condition is satisfied when

$$\pi\mu^2\lambda = 2N\pi \quad \text{for } N = 0, 1, 2, 3, \ldots \tag{9.69}$$

The planes at which this condition is satisfied are the Talbot planes. However, when N takes half-integer values, we still obtain the transmittance function of a sinusoidal grating, but it is phase-shifted by π. Thus, the Talbot planes are separated by $z_T = 1/\mu^2\lambda$. In the case of collimated illumination, the Talbot planes are equispaced.

9.14.2 Cut-Off Distance

The Talbot images are formed as a result of constructive interference among the diffracted waves at successive Talbot planes. For an infinite grating illuminated by an infinitely large beam, the various diffracted waves will continue to overlap, irrespective of distance, thus producing an infinite number of Talbot images. In a practical situation, both the grating and the beam are of finite dimensions. Therefore, the diffracted waves will cease to overlap after a certain distance; consequently, no Talbot images are formed after this distance. Let us consider a grating of linear dimension D and spatial frequency μ illuminated by a beam of size D. The cut-off distance z_c is defined as the distance from the grating over which the first-order diffracted beam deviates by half the beam size. This is given by $z_c = D/2\mu\lambda$.

9.14.3 Talbot Effect in Noncollimated Illumination

Let us assume that a point source is placed at $(0, 0)$ and that the grating is at the plane $z = R$. The grating is illuminated by a divergent spherical wave from the point source. The amplitude of the wave just behind the grating is

$$u(x, R) = \frac{A}{2R} e^{ikR}\, e^{i(k/2R)x^2}(1 + \cos 2\pi\mu x) \tag{9.70}$$

This is valid under the paraxial approximation. Using the Fresnel diffraction approximation, the amplitude at a plane distant z from the grating is

$$u(x_1, R + z) = \frac{A}{2(R+z)} e^{ik(R+z)} \exp\left[i\frac{k}{2(R+z)}x_1^2\right] \times \left[1 + \exp\left(-i\pi\frac{Rz}{R+z}\mu^2\lambda\right)\cos\left(2\pi\mu\frac{R}{R+z}x_1\right)\right] \tag{9.71}$$

This expression represents a transmittance function of the grating of spatial frequency $\mu' = \mu R/(R+z)$, provided that

$$\exp\left(-i\pi\frac{Rz}{R+z}\mu^2\lambda\right) = 1$$

This yields the self-image planes distances $(z_T)_s$ as

$$(z_T)_s = \frac{2N}{\mu^2\lambda - \dfrac{2N}{R}} \quad \text{for } N = 1, 2, 3, \tag{9.72}$$

The spacing between successive Talbot planes increases with the order N. The period of the grating also increases as if it were geometrically projected. Similarly, when the grating is illuminated by a convergent spherical wave, the successive Talbot planes come closer and the spatial frequency increases. This is valid until some distance from the point of convergence.

9.14.4 Talbot Effect for Measurement

Instead of using a projection system for projecting a grating on the object surface for either shadow moiré or projection moiré work, the projection can be done without a lens but using the Talbot effect. This, however, necessitates the use of laser radiation. The additive moiré method can then be applied to measure out-of-plane deformation. The shape of the object can also be obtained by imaging the Talbot grating on another identical grating.

When a grating of higher frequency is used, several Talbot planes may intersect the object surface. A moiré pattern of high contrast will be formed at the Talbot planes. Therefore, objects of large depth can be topographically examined using this technique.

9.14.4.1 Temperature Measurement

Several optical methods, such as holographic interferometry, moiré deflectometry, schlieren photography, and laser speckle photography, have been used for the measurement of flame temperature. These are noncontact methods. Talbot interferometry with a circular grating has been proposed as another method of temperature profiling of flames. Two identical circular gratings (containing equispaced concentric circles) are used in the experiment. They are aligned with their centers collinear with the beam direction and are placed such that the second grating lies on the Talbot image of the first grating. The flame is placed between the two gratings. For a centrosymmetric flame, the center of the flame lies on the line joining the centers of the gratings. In such a case, the rays bend away from the center: the Talbot image of the circular grating takes an elliptical shape, which forms a moiré pattern with the second grating. The angle of deflection is calculated from the moiré pattern. Using the inverse Abel transform, the change in refractive index, $(n - n_0)/n_0$, is obtained, where n and n_0 are the refractive indices at a point in the flame and of the ambient atmosphere, respectively. This value is then used to calculate the temperature using the formula

$$T = \frac{T_0}{\left(\dfrac{n - n_0}{n_0}\right)\left(\dfrac{3PA + 2RT_0}{3PA}\right) + 1}$$

where T_0 is the temperature at the ambient condition for which the refractive index is n_0, P is the pressure, R is the gas constant, and A is the molar refractivity of air at the wavelength of the light used in the experiment. Temperatures measured at various locations in the flame with Talbot interferometry and with thermocouples are in good agreement.

The sensitivity of the method depends nonlinearly on the pitch of the circular grating.

9.14.4.2 Measurement of the Focal Length of a Lens and the Long Radius of Curvature of a Surface

Talbot interferometry has been used for measuring the focal length of a positive lens and power variations for a multifocus lens. It is a convenient technique to the measure radii of curvature of shallower concave surfaces.

BIBLIOGRAPHY

1. K. J. Gasvik, *Optical Metrology*, Wiley, Chichester, 1987.
2. O. D. D. Soares (Ed.), *Optical Metrology*, Martinus Nijhoff, Dordrecht, 1987.
3. A. Lagarde (Ed.), *IUTAM Symposium on Advanced Optical Methods and Applications in Solids Mechanics* (Poitiers, France, 31 August–4 September, 1998), Springer Vienna, 2000.
4. J.F. Doyle and J. W. Phillips (Eds.), *Manual of Experimental Stress Analysis*, Society for Experimental Stress Analysis, Brookfield, CT, 1989.
5. O. Kafri and I. Glatt, *The Physics of Moiré Metrology*, Wiley, New York, 1990.
6. G. Indebetouw and R. Czarnek (Eds.), *Selected Papers on Optical Moiré and Applications*, MS64, SPIE, Bellingham, WA, 1992.
7. K. Patorski and M. Kujawińska, *Handbook of the Moiré Fringe Technique*, Elsevier, Amsterdam, 1993.
8. D. Post, B. Han, and P. Ifju, *High Sensitivity Moiré*, Springer-Verlag, New York, 1994.
9. G. L. Cloud, *Optical Methods of Engineering Analysis*, Cambridge University Press, Cambridge, 1995.
10. N. H. Abramson, *Light in Flight or The Holodiagram*; *The Columbi Egg of Optics*, SPIE, Bellingham, WA, 1997.
11. I. Amidror, *The Theory of the Moire Phenomenon*, Springer Verlag, 2000.
12. C. A. Walker (Ed.), *Handbook of Moiré Measurement*, Taylor & Francis, New York, 2003.

ADDITIONAL READING

1. F. K. Ligtenberg, The moiré method, a new experimental method for the determination of moments in small slab models, *Proc. Soc. Exp. Stress Anal.*, 12, 83–98, 1954/55.
2. G. L. Rogers, A simple method of calculating moiré patterns, *Proc. Phys. Soc.*, 73, 142–144, 1959.
3. D. L. A. Barber and M. P. Atkinson, Method of measuring displacement using optical gratings, *J. Sci. Instrum.*, 36, 501–504, 1959.
4. J. M. Burch, The metrological applications of diffraction gratings, In *Progress in Optics* (ed. E. Wolf), Vol. 2, 73–108, North-Holland Amsterdam, 1963.
5. G. Oster and Y. Nishijima, Moiré patterns, *Sci. Am.*, 208, 54–63, 1963.

6. P. S. Theocaris, Isopachic patterns by the moiré method, *Exp. Mech.*, 4, 153–159, 1964.
7. G. Rieder and R. Ritter, Kruemmungsmessung an belasteter Platten nach dem Ligtenbergschen Moiré-Verfahren, *Forsch. Ing.-Wes.*, 31, 33–44, 1965.
8. C. A. Sciammarella, Basic optical law in the interpretation of moiré patterns applied to the analysis of strains, Part 1, *Exp. Mech.*, 5, 154–160, 1965.
9. B. E. Ross, C. A. Sciammarella, and D. E. Sturgeon, Basic law in the interpretation of moiré patterns applied to the analysis of strains, Part 2, *Exp. Mech.*, 5, 161–166, 1965.
10. D. Post, Analysis of moiré fringe multiplication phenomenon, *Appl. Opt.*, 6, 1938–1942, 1967.
11. U. Heise, A moiré method for measuring plate curvature, *Exp. Mech.*, 7, 47–48, 1967.
12. F. P. Chiang, Techniques of optical spatial filtering applied to processing of moiré fringe patterns, *Exp. Mech.*, 9, 523–526, 1969.
13. H. H. M. Chau, Moiré pattern resulting from the superposition of two zone plates, *Appl. Opt.*, 8, 1707–1712, 1969.
14. D. W. Meadows, W. O. Johnson and J. B. Allen, Generation of surface contours by moiré patterns, *Appl. Opt.*, 9, 942–947, 1970.
15. H. Takasaki, Moiré topography, *Appl. Opt.*, 9, 1467–1472, 1970.
16. J. der Hovanesian and Y. Y. Hung, Moiré contour-sum contour-difference and vibration analysis of arbitrary objects, *Appl. Opt.*, 10, 2734–2738, 1971.
17. H. Takasaki, Moiré topography, *Appl. Opt.*, 12, 845–850, 1973.
18. F. P. Chiang and G. Jaisingh, Dynamic moiré methods for the bending of plates, *Exp. Mech.*, 13, 168–171, 1973.
19. K. Matsumoto and M. Takashima, Improvement on moiré technique for in-plane deformation measurements, *Appl. Opt.*, 12, 858–864, 1973.
20. O. Bryngdahl, Moiré: Formation and interpretation, *J. Opt. Soc. Am.*, 64, 1287–1294, 1974.
21. J. M. Burch and C. Forno, A high sensitivity moiré grid technique for studying deformation in large objects, *Opt. Eng.*, 14, 178–185, 1975.
22. C. Chiang, Moiré topography, *Appl. Opt.*, 14, 177–179, 1975.
23. J. M. Walls and H. N. Southworth, The moiré pattern formed on superposing a zone plate with a grating or grid, *Opt. Acta*, 22, 591–601, 1975.
24. J. Motycka, A grazing-incidence moiré interferometer for displacement and planeness measurement, *Exp. Mech.*, 15, 279–281, 1975.
25. A. L. Browne and S. M. Rhode, Moiré contouring: Mathematical analysis, *Appl. Opt.*, 14, 1260–1261, 1975.
26. F. P. Chiang, A shadow moiré method with two discrete sensitivities, *Exp. Mech.*, 15, 382–385, 1975.
27. J. Marasco, Use of curved grating in shadow moiré, *Exp. Mech.*, 15, 464–470, 1975.
28. Y. Yoshino and H. Takasaki, Doubling and visibility enhancement of moiré fringes of the summation type, *Appl. Opt.*, 15, 1124–1126, 1976.
29. Y. Y. Hung, C. Y. Liang, J. D. Hovanesian and A. J. Durelli, Time-averaged shadow-moiré method for studying vibrations, *Appl. Opt.*, 16, 1717–1719, 1977.
30. M. Idesawa, T. Yatagai and T. Soma, Scanning moiré method and automatic measurement of 3-D shapes, *Appl. Opt.*, 16, 2152–2162, 1977.
31. A. Assa, A. A. Betser and J. Politch, Recording slope and curvature contours of flexed plates using a grating shearing interferometer, *Appl. Opt.*, 16, 2504–2513, 1977.
32. C. Forno, Welds at high temperatures studied by moiré photography, *Weld Met. Fabr.*, 46, 661–667, 1978.

33. A. Assa, J. Politch, and A. A. Betser, Slope and curvature measurement by a double-frequency-grating shearing interferometer, *Exp. Mech.*, 20, 129–137, 1979.
34. D. T. Moore and B. E. Truax, Phase locked moiré fringe analysis for automated contouring of diffuse surfaces, *Appl. Opt.*, 18, 91–97, 1979.
35. J. C. Perrin and A. Thomas, Electronic processing of moiré fringes: Application to moiré topography and comparison with photogrammetry, *Appl. Opt.*, 18, 563–574, 1979.
36. S. Yokozeki and S. Mihara, Moiré interferometry, *Appl. Opt.*, 18, 1275–1280, 1979.
37. Y. Y. Hung, C. Y. Liang, J. D. Hovanesian and A. J. Durelli, Time averaged shadow moiré method for studying vibrations, *Appl. Opt.*, 18, 1717–1719, 1979.
38. R. E. Rowlands and P. O. Lemens, Moiré strain analysis in cryogenic environments, *Appl. Opt.*, 18, 1886–1887, 1979.
39. F. P. Chiang, Moiré methods of strain analysis, *Exp. Mech.*, 19, 290–308, 1979.
40. M. L. Basehore and D. Post, Moiré method for in-plane and out-of-plane displacement measurement, *Exp. Mech.*, 21, 321–328, 1981.
41. C. A. Sciammarella, Holographic moiré an optical tool for the determination of displacements, strains, contours and slopes of surfaces, *Opt. Eng.*, 21, 447–457, 1982.
42. J. M. Burch and C. Forno, High resolution moiré photography, *Opt. Eng.*, 21, 602–614, 1982.
43. L. Pirodda, Shadow and projection moiré techniques for absolute or relative mapping of surface shapes, *Opt. Eng.*, 21, 640–649, 1982.
44. D. R. Andrews, Shadow moiré contouring of impact craters, *Opt. Eng.*, 21, 650–654, 1982.
45. T. Y. Kao and F. P. Chiang, Family of grating techniques of slope and curvature measurements for static and dynamic flexure of plates, *Opt. Eng.*, 21, 721–742, 1982.
46. S. Yokozeki, Moiré fringes, *Opt. Lasers Eng.*, 3, 15–27, 1982.
47. C. A. Sciammarella, The moiré method—A Review, *Exp. Mech.*, 22, 418–433, 1982.
48. T. Yatagai and M. Idesawa, Automatic fringe analysis for moiré topography, *Opt. Lasers Eng.*, 3, 73–83, 1982.
49. R. Ritter, Reflection moiré methods for plate bending studies, *Opt. Eng.*, 21, 663–671, 1982.
50. M. L. Basehore and D. Post, Displacement fields (U, W) obtained simultaneously by moiré interferometry, *Appl. Opt.*, 21, 2558–2562, 1982.
51. J. Fujimoto, Determination of the vibrating phase by a time-averaged shadow moiré method, *Appl. Opt.*, 21, 4373–4376, 1982.
52. T. Y. Kao and F. P. Chiang, Family of grating techniques of slope and curvature measurements for static and dynamic flexure of plates, *Opt. Eng.*, 21, 721–742, 1982.
53. R. Ritter and R. Schettler-Koehler, Curvature measurement by moiré effect, *Exp. Mech.*, 23, 165–170, 1983.
54. G. Subramanian and S. Krishnakumar, On a curvature based non-destructive moiré technique for detecting defects in plates, *NDT Int. (UK)*, 16, 271–274, 1983.
55. O. Kafri, A. Livnat, and E. Keren, Optical second differentiation by shearing moiré deflectometry, *Appl. Opt.*, 22, 650–652, 1983.
56. M. Halioua, R. S. Krishnamurthy, H. Liu, and F. P. Chiang, Projection moiré with moving grating for automated 3-D topography, *Appl. Opt.*, 22, 850–855, 1983.
57. G. T. Reid, Moiré fringes in metrology, *Opt. Lasers Eng.*, 5, 63–93, 1984.
58. H. E. Cline, W. E. Lorensen, and A. S. Holik, Automatic moiré contouring, *Appl. Opt.*, 23, 1454–1459, 1984.

59. G. T. Reid, R. C. Rixon, and H. I. Messer, Absolute and comparative measurements of three dimensional shape by phase measuring moiré topography, *Opt. Laser Technol.*, 16, 315–319, 1984.
60. Y. Morimoto, T. Hayashi, and N. Yamaguchi, Strain measurement by scanning—Moiré method, *Bull. JSME*, 27, 2347–2352, 1984.
61. E. M. Weissman, D. Post, and A. Asundi, Wholefield strain determination by moiré shearing interferometry, *J. Strain Anal.*, 19, 77–80, 1984.
62. Y. Nakano and K. Murata, Measurement of phase objects using Talbot effect and moiré techniques, *Appl. Opt.*, 23, 2296–2299, 1984.
63. L. Pirroda, Strain analysis by grating interferometry, *Opt. Lasers Eng.*, 5, 7, 1984.
64. J. Jahns, A. W. Lohmann, and J. Ojeda-Castaneda, Talbot and Lau effects, a parageometrical approach, *Opt. Acta*, 31, 313–324, 1984.
65. D. Post, R. Czarnek, and D. Joh, Shear strain contours from moiré interferometry, *Exp. Mech.*, 25, 282–287, 1985.
66. G. Subramanian and S. M. Nair, Direct determination of curvatures of bent plates using double-glass plate shearing interferometer, *Exp. Mech.*, 25, 376–380, 1985.
67. K. Patorski and M. Kujawińska, Optical differentiation of displacement patterns using moiré interferometry, *Appl. Opt.*, 24, 3041–3048, 1985.
68. A. Asundi, Variable sensitivity shadow moiré method, *J. Strain Anal.*, 20, 59–61, 1985.
69. K. Patorski, Generation of derivatives of out-of-plane displacement using conjugate shear and moiré interferometry, *Appl. Opt.*, 25, 3146–3151, 1986.
70. O. J. Løkberg and J. T. Malmo, Deformation measurements at very high temperatures by ESPI and moiré methods, *Proc. SPIE*, 661, 62–68, 1986.
71. A. S. Voloshin, C. P. Burger, R. E. Rowlands, and T. S. Richard, Fractional moiré strain analysis using digital image techniques, *Exp. Mech.*, 26, 254–258, 1986.
72. K. J. Gasvik and M. E. Fourney, Projection moiré using digital video processing. A technique for improving the accuracy and sensitivity, *Trans. ASME*, 108, 652–656, 1986.
73. A. Asundi and M. T. Cheung, Moiré interferometry for out-of-plane displacement measurement, *J. Strain Anal.*, 21, 51–54, 1986.
74. K. Andresen and D. Klassen, The phase-shift method applied to cross grating moiré measurements, *Opt. Lasers Eng.*, 7, 101–114, 1986/87.
75. K. Patorski, D. Post, R. Czarnek, and Y. Guo, Real-time optical differentiation for moiré interferometry, *Appl. Opt.*, 26, 1977–1982, 1987.
76. Y. Marimoto, Y. Seguchi, and T. Higashi, Application of moiré analysis of strain using Fourier transform, *Proc. SPIE*, 814, 295–302, 1987.
77. A. Asundi and M. T. Cheung, Three-dimensional measurement using moiré interferometry, *Strain*, 24, 25–26, 1988.
78. M. Suzuki and M. Kanaya, Applications of moiré topography measurement methods in industry, *Opt. Lasers Eng.*, 8, 171–188, 1988.
79. C. Forno, Deformation measurement using high resolution moiré photography, *Opt. Lasers Eng.*, 8, 189–212, 1988.
80. Y. Morimoto, Y. Seguchi, and T. Higashi, Application of moiré analysis of strain using Fourier Transform, *Opt. Eng.*, 27, 650–656, 1988.
81. I. Glatt and O. Kafri, Moiré deflectometry–ray tracing interferometry, *Opt. Lasers Eng.*, 8, 277–320, 1988.
82. J. M. Huntley and J. E. Field, High resolution moiré photography: Application to dynamic stress analysis, *Opt. Eng.*, 28, 926–933, 1989.
83. A. Asundi, Moiré interferometry for deformation measurement, *Opt. Lasers Eng.*, 11, 281–292, 1989.

84. Y. Guo, D. Post, and R. Czarnek, The magic of carrier fringes in moiré interferometry, *Exp. Mech.*, 29, 169–173, 1989.
85. D. Post, Determination of the thermal strains by moiré interferometry, *Exp. Mech.*, 29, 318–323, 1989.
86. J. S. Lim, J. Kim and M. S. Chung, Additive type moiré with computer image processing, *Appl. Opt.*, 28, 2677–2680, 1989.
87. M. E. Tuttle and D. L. Graesser, Compression creep of graphite/epoxy laminates monitored using moiré interferometry, *Opt. Lasers Eng.*, 12, 151–171, 1990.
88. L. Salbut and K. Patorski, Polarization phase-shifting method for moiré interferometry and flatness testing, *Appl. Opt.*, 29, 1471–1473, 1990.
89. J. J. J. Dirckx and W. F. Decraemer, Automatic calibration method for phase shift shadow moiré interferometry, *Appl. Opt.*, 29, 1474–1476, 1990.
90. F.-L. Dai, J. McKelvie, and D. Post, An interpretation of moiré interferometry from wavefront interference theory, *Opt. Lasers Eng.*, 12, 101–118, 1990.
91. Y. Sogabe, Y. Morimoto, and S. Murata, Displacement measurement of track on magnetic tape by Fourier transform grid method, *Proc. SPIE*, 1554B, 289–297, 1991.
92. C. G. Woychik and Y. Guo, Thermal strain measurements of solder joints in electronic packaging using moiré interferometry, *Proc. SPIE*, 1554B, 461–470, 1991.
93. Y. Morimoto, D. Post, and H. E. Gascoigne, Carrier pattern analysis of moiré interferometry using Fourier transform, *Proc. SPIE*, 1554B, 493–502, 1991.
94. M. Kujawińska, L. Salbut, and K. Patorski, 3-channel phase stepped system for moiré interferometry, *Appl. Opt.*, 30, 1633–1635, 1991.
95. M. Suganuma and T. Yoshizawa, Three-dimensional shape analysis by the use of a projected grating image, *Opt. Eng.*, 30, 1529–1533, 1991.
96. Y. Z. Dai and F. P. Chiang, Contouring by moiré interferometry, *Exp. Mech.*, 31, 76–81, 1991.
97. Q. Yu, K. Andresen, and D. S. Zhang, Digital pure shear-strain moiré patterns, *Appl. Opt.*, 31, 1813–1817, 1992.
98. K. V. Sriram, M. P. Kothiyal, and R. S. Sirohi, Talbot interferometry in noncollimated illumination for curvature and focal length measurement, *Appl. Opt.*, 31, 75–79, 1992.
99. B. Han and D. Post, Immersion interferometer for microscopic moiré interferometry, *Exp. Mech.*, 32, 38–41, 1992.
100. B. Han, Higher sensitivity moiré interferometry for micromechanics studies, *Opt. Eng.*, 31, 1517–1526, 1992.
101. K. J. Gasvik and G. K. Robbersmyr, Autocorrelation of the setup for the projected moiré method, *Exp. Tech.*, 17, 41–44, 1993.
102. J. Kato, I. Yamaguchi, and S. Kuwashima, Real-time fringe analysis based on electronic moiré and its applications, In *Fringe '93* (ed. W. Jüptner and W. Osten), 66–71. Akademie Verlag, Berlin, 1993.
103. M. Priga, A. Kozlowska, and M. Kujawińska, Generalization of the scaling problem for the automatic moiré and fringe projection shape measurement systems, In *Fringe '93* (ed. W. Jüptner and W. Osten), 188–193, Akademie Verlag, Berlin, 1993.
104. C. Y. Poon, M. Kujawiñska, and C. Ruiz, Automated fringe pattern analysis for moiré interferometry, *Exp. Mech.*, 33, 234–241, 1993.
105. J. T. Atkinson and M. J. Lalor, Range measurement using moiré contouring, *Proc. SPIE*, 2340, 202–210, 1994.
106. M. B. Whitworth and J. M. Huntley, Dynamic stress analysis by high-resolution reflection moiré, *Opt. Eng.*, 33, 924–931, 1994.
107. P. K. Rastogi, Visualization of contours of equal slope of an arbitrarily shaped object using holographic moiré, *Opt. Eng.*, 33, 2373–2377, 1994.

108. Y. Morimoto, H. E. Gascoigne, and D. Post, Carrier pattern analysis of moiré interferometry using the Fourier transform method, *Opt. Eng.*, 33, 2646–2653, 1994.
109. Y. Wang and F. P. Chiang, New moiré interferometry for measuring three-dimensional displacements, *Opt. Eng.*, 33, 2654–2658, 1994.
110. A. K. Asundi, Chi-Shing Chan and M. R. Sajan, 360° profilometry: New techniques for display and acquisition, *Opt. Eng.*, 33, 2760–2769, 1994.
111. C. A. Walker, A historical review of moiré interferometry, *Exp. Mech.*, 34, 281–299, 1994.
112. S. K. Bhadra, S. K. Sarkar, R. N. Chakraborty, and A. Basuray, Coherent moiré technique for obtaining slope and curvature of stress patterns, *Opt. Eng.*, 33, 3359–3363, 1994.
113. C. Shakher and A. J. Pramila Daniel, Talbot interferometer with circular gratings for the measurement of temperature in axisymmetric gaseous flames, *Appl. Opt.*, 33, 6068–6072, 1994.
114. A. J. Moore and J. R. Tyre, Phase-stepped ESPI and moiré interferometry for measuring stress-intensity factor and J integral, *Exp. Mech.*, 35, 306–314, 1995.
115. T. Matsumoto, Y. Kitagawa, and T. Minemoto, Sensitivity-variable moiré topography with a phase-shift method, *Opt. Eng.*, 35, 1754–1760, 1996.
116. M. Wang, Projection moiré deflectometry for the automatic measurement of phase objects, *Opt. Eng.*, 35, 2005–2011, 1996.
117. M. K. Kalms, W. Jüptner and W. Osten, Projected-fringe-technique with automatic pattern adoption using a programmable LCD-projector, In *Fringe '97* (ed. W. Jüptner and W. Osten), 231–236 Akademie Verlag, Berlin, 1997.
118. D. Malacara-Doblado, Measuring the curvature of spherical wavefronts with Talbot interferometry, *Opt. Eng.*, 36, 2016–2024, 1997.
119. X. He, D. Zou, S. Liu, and Y. Guo, Phase-shifting analysis in moiré interferometry and its applications in electronic packaging, *Opt. Eng.*, 37, 1410–1419, 1998.
120. B. Han, Recent advancements of moiré and microscopic moiré interferometry for thermal deformation analyses of microelectronics devices, *Exp. Mech.*, 38, 278–288, 1998.
121. S. De Nicola, P. Ferraro, and S. De Nicola, Fourier transform method of fringe analysis for moiré interferometry, *J. Opt. A: Pure Appl. Opt.*, 2, 228–233, 2000.
122. J. Villa, J. A. Quiroga, and M. Servin, Improved regularized phase-tracking technique for the processing of square-grating deflectograms, *Appl. Opt.*, 39, 502–508, 2000.
123. M. R. Miller, I. Mohammed, and P.S. Ho, Quantitative strain analysis of flip-chip electronic packages using phase-shifting moiré interferometry, *Opt. Lasers Eng.*, 36, 127–139, 2001.
124. S.-T. Lin, Three-dimensional displacement measurement using a newly designed moiré interferometer, *Opt. Eng.*, 40, 822–826, 2001.
125. A. Martínez, R.Rodríguez-Vera, J. A. Rayas, and H. J. Puga, Fracture detection by grating moiré and in-plane ESPI techniques, *Opt. Lasers Eng.*, 39, 525–536, 2003.
126. K. Kadooka, K. Kunoo, N. Uda, K. Ono, and T. Nagayasu, Strain analysis for moiré interferometry using the two-dimensional continuous wavelet transform, *Exp. Mech.*, 43, 45–51, 2003.
127. Q. Kemao, S. H. Soon, and A. Asundi, Instantaneous frequency and its application to strain extraction in moiré interferometry, *Appl. Opt.*, 42, 6504–6513, 2003.
128. R. R. Cordero and I. Lira, Uncertainty analysis of displacements measured by phase-shifting moiré interferometry, *Opt. Commun.*, 237, 25–36, 2004.

129. C. J. Tay, C. Quan, Y. Fu, and Y. Huang, Instantaneous velocity, displacement and contour measurement by use of shadow moiré and temporal wavelet analysis, *Appl. Opt.*, 43, 4164–4171, 2004.
130. L. J. Fellows and D. Nowell, Measurement of crack closure after the application of an overload cycle, using moiré interferometry, *Int. J. Fatigue*, 27, 1453–1462, 2005.
131. F. Labbe and R. R. Cordero, Monitoring the plastic deformation progression of a specimen undergoing tensile deformation by moiré interferometry, *Meas. Sci. Technol.*, 16, 1469–1476, 2005.
132. D. Post, Moiré interferometry for engineering and science, *Proc. SPIE*, 5776, 29–43, 2005.
133. A. Kishen, K. Tan, and A. Asundi, Digital moiré interferometric investigations on the deformation gradients of enamel and dentine: An insight into non-carious cervical lesions, *J. Dentistry*, 34, 12–18, 2006.
134. Y. Yamamoto, Y. Morimoto, and M. Fusigaki, Two-direction phase-shifting moiré interferometry and its application to thermal deformation measurement of an electronic device, *Meas. Sci. Technol.*, 18, 561–566, 2007.

Index

Printed and bound by CPI Group (UK) Ltd, Croydon, CR0 4YY
21/10/2024
01777107-0007